Pitman Research Notes in Mathema Series

Submission of proposals for consideration
Suggestions for publication, in the form of outlines and representative samples, are invited by the Editorial Board for assessment. Intending authors should approach one of the main editors or another member of the Editorial Board, citing the relevant AMS subject classifications. Alternatively, outlines may be sent directly to the publisher's offices. Refereeing is by members of the board and other mathematical authorities in the topic concerned, throughout the world.

Preparation of accepted manuscripts
On acceptance of a proposal, the publisher will supply full instructions for the preparation of manuscripts in a form suitable for direct photo-lithographic reproduction. Specially printed grid sheets can be provided and a contribution is offered by the publisher towards the cost of typing. Word processor output, subject to the publisher's approval, is also acceptable.

Illustrations should be prepared by the authors, ready for direct reproduction without further improvement. The use of hand-drawn symbols should be avoided wherever possible, in order to maintain maximum clarity of the text.

The publisher will be pleased to give any guidance necessary during the preparation of a typescript, and will be happy to answer any queries.

Important note
In order to avoid later retyping, intending authors are strongly urged not to begin final preparation of a typescript before receiving the publisher's guidelines. In this way it is hoped to preserve the uniform appearance of the series.

Addison Wesley Longman Ltd
Edinburgh Gate
Harlow, Essex, CM20 2JE
UK
(Telephone (0) 1279 623623)

Titles in this series. A full list is available from the publisher on request.

151 A stochastic maximum principle for optimal control of diffusions
U G Haussmann
152 Semigroups, theory and applications. Volume II
H Brezis, M G Crandall and F Kappel
153 A general theory of integration in function spaces
P Muldowney
154 Oakland Conference on partial differential equations and applied mathematics
L R Bragg and J W Dettman
155 Contributions to nonlinear partial differential equations. Volume II
J I Díaz and P L Lions
156 Semigroups of linear operators: an introduction
A C McBride
157 Ordinary and partial differential equations
B D Sleeman and R J Jarvis
158 Hyperbolic equations
F Colombini and M K V Murthy
159 Linear topologies on a ring: an overview
J S Golan
160 Dynamical systems and bifurcation theory
M I Camacho, M J Pacifico and F Takens
161 Branched coverings and algebraic functions
M Namba
162 Perturbation bounds for matrix eigenvalues
R Bhatia
163 Defect minimization in operator equations: theory and applications
R Reemtsen
164 Multidimensional Brownian excursions and potential theory
K Burdzy
165 Viscosity solutions and optimal control
R J Elliott
166 Nonlinear partial differential equations and their applications: Collège de France Seminar. Volume VIII
H Brezis and J L Lions
167 Theory and applications of inverse problems
H Haario
168 Energy stability and convection
G P Galdi and B Straughan
169 Additive groups of rings. Volume II
S Feigelstock
170 Numerical analysis 1987
D F Griffiths and G A Watson
171 Surveys of some recent results in operator theory. Volume I
J B Conway and B B Morrel
172 Amenable Banach algebras
J-P Pier
173 Pseudo-orbits of contact forms
A Bahri
174 Poisson algebras and Poisson manifolds
K H Bhaskara and K Viswanath
175 Maximum principles and eigenvalue problems in partial differential equations
P W Schaefer
176 Mathematical analysis of nonlinear, dynamic processes
K U Grusa
177 Cordes' two-parameter spectral representation theory
D F McGhee and R H Picard
178 Equivariant K-theory for proper actions
N C Phillips
179 Elliptic operators, topology and asymptotic methods
J Roe
180 Nonlinear evolution equations
J K Engelbrecht, V E Fridman and E N Pelinovski
181 Nonlinear partial differential equations and their applications: Collège de France Seminar. Volume IX
H Brezis and J L Lions
182 Critical points at infinity in some variational problems
A Bahri
183 Recent developments in hyperbolic equations
L Cattabriga, F Colombini, M K V Murthy and S Spagnolo
184 Optimization and identification of systems governed by evolution equations on Banach space
N U Ahmed
185 Free boundary problems: theory and applications. Volume I
K H Hoffmann and J Sprekels
186 Free boundary problems: theory and applications. Volume II
K H Hoffmann and J Sprekels
187 An introduction to intersection homology theory
F Kirwan
188 Derivatives, nuclei and dimensions on the frame of torsion theories
J S Golan and H Simmons
189 Theory of reproducing kernels and its applications
S Saitoh
190 Volterra integrodifferential equations in Banach spaces and applications
G Da Prato and M Iannelli
191 Nest algebras
K R Davidson
192 Surveys of some recent results in operator theory. Volume II
J B Conway and B B Morrel
193 Nonlinear variational problems. Volume II
A Marino and M K V Murthy
194 Stochastic processes with multidimensional parameter
M E Dozzi
195 Prestressed bodies
D Iesan
196 Hilbert space approach to some classical transforms
R H Picard
197 Stochastic calculus in application
J R Norris
198 Radical theory
B J Gardner
199 The C^*-algebras of a class of solvable Lie groups
X Wang
200 Stochastic analysis, path integration and dynamics
K D Elworthy and J C Zambrini

201 Riemannian geometry and holonomy groups
S Salamon
202 Strong asymptotics for extremal errors and polynomials associated with Erdös type weights
D S Lubinsky
203 Optimal control of diffusion processes
V S Borkar
204 Rings, modules and radicals
B J Gardner
205 Two-parameter eigenvalue problems in ordinary differential equations
M Faierman
206 Distributions and analytic functions
R D Carmichael and D Mitrovic
207 Semicontinuity, relaxation and integral representation in the calculus of variations
G Buttazzo
208 Recent advances in nonlinear elliptic and parabolic problems
P Bénilan, M Chipot, L Evans and M Pierre
209 Model completions, ring representations and the topology of the Pierce sheaf
A Carson
210 Retarded dynamical systems
G Stepan
211 Function spaces, differential operators and nonlinear analysis
L Paivarinta
212 Analytic function theory of one complex variable
C C Yang, Y Komatu and K Niino
213 Elements of stability of visco-elastic fluids
J Dunwoody
214 Jordan decomposition of generalized vector measures
K D Schmidt
215 A mathematical analysis of bending of plates with transverse shear deformation
C Constanda
216 Ordinary and partial differential equations. Volume II
B D Sleeman and R J Jarvis
217 Hilbert modules over function algebras
R G Douglas and V I Paulsen
218 Graph colourings
R Wilson and R Nelson
219 Hardy-type inequalities
A Kufner and B Opic
220 Nonlinear partial differential equations and their applications: Collège de France Seminar. Volume X
H Brezis and J L Lions
221 Workshop on dynamical systems
E Shiels and Z Coelho
222 Geometry and analysis in nonlinear dynamics
H W Broer and F Takens
223 Fluid dynamical aspects of combustion theory
M Onofri and A Tesei
224 Approximation of Hilbert space operators. Volume I. 2nd edition
D Herrero
225 Operator theory: proceedings of the 1988 GPOTS–Wabash conference
J B Conway and B B Morrel
226 Local cohomology and localization
J L Bueso Montero, B Torrecillas Jover and A Verschoren
227 Nonlinear waves and dissipative effects
D Fusco and A Jeffrey
228 Numerical analysis 1989
D F Griffiths and G A Watson
229 Recent developments in structured continua. Volume II
D De Kee and P Kaloni
230 Boolean methods in interpolation and approximation
F J Delvos and W Schempp
231 Further advances in twistor theory. Volume I
L J Mason and L P Hughston
232 Further advances in twistor theory. Volume II
L J Mason, L P Hughston and P Z Kobak
233 Geometry in the neighborhood of invariant manifolds of maps and flows and linearization
U Kirchgraber and K Palmer
234 Quantales and their applications
K I Rosenthal
235 Integral equations and inverse problems
V Petkov and R Lazarov
236 Pseudo-differential operators
S R Simanca
237 A functional analytic approach to statistical experiments
I M Bomze
238 Quantum mechanics, algebras and distributions
D Dubin and M Hennings
239 Hamilton flows and evolution semigroups
J Gzyl
240 Topics in controlled Markov chains
V S Borkar
241 Invariant manifold theory for hydrodynamic transition
S Sritharan
242 Lectures on the spectrum of $L^2(\Gamma\backslash G)$
F L Williams
243 Progress in variational methods in Hamiltonian systems and elliptic equations
M Girardi, M Matzeu and F Pacella
244 Optimization and nonlinear analysis
A Ioffe, M Marcus and S Reich
245 Inverse problems and imaging
G F Roach
246 Semigroup theory with applications to systems and control
N U Ahmed
247 Periodic-parabolic boundary value problems and positivity
P Hess
248 Distributions and pseudo-differential operators
S Zaidman
249 Progress in partial differential equations: the Metz surveys
M Chipot and J Saint Jean Paulin
250 Differential equations and control theory
V Barbu

251 Stability of stochastic differential equations with respect to semimartingales
X Mao
252 Fixed point theory and applications
J Baillon and M Théra
253 Nonlinear hyperbolic equations and field theory
M K V Murthy and S Spagnolo
254 Ordinary and partial differential equations. Volume III
B D Sleeman and R J Jarvis
255 Harmonic maps into homogeneous spaces
M Black
256 Boundary value and initial value problems in complex analysis: studies in complex analysis and its applications to PDEs 1
R Kühnau and W Tutschke
257 Geometric function theory and applications of complex analysis in mechanics: studies in complex analysis and its applications to PDEs 2
R Kühnau and W Tutschke
258 The development of statistics: recent contributions from China
X R Chen, K T Fang and C C Yang
259 Multiplication of distributions and applications to partial differential equations
M Oberguggenberger
260 Numerical analysis 1991
D F Griffiths and G A Watson
261 Schur's algorithm and several applications
M Bakonyi and T Constantinescu
262 Partial differential equations with complex analysis
H Begehr and A Jeffrey
263 Partial differential equations with real analysis
H Begehr and A Jeffrey
264 Solvability and bifurcations of nonlinear equations
P Drábek
265 Orientational averaging in mechanics of solids
A Lagzdins, V Tamuzs, G Teters and A Kregers
266 Progress in partial differential equations: elliptic and parabolic problems
C Bandle, J Bemelmans, M Chipot, M Grüter and J Saint Jean Paulin
267 Progress in partial differential equations: calculus of variations, applications
C Bandle, J Bemelmans, M Chipot, M Grüter and J Saint Jean Paulin
268 Stochastic partial differential equations and applications
G Da Prato and L Tubaro
269 Partial differential equations and related subjects
M Miranda
270 Operator algebras and topology
W B Arveson, A S Mishchenko, M Putinar, M A Rieffel and S Stratila
271 Operator algebras and operator theory
W B Arveson, A S Mishchenko, M Putinar, M A Rieffel and S Stratila
272 Ordinary and delay differential equations
J Wiener and J K Hale
273 Partial differential equations
J Wiener and J K Hale
274 Mathematical topics in fluid mechanics
J F Rodrigues and A Sequeira
275 Green functions for second order parabolic integro-differential problems
M G Garroni and J F Menaldi
276 Riemann waves and their applications
M W Kalinowski
277 Banach C(K)-modules and operators preserving disjointness
Y A Abramovich, E L Arenson and A K Kitover
278 Limit algebras: an introduction to subalgebras of C*-algebras
S C Power
279 Abstract evolution equations, periodic problems and applications
D Daners and P Koch Medina
280 Emerging applications in free boundary problems
J Chadam and H Rasmussen
281 Free boundary problems involving solids
J Chadam and H Rasmussen
282 Free boundary problems in fluid flow with applications
J Chadam and H Rasmussen
283 Asymptotic problems in probability theory: stochastic models and diffusions on fractals
K D Elworthy and N Ikeda
284 Asymptotic problems in probability theory: Wiener functionals and asymptotics
K D Elworthy and N Ikeda
285 Dynamical systems
R Bamon, R Labarca, J Lewowicz and J Palis
286 Models of hysteresis
A Visintin
287 Moments in probability and approximation theory
G A Anastassiou
288 Mathematical aspects of penetrative convection
B Straughan
289 Ordinary and partial differential equations. Volume IV
B D Sleeman and R J Jarvis
290 *K*-theory for real *C**-algebras
H Schröder
291 Recent developments in theoretical fluid mechanics
G P Galdi and J Necas
292 Propagation of a curved shock and nonlinear ray theory
P Prasad
293 Non-classical elastic solids
M Ciarletta and D Ieşan
294 Multigrid methods
J Bramble
295 Entropy and partial differential equations
W A Day
296 Progress in partial differential equations: the Metz surveys 2
M Chipot
297 Nonstandard methods in the calculus of variations
C Tuckey
298 Barrelledness, Baire-like- and (LF)-spaces
M Kunzinger
299 Nonlinear partial differential equations and their applications. Collège de France Seminar. Volume XI
H Brezis and J L Lions

300 Introduction to operator theory
T Yoshino
301 Generalized fractional calculus and applications
V Kiryakova
302 Nonlinear partial differential equations and their applications. Collège de France Seminar Volume XII
H Brezis and J L Lions
303 Numerical analysis 1993
D F Griffiths and G A Watson
304 Topics in abstract differential equations
S Zaidman
305 Complex analysis and its applications
C C Yang, G C Wen, K Y Li and Y M Chiang
306 Computational methods for fluid-structure interaction
J M Crolet and R Ohayon
307 Random geometrically graph directed self-similar multifractals
L Olsen
308 Progress in theoretical and computational fluid mechanics
G P Galdi, J Málek and J Necas
309 Variational methods in Lorentzian geometry
A Masiello
310 Stochastic analysis on infinite dimensional spaces
H Kunita and H-H Kuo
311 Representations of Lie groups and quantum groups
V Baldoni and M Picardello
312 Common zeros of polynomials in several variables and higher dimensional quadrature
Y Xu
313 Extending modules
N V Dung, D van Huynh, P F Smith and R Wisbauer
314 Progress in partial differential equations: the Metz surveys 3
M Chipot, J Saint Jean Paulin and I Shafrir
315 Refined large deviation limit theorems
V Vinogradov
316 Topological vector spaces, algebras and related areas
A Lau and I Tweddle
317 Integral methods in science and engineering
C Constanda
318 A method for computing unsteady flows in porous media
R Raghavan and E Ozkan
319 Asymptotic theories for plates and shells
R P Gilbert and K Hackl
320 Nonlinear variational problems and partial differential equations
A Marino and M K V Murthy
321 Topics in abstract differential equations II
S Zaidman
322 Diffraction by wedges
B Budaev
323 Free boundary problems: theory and applications
J I Diaz, M A Herrero, A Liñan and J L Vazquez
324 Recent developments in evolution equations
A C McBride and G F Roach
325 Elliptic and parabolic problems: Pont-à-Mousson 1994
C Bandle, J Bemelmans, M Chipot, J Saint Jean Paulin and I Shafrir
326 Calculus of variations, applications and computations: Pont-à-Mousson 1994
C Bandle, J Bemelmans, M Chipot, J Saint Jean Paulin and I Shafrir
327 Conjugate gradient type methods for ill-posed problems
M Hanke
328 A survey of preconditioned iterative methods
A M Bruaset
329 A generalized Taylor's formula for functions of several variables and certain of its applications
J-A Riestra
330 Semigroups of operators and spectral theory
S Kantorovitz
331 Boundary-field equation methods for a class of nonlinear problems
G N Gatica and G C Hsiao
332 Metrizable barrelled spaces
J C Ferrando, M López Pellicer and L M Sánchez Ruiz
333 Real and complex singularities
W L Marar
334 Hyperbolic sets, shadowing and persistence for noninvertible mappings in Banach spaces
B Lani-Wayda
335 Nonlinear dynamics and pattern formation in the natural environment
A Doelman and A van Harten
336 Developments in nonstandard mathematics
N J Cutland, V Neves, F Oliveira and J Sousa-Pinto
337 Topological circle planes and topological quadrangles
A E Schroth
338 Graph dynamics
E Prisner
339 Localization and sheaves: a relative point of view
P Jara, A Verschoren and C Vidal
340 Mathematical problems in semiconductor physics
P Marcati, P A Markowich and R Natalini
341 Surveying a dynamical system: a study of the Gray–Scott reaction in a two-phase reactor
K Alhumaizi and R Aris
342 Solution sets of differential equations in abstract spaces
R Dragoni, J W Macki, P Nistri and P Zecca
343 Nonlinear partial differential equations
A Benkirane and J-P Gossez
344 Numerical analysis 1995
D F Griffiths and G A Watson
345 Progress in partial differential equations: the Metz surveys 4
M Chipot and I Shafrir
346 Rings and radicals
B J Gardner, Liu Shaoxue and R Wiegandt
347 Complex analysis, harmonic analysis and applications
R Deville, J Esterle, V Petkov, A Sebbar and A Yger
348 The theory of quantaloids
K I Rosenthal
349 General theory of partial differential equations and microlocal analysis
Qi Min-you and L Rodino

350 Progress in elliptic and parabolic partial differential equations
A Alvino, P Buonocore, V Ferone, E Giarrusso, S Matarasso, R Toscano and G Trombetti
351 Integral representations for spatial models of mathematical physics
V V Kravchenko and M V Shapiro
352 Dynamics of nonlinear waves in dissipative systems: reduction, bifurcation and stability
G Dangelmayr, B Fiedler, K Kirchgässner and A Mielke
353 Singularities of solutions of second order quasilinear equations
L Véron
354 Mathematical theory in fluid mechanics
G P Galdi, J Málek and J Necas
355 Eigenfunction expansions, operator algebras and symmetric spaces
R M Kauffman
356 Lectures on bifurcations, dynamics and symmetry
M Field
357 Noncoercive variational problems and related results
D Goeleven
358 Generalised optimal stopping problems and financial markets
D Wong
359 Topics in pseudo-differential operators
S Zaidman
360 The Dirichlet problem for the Laplacian in bounded and unbounded domains
C G Simader and H Sohr
361 Direct and inverse electromagnetic scattering
A H Serbest and S R Cloude
362 International conference on dynamical systems
F Ledrappier, J Lewowicz and S Newhouse
363 Free boundary problems, theory and applications
M Niezgódka and P Strzelecki
364 Backward stochastic differential equations
N El Karoui and L Mazliak
365 Topological and variational methods for nonlinear boundary value problems
P Drábek
366 Complex analysis and geometry
V Ancona, E Ballico, R M Mirò-Roig and A Silva
367 Integral expansions related to Mehler–Fock type transforms
B N Mandal and N Mandal
368 Elliptic boundary value problems with indefinite weights: variational formulations of the principal eigenvalue and applications
F Belgacem
369 Integral transforms, reproducing kernels and their applications
S Saitoh
370 Ordinary and partial differential equations. Volume V
P D Smith and R J Jarvis
371 Numerical methods in mechanics
C Conca and G N Gatica
372 Generalized manifolds
K-G Schlesinger
373 Independent axioms for Minkowski space-time
J W Schutz
374 Integral methods in science and engineering Volume one: analytic methods
C Constanda, J Saranen and S Seikkala
375 Integral methods in science and engineering Volume two: approximation methods
C Constanda, J Saranen and S Seikkala

C Constanda

University of Strathclyde

J Saranen and S Seikkala

University of Oulu

(Editors)

Integral methods in science and engineering

Volume one: analytic methods

LONGMAN

Addison Wesley Longman Limited
Edinburgh Gate, Harlow
Essex CM20 2JE, England
and Associated Companies throughout the world.

*Published in the United States of America
by Addison Wesley Longman Inc.*

First published 1997

AMS Subject Classifications (Main) 35-XX, 45-XX, 65-XX
(Subsidiary) 73-XX, 76-XX, 34-XX

ISSN 0269-3674

ISBN 0 582 30406 7

British Library Cataloguing in Publication Data

A catalogue record for this book is
available from the British Library

Printed and bound in Great Britain
by Biddles Ltd, Guildford and King's Lynn

Contents

Preface

The aim of the international conferences on Integral Methods in Science and Engineering (IMSE) is twofold: to bring together specialists from various research fields who employ integration as an essential tool in their work, and to promote and consolidate the use by the world scientific community of such elegant, powerful and far-reaching techniques.

IMSE85 and IMSE90 were organised by the University of Texas at Arlington, USA, IMSE93 was hosted by Tohoku University, Sendai, Japan, and IMSE96 took place at the University of Oulu, Finland. The last meeting, attended by participants from 26 countries on 4 continents, has confirmed IMSE as an important event on the international conference circuit, which gives scientists and engineers from quite varied research backgrounds the opportunity to get together and discuss advances in a large class of mathematical procedures.

The organisation of IMSE96 was, as expected, of a high standard; acknowledging this fact, the participants wish to thank the sponsoring agencies and the host institution for their generosity. Special thanks are due, in particular, to the members of the Local Committee:

J. Saranen (Department of Mathematical Sciences), Chairman,
S. Heikkilä (Department of Mathematical Sciences),
A. Pramila (Department of Mechanical Engineering),
K. Ruotsalainen (Division of Mathematics, Faculty of Technology),
S. Seikkala (Division of Mathematics, Faculty of Technology).

Advice and general guidance was provided by the International Steering Committee.

The next two conferences will be held on the American continent: IMSE98 at Michigan Technological University, Houghton, USA, and IMSE2000 in Banff, Canada (organised by the University of Alberta at Edmonton).

These two volumes contain 87 refereed papers from among those presented in Oulu. Owing to logistic problems, the original intention of grouping the material into 6 more sharply delineated categories had to be abandoned in favour of the more pragmatic choice of only two: analytic and approximation methods. In each volume the invited papers precede the contributed ones, which have been arranged in alphabetical order by (first) author's surname. It is possible that some of the authors may disagree with the choice of volume for their papers, but the editors' decision was heavily influenced by the need to balance page numbers in order to meet the publisher's requirements.

The editors are indebted to all those who helped with the refereeing process, to Mary Doherty and Gavin Thomson for their sub-editing efforts, and to the staff of Addison Wesley Longman for their efficient handling of the entire publication business.

Information on IMSE98 can be obtained on the Internet at the MTU site http://www.math.mtu.edu/home/math/imse/Home.html or through the link from the AMS Calendar page http://www.ams.org/mathcal/ (where IMSE2000 is also announced).

Christian Constanda
Chairman, IMSE96

The new International Steering Committee of IMSE:

C. Constanda (University of Strathclyde, Glasgow), Chairman,
B. Bertram (Michigan Technological University),
H.H. Chiu (National Chen Kung University, Tainan),
C. Corduneanu (University of Texas at Arlington),
C.A.J. Fletcher (University of New South Wales),
R.P. Gilbert (University of Delaware),
A. Haji-Sheikh (University of Texas at Arlington),
V.P. Korobeinikov (Institute for Computer Aided Design, Moscow),
A. Nastase (Reinisch-Westfälische Technische Hochschule, Aachen),
K. Oshima (Japan Society of Computational Fluid Dynamics, Tokyo),
F.R. Payne (University of Texas at Arlington),
K. Ruotsalainen (University of Oulu),
J. Saranen (University of Oulu),
P. Schiavone (University of Alberta, Edmonton),
S. Seikkala (University of Oulu),
H. Wu (Computing Centre, Beijing),
K.S. Yajnik (National Aeronautical Laboratory, Bangalore),
D.W. Yarbrough (Tennessee Technological University, Cookeville),
F.-G. Zhuang (Chinese Aerodynamics Research Society, Beijing).

Invited papers

D. COLTON

A brief introduction to inverse scattering theory

1. Introduction

The purpose of this short note is to do exactly what the title says: to provide the reader who is uninformed about inverse scattering theory with a view of what the field is about and a few of the recent advances that have been made. In order to do this in a few pages we obviously can only focus on a special, albeit representative, problem, and this choice has of course been strongly influenced by our own research interests. Hence, far from being encyclopaedic, our presentation will be very specific. Nevertheless, by learning about this specific problem, our hope is that the reader will get a glimpse of the fascination of inverse scattering theory. For a detailed and comprehensive presentation of our view of the field of inverse scattering, we refer the reader to our monograph [1].

2. Scattering by an infinite cylinder

We consider the scattering of a time-harmonic electromagnetic plane wave propagating in a direction d by an infinite cylinder of cross section D, where D is assumed to have a smooth boundary ∂D with unit outward normal ν. Assuming that ∂D has (constant) surface impedance $\lambda \geq 0$ and letting $k > 0$ be the wave number, under appropriate assumptions the direct scattering problem can be formulated as finding $u \in C^2(\mathbb{R}^2 \setminus \overline{D}) \cap C^1(\mathbb{R}^2 \setminus D)$ such that

$$\Delta_2 u + k^2 u = 0 \quad \text{in } \mathbb{R}^2 \setminus \overline{D}, \tag{2.1}$$

$$u = u^i + u^s, \tag{2.2}$$

$$\frac{\partial u}{\partial \nu} + ik\lambda u = 0 \quad \text{on } \partial D, \tag{2.3}$$

$$\lim_{r \to \infty} \sqrt{r} \left(\frac{\partial u^s}{\partial r} - iku^s \right) = 0, \tag{2.4}$$

where the *incident field* $u^i(x) = e^{ikx \cdot d}$, $x \in \mathbb{R}^2$, and $r = |x|$. The *Sommerfeld radiation condition* (2.4) for the *scattered field* u^s is assumed to hold uniformly for all directions $\hat{x} = x/|x|$.

We now note that a fundamental solution to the Helmholtz equation (2.1) satisfying the Sommerfeld radiation condition is given by

$$\Phi(x, y) = \frac{i}{4} H_0^{(1)}(k|x - y|), \tag{2.5}$$

This work was supported in part by a grant from the Air Force Office of Scientific Research.

where $H_0^{(1)}$ denotes the Hankel function of the first kind of order zero. Using Green's identities, it is now possible to show [1] that for $x \in \mathbb{R}^2 \setminus \overline{D}$

$$u^s(x) = \int_{\partial D} \left(u^s(y) \frac{\partial}{\partial \nu(y)} \Phi(x,y) - \frac{\partial u^s}{\partial \nu}(y) \Phi(x,y) \right) ds(y), \tag{2.6}$$

and hence from the asymptotic behaviour of the Hankel function we have that

$$u^s(x) = \frac{e^{ikr}}{\sqrt{r}} u_\infty(\hat{x}, d) + O\left(r^{-3/2}\right) \tag{2.7}$$

as $r \to \infty$, where

$$u_\infty(\hat{x}, d) = \frac{e^{i\pi/4}}{\sqrt{8\pi k}} \int_{\partial D} \left(u^s(y) \frac{\partial}{\partial \nu(y)} e^{-ik\hat{x}\cdot y} - \frac{\partial u^s}{\partial \nu}(y) e^{-ik\hat{x}\cdot y} \right) ds(y) \tag{2.8}$$

is the *far field pattern* of the scattered field u^s. The basic problem in inverse scattering theory is to obtain information about ∂D from a knowledge of u_∞.

Having obtained the far field pattern, we can now define the far field operator which plays a central role in inverse scattering theory. Let Ω denote the unit circle. Then the *far field operator* $F : L^2(\Omega) \to L^2(\Omega)$ is defined by

$$(Fg)(\hat{x}) := \int_{\Omega} u_\infty(\hat{x}, d)\, g(d) ds(d) \tag{2.9}$$

for $\hat{x} \in \Omega$. We remark that F is a compact operator on $L^2(\Omega)$. Closely connected to the far field operator are solutions to the Helmholtz equation in $\mathbb{R}^2$ of the form

$$v(x) := \int_{\Omega} e^{ikx\cdot d} g(d) ds(d), \quad x \in \mathbb{R}^2, \tag{2.10}$$

where $g \in L^2(\Omega)$. Such a function is called a *Herglotz wave function* with kernel g. Note that $(Fg)(\hat{x})$ is the far field pattern corresponding to the incident field $v(x)$. The following theorem is due to Colton and Kirsch (see [1] and [2]), where $N(F)$ denotes the null space of F.

Theorem. *Let $\lambda > 0$. Then the far field operator F is injective and has dense range. If $\lambda = 0$, then F is injective with dense range if and only if there does not exist a Neumann eigenfunction for D which is a Herglotz wave function. If* dim $N(F)$ *is not zero, then* dim $N(F)$ *equals the number of linearly independent Neumann eigenfunctions which are Herglotz wave functions.*

3. Spectral theory of the far field operator

We now ask the question if the compact operator F has eigenvalues and, if so, what do they say about ∂D? We will assume in this section that λ is known.

Definition. An operator $T : L^2(\Omega) \to L^2(\Omega)$ is a *trace class operator* if there exists a sequence of operators $\{T_n\}$ having finite rank not greater than n such that

$$\sum_{n=1}^{\infty} \|T - T_n\| < \infty.$$

From the representation (2.8) it can easily be shown that the far field operator F is a trace class operator [3]. The main result about trace class operators that we will need is the following theorem due to Lidskii [4].

Lidskii's Theorem. *Let T be a trace class operator such that T has finite dimensional null space and* $\mathrm{Im}(Tg, g) \geq 0$ *for every $g \in L^2(\Omega)$. Then T has an infinite number of eigenvalues.*

We remark that by the theorem of Colton and Kirsch given in the previous section, the far field operator F has a finite-dimensional null space.

The following lemma, due to Colton and Kress [3], is fundamental to the spectral analysis of F.

Lemma. *Let u_g and u_h be the solutions of* (2.1)–(2.4) *with u^i equal to the Herglotz wave function with kernels g and h, respectively. Then*

$$k\lambda \int_{\partial D} u_g \bar{u}_h ds = \frac{\sqrt{8\pi k}}{2i}\left[(e^{-i\pi/4}Fg, h) - (g, e^{-i\pi/4}Fh)\right] - k(Fg, Fh).$$

We will now use this lemma to deduce some spectral properties of F. Our first result follows from the lemma by setting $\lambda = 0$, $g = h$, $Fg = \mu g$, and using the *reciprocity relation* $u_\infty(\hat{x}, d) = u_\infty(-d, -\hat{x})$ [1] (details can be found in [5]).

Theorem. *Suppose that $\lambda = 0$. Then F is normal, and hence there exist an infinite number of eigenvalues of F, all of which lie on the circle $\sqrt{8\pi k}\,\mathrm{Im}(e^{-i\pi/4}\mu) - k|\mu|^2 = 0$.*

The case when $\lambda > 0$ is handled by Lidskii's theorem.

Theorem. *Suppose that $\lambda > 0$. Then F has an infinite number of eigenvalues, all of which lie inside the disk $\sqrt{8\pi k}\,\mathrm{Im}\left(e^{-i\pi/4}\mu\right) - k|\mu|^2 > 0$.*

Proof. Setting $g = h$, the lemma implies that $\mathrm{Im}(e^{-i\pi/4}Fg, g) \geq 0$. Lidskii's theorem implies that $e^{-i\pi/4}F$, and hence F has an infinite number of eigenvalues. The equation for the disk again follows from the lemma by setting $g = h$ and $Fg = \mu g$.

The above theorem shows that if $\lambda > 0$, then the eigenvalues of F move off the circle $\sqrt{8\pi k}\,\mathrm{Im}\,(e^{-i\pi/4}\mu) - k|\mu|^2 = 0$ into its interior. How far inside they move depends on the arc length $|\partial D|$ of ∂D [3].

Theorem. *Suppose that $\lambda > 0$, and let $Fg = \mu g$. Then*

$$\frac{4\lambda|\mu|^2}{|\partial D|\,(\lambda^2+1)} \leq \sqrt{8\pi k}\,\mathrm{Im}\left(e^{-i\pi/4}\mu\right) - k|\mu|^2.$$

The proof of this theorem follows from the lemma and obtaining a lower bound for $\int\limits_{\partial D} |u_g|^2 ds$ [3]. Suppose that ρ is the radius of the smallest circle with center on $\mathrm{Im}\,\mu = -\mathrm{Re}\,\mu$, $\mathrm{Im}\,\mu > 0$, and passing through the origin which contains all the eigenvalues of F. Then, knowing ρ and λ, the theorem yields a lower bound for $|\partial D|$, that is, we now have a partial solution to the inverse scattering problem of determining ∂D from $u_\infty(\hat{x}, d)$.

4. The inverse scattering problem

In this final section we will consider in more detail the inverse scattering problem of determining ∂D from $u_\infty(\hat{x}, d)$. We first state a uniqueness result due to Kirsch and Kress (see [1] and [6]). (Their proof for $\lambda = 0$ carries over immediately to the case when $\lambda \geq 0$). We note in passing that uniqueness theorems for inverse scattering problems have been a major area of research in recent years; for a sample of some of this research we refer the reader to [7]–[9] and the references contained therein.

Theorem. *∂D is uniquely determined by $u_\infty(\hat{x}, d)$ for $\hat{x}, d \in \Omega$.*

Unfortunately, from the point of view of applications, the inverse scattering problem is *improperly posed* in the sense that the solution does not depend continuously on the data! Furthermore, the problem is *non-linear* (for example, if ∂D is described by $x = r(\hat{x})\hat{x}$, $r(\hat{x}) > 0$, then u_∞ is not a linear function of $r(\hat{x})$). Hence, the usual approach to solving the inverse scattering problem is through non-linear optimisation methods using regularisation techniques (cf. [1] and [10]). Here we will briefly describe a new method, which avoids non-linear optimisation techniques [11]. (For a somewhat related method, see [12]). In what follows, we need only assume that $\lambda \geq 0$, and do not need to know λ precisely, i.e., in contrast to nonlinear optimization methods we do not need to know the boundary condition! The method we propose is based on the following theorem, where we write $\hat{x} = (\cos\theta, \sin\theta)$, $d = (\cos\alpha, \sin\alpha)$, $u_\infty(\hat{x}, d) = u_\infty(\theta, \alpha)$, $g(d) = g(\alpha) \in L^2(-\pi, \pi)$, and $y_0 = (\rho\cos\varphi,\ \rho\sin\varphi)$ is assumed to be a point in D.

Theorem. *Suppose that $\lambda > 0$, or , if $\lambda = 0$, that k^2 is not a Neumann eigenvalue. Then for every $\epsilon > 0$ there exists a function $g \in L^2(-\pi, \pi)$ such that*

(i) $\left\| \int\limits_{-\pi}^{\pi} u_\infty(\theta, \alpha) g(\alpha) d\alpha - e^{-ik\rho\cos(\theta-\varphi)} \right\|_{L^2(-\pi,\pi)} < \epsilon$, *where $\|\cdot\|$ is with respect to θ;*

(ii) $\lim\limits_{y_0 \to \partial D} \|g\| = \infty$;

(iii) *if v is the Herglotz wave function with kernel g, then*

$$\lim_{y_0 \to \partial D} \max_{x \in \partial D} |v(x)| = \infty.$$

We remark that the third condition implies that $\|g\|$ becoming large is not due to wild oscillations of g, whereas the assumption that k^2 is not a Neumann eigenvalue implies that the far field operator is injective, by the theorem stated in Section 2.

The above theorem suggests the following method for determining ∂D (assuming that $\lambda > 0$ or k^2 is not a Neumann eigenvalue): find a solution of the *far field equation*

$$\int_{-\pi}^{\pi} u_\infty(\theta, \alpha) g(\alpha) d\alpha = e^{-ik\rho \cos(\theta - \varphi)} \tag{4.1}$$

for $y_0 = (\rho \cos \varphi, \ \rho \sin \varphi)$ on some partition P of a rectangle known a priori to contain D; having found $g(\alpha) = g(\alpha; y_0)$, ∂D is determined by those points $y_0 \in P$ where $\|g\|$ achieves its maximum (approaching ∂D from inside D). For numerical examples using this approach we refer the reader to [11]. We note in closing that, in general, the far field equation requires regularisation. This must be done with respect to $\|g'\|$, not $\|g\|$! In particular, we are looking for an *unbounded* solution of the far field equation, that is, we are making very explicit use of the fact that the inverse scattering problem is improperly posed!

References

1. D. Colton and R. Kress, *Inverse acoustic and electromagnetic scattering theory*, Springer-Verlag, Berlin, 1992.
2. D. Colton and A. Kirsch, Dense sets and far field patterns in acoustic wave propagation, *SIAM J. Math. Anal.* **15** (1984), 996–1006.
3. D. Colton and R. Kress, Eigenvalues of the far field operator for the Helmholtz equation in an absorbing medium, *SIAM J. Appl. Math.* **55** (1995), 1724–1735.
4. J. Ringrose, *Compact non-selfadjoint operators*, Van Nostrand Reinhold, London, 1971.
5. D. Colton and R. Kress, Eigenvalues of the far field operator and inverse scattering theory, *SIAM J. Math. Anal.* **26** (1995), 601–615.
6. A. Kirsch and R. Kress, Uniqueness in inverse obstacle scattering, *Inverse Problems* **9** (1993), 285–299.
7. A. Nachman, Reconstructions from boundary measurements, *Ann. of Math.* **128** (1988), 531–576.
8. P. Ola, L. Päivärinta and E. Somersalo, An inverse boundary value problem in electrodynamics, *Duke Math. J.* **70** (1993), 617–653.
9. Z. Sun and G. Uhlmann, Recovery of singularities for formally determined inverse problems, *Comm. Math. Phys.* **153** (1993), 431–445.

10. T.S. Angell, R.E. Kleinman and G.F. Roach, An inverse transmission problem for the Helmholtz equation, *Inverse Problems* **3** (1987), 149–180.

11. D. Colton and A. Kirsch, A simple method for solving inverse scattering problems in the resonance region, *Inverse Problems* (to appear).

12. R. Potthast, A fast new method to solve inverse scattering problems, *Inverse Problems* (to appear).

Department of Mathematical Sciences, University of Delaware, Newark, DE 19716, USA

M. PAUKSHTO

On some applications of integral equations in elasticity

1. Introduction

This paper discusses applications of the Cosserat spectrum theory in thermoelasticity, coupled non-linear theory of diffusion in elastic media and some problems of Helmholtz decomposition.

Cosserat spectrum theory was introduced and extensively studied by S. Mikhlin during the Sixties and Seventies, being inspired by the ideas of E. and F. Cosserat, who published nine papers on this topic between 1898 and 1901. Some results were established jointly by V.G. Maz'ya and S.G. Mikhlin (1967). Further important results belong to A.N. Kozhevnikov (1988).

Let us define the term "Cosserat spectrum". The homogeneous Navier equations

$$\Delta u + \omega \operatorname{grad} \operatorname{div} u = 0, \quad x \in \Omega, \quad \omega = \frac{1}{1-2\nu} \tag{1}$$

with homogeneous boundary conditions of displacement

$$u|_{\partial\Omega} = 0, \quad x \in \partial\Omega, \tag{2}$$

or traction

$$t_i(u) = \frac{1}{\mu}\sigma_{ij}(u)n_j = (\omega - 1)n_i \operatorname{div} u + (u_{i,j} + u_{j,i})n_j = 0, \quad x \in \partial\Omega, \tag{3}$$

admit non-trivial solutions when ω takes values in a set of points (lying, of course, outside the physical range for Poisson's ratio ν) called the Cosserat spectrum.

For a very long time, the physical sense of the eigenvalues ω_k and corresponding eigenvectors u_k have been unknown. Therefore, the field of applications of the Cosserat spectrum was not very wide.

Last year it was found by X. Markenscoff and M. Paukshto [1] that any eigenvector u_k of the above eigenvalue problem may be visualized as the displacement in an equilibrium thermoelastic problem with temperature distribution $T \sim \operatorname{div} u_k$, and can be characterized as a stationary point of the elastic energy under a constraint on the "thermal energy". This principle lies at the foundation of our subsequent discussion.

2. Equivalent formulations

Let Ω be a bounded (or unbounded) domain with a smooth compact boundary $\partial\Omega$. The problem of the Cosserat spectrum may be reduced to that of the spectrum of

This research was supported by the Russian Ministry of Science and Technology through grant 216/67/2-1, and by the RFFI through grant 9601–01334.

the Mikhlin integral equation, as given here through the example of the problem (1), (2). Let

$$G(x,y) = \frac{1}{4\pi}\left(\frac{1}{r} + g(x,y)\right), \quad r = |x-y|,$$

be the Green's function of the Dirichlet problem for the Laplace operator in the domain Ω. Equations (1) and (2) yield the integral equation [2]

$$u(x) - \frac{3\omega}{3+\omega}\int_{\Omega} \operatorname{grad}_y \operatorname{grad}_y G(x,y)u(y)\,dy = 0, \quad x \in \Omega. \tag{4}$$

Clearly, the Cosserat spectrum coincides with the spectrum of this system. Every element of the matrix of the kernel of (4) is a sum of two terms. The first term is the singular kernel $\partial^2 r^{-1}/\partial y_j \partial y_i$. The second term $\partial^2 g/\partial y_j \partial y_i$ is continuous if one of the points x or y lies inside Ω; however, if both points lie on $\partial\Omega$, then, if they coincide, this term becomes infinite, so it is not summable in Ω. The theory of such equations has not been developed enough.

Another integral equation which has the same spectrum as the Cosserat spectrum is the Lichtenshtein integral equation. In the case of the first boundary value problem (1), (2) this has the form

$$\theta(x) + \frac{\omega}{\omega+2}\int_{\partial\Omega} \rho\,\frac{\partial^2 G(x,y)}{\partial\rho\partial n_y}\,\theta(y)\,d_y\Gamma = 0, \quad x \in \Gamma = \partial\Omega, \tag{5}$$

where $\theta = \operatorname{div} u$, n is the external normal to Γ and ρ is the radius vector from the point x to the point y. An analogous equation for the second boundary value problem (1), (3) has been studied by Kozhevnikov [3].

A system of boundary pseudodifferential equations considered by Goldstein [4] may be written in the operator bundle form

$$\Delta P + \frac{\omega-1}{\omega+1}\operatorname{grad}\operatorname{div} P = -2\frac{t}{\mu}, \quad P = (P_1, P_2) = \gamma_G \Lambda^{-1}[u],$$

where t/μ is the traction, Λ^{-1} is a pseudodifferential operator with symbol $|\xi|^{-1}$, γ_G is the restriction to the "surface of a crack" $G \subset \mathbb{R}^2$ and $[u] = ([u_1],[u_2])$ is the displacement jump.

3. The main results of Cosserat spectrum theory

We state in brief some results on the spectrum of the above problems [2]. We introduce the operators

$$N_0 u = (-\Delta u, \gamma_\Gamma u), \quad N_1 u = (\operatorname{grad}\operatorname{div} u,\ \gamma_\Gamma 0),$$
$$N_3 u = (-\operatorname{grad}\operatorname{div} u,\ n\operatorname{div} u|_\Gamma),$$
$$N_2 u = (-\Delta u - \operatorname{grad}\operatorname{div} u, \delta_\Gamma u), \quad (\delta_\Gamma u)_i = (u_{j,i} + u_{i,j})n_j|_\Gamma.$$

Then the spectrum problems (1), (2) and (1), (3) are equivalent, respectively, to the problems

$$u - \omega N_0^{-1} N_1 u = 0, \tag{6}$$

$$u + (\omega - 1) N_2^{-1} N_3 u = 0. \tag{7}$$

The first boundary value problem (6). The corresponding eigenvalues of the Cosserat spectrum are located on the segment $-\infty \leq \omega \leq -1$ and may accumulate only at the point $\omega = -2$. The numbers $\omega = -1$ and $\omega = \infty$ are characteristic values of infinite multiplicity. Let $\{u_n\}$ be the Cosserat eigenfunctions, that is,

$$\Delta u_n + \omega_n \operatorname{grad} \operatorname{div} u_n = 0,$$
$$u_n|_\Gamma = 0, \quad x \in \Omega, \quad \omega_n \neq -1, \infty,$$

and $\{u_n^{(-1)}\}, \{u_n^{(\infty)}\}$ any bases for the subspaces $H_1^{(-1)} \subset H_1$ and $H_1^{(\infty)} \subset H_1$, respectively. Here $H_1 = \mathring{W}_2^1(\Omega)$ is the Sobolev space with the inner product

$$[u, v] = \int_\Omega \nabla u \, \nabla v \, dx$$

and

$$H_1^{(-1)} = \{u \in H_1 : \operatorname{curl} u = 0\},$$
$$H_1^{(\infty)} = \{u \in H_1 : \operatorname{div} u = 0\}.$$

All the systems $\{u_n\}$, $\{u_n^{(-1)}\}$ and $\{u_n^{(\infty)}\}$ are orthogonal in H_1. The system $\{u_n\} \cup \{u_n^{(-1)}\} \cup \{u_n^{(\infty)}\}$ is complete in each of the spaces H_1 and $L_2(\Omega)$. The system $\{\operatorname{div} u_n\}$, augmented by a constant, is complete in the space $G_2(\Omega)$ of functions that are harmonic and square-summable in Ω. The following equalities hold:

$$(\operatorname{div} u_n, \operatorname{div} u_k)_{L_2(\Omega)} = -\frac{1}{\omega_n}\delta_{nk}, \quad (\operatorname{curl} u_n, \operatorname{curl} u_k)_{L_2(\Omega)} = \frac{\omega_n + 1}{\omega_n}\delta_{nk}.$$

This result carries over to the plane problem. In that case the Cosserat spectrum is located in the set $[-\infty, -2] \cup \{-1\}$, as was proved by Sherman (1938).

The second boundary value problem (7). The corresponding eigenvalues of the Cosserat spectrum are located on the segment $-\infty \leq \omega \leq 1/3$ and may accumulate only at the point $\omega = 0$. The numbers $\omega = -1$ and $\omega = \infty$ are characteristic values of infinite multiplicity. Let $\{\tilde{u}_n\}$ be the Cosserat eigenfunctions, that is,

$$\Delta \tilde{u}_n + \tilde{\omega}_n \operatorname{grad} \operatorname{div} \tilde{u}_n = 0, \quad x \in \Omega, \quad \tilde{\omega}_n \neq -1, \infty, \quad \sigma_{ij}(\tilde{u}_n) n_j|_\Gamma = 0,$$

and $\{\tilde{u}_n^{(-1)}\}$, $\{\tilde{u}_n^{(\infty)}\}$ any bases for the subspaces $\tilde{H}_1^{(-1)} \subset \tilde{H}_1$ and $\tilde{H}_1^{(\infty)} \subset \tilde{H}_1$, respectively. Here $\tilde{H}_1$ is the Sobolev space of the functions $u \in W_2^1(\Omega)$ such that

$$\int_\Omega uh\,dx = 0,$$

where h is a rigid displacement. The inner product in $\tilde{H}_1$ is

$$(u,v)_{\tilde{H}_1} = 2\int_\Omega \varepsilon_{ij}(u)\,\varepsilon_{ij}(v)\,dx,$$

and

$$\tilde{H}_1^{(-1)} = \{u \in \tilde{H}_1 : \operatorname{curl}\operatorname{curl} u = 0 \text{ on } \Omega,\ -2n_j \operatorname{div} u + 2\varepsilon_{ij}(u)n_i = 0 \text{ on } \Gamma\},$$
$$\tilde{H}_1^{(\infty)} = \{u \in \tilde{H}_1 : \operatorname{div} u = 0\}.$$

All the systems $\{\tilde{u}_n\}$, $\{\tilde{u}_n^{(-1)}\}$ and $\{\tilde{u}_n^{(\infty)}\}$ are orthogonal in $\tilde{H}_1$. The system $\{\tilde{u}_n\} \cup \{\tilde{u}_n^{(-1)}\} \cup \{\tilde{u}_n^{(\infty)}\}$ is complete in each of the spaces $\tilde{H}_1$ and $\tilde{L}_2(\Omega)$, where

$$\tilde{L}_2(\Omega) = \left\{u \in L_2(\Omega) : \int_\Omega uh\,dx = 0,\ h \text{ is a rigid displacement}\right\}.$$

The following equalities hold:

$$(\operatorname{div}\tilde{u}_n, \operatorname{div}\tilde{u}_k)_{L_2(\Omega)} = \frac{1}{1-\tilde{\omega}_n}\,\delta_{nk}, \qquad \Delta \operatorname{div}\tilde{u}_n = 0.$$

This result also carries over to the plane problem. In that case the Cosserat spectrum is located in the set $\{-\infty, -1, 0\}$.

4. A new variational principle in thermoelasticity

Consider an isotropic elastic solid Ω bounded by a smooth surface Γ and subjected to a harmonic temperature field T, which induces a stress distribution of Cartesian components σ_{ij}. If F is the body force, then the stresses σ_{ij}, strains ε_{ij} and displacements u_i satisfy the stress equilibrium equations, the Duhamel-Neumann law and the strain-displacement relations; that is,

$$\partial_j \sigma_{ij} + F_i = 0,$$
$$\sigma_{ij} = \mu(\omega - 1)\varepsilon_{kk}\delta_{ij} + 2\mu\varepsilon_{ij} - (3\omega - 1)\alpha\mu T\delta_{ij}, \quad \varepsilon_{ij} = (\partial_j u_i + \partial_i u_j)/2,$$

where $\omega = 1/(1-2\nu)$, μ is the shear modulus, ν is the Poisson ratio and α is the coefficient of thermal expansion.

The total strain energy

$$E(u) = \frac{1}{2} \int\limits_{\Omega} \left[\mu(\omega - 1)\varepsilon_{kk}^2 + 2\mu\varepsilon_{ij}\varepsilon_{ij}\right] dx$$

has the form

$$E(u) = \frac{1}{2}(3\omega - 1)\alpha\mu \int\limits_{\Omega} T\varepsilon_{kk}\, dx.$$

We define a bounded self-adjoint operator in $L_2(\Omega)$ by the formula $B\,T = \varepsilon_{kk}$. Then

$$E(u) = \frac{1}{2}(3\omega - 1)\mu\alpha \int\limits_{\Omega} TBT\, dx.$$

Thus, we see that the total strain energy is a quadratic form in the temperature T. Under the circumstances, it is natural to ask the following question:

For a specific value of the total "thermal energy"

$$\Theta(T) = \int\limits_{\Omega} T^2\, dx,$$

say equal to 1, what is the temperature distribution that yields the maximum strain energy or the minimum strain energy, or, in general, renders the strain energy stationary:

$$\inf E(u) = ? \qquad \Theta(T) = 1.$$

This question is equivalent [1] to making the functional

$$\int\limits_{\Omega} TBT\, dx \tag{8}$$

stationary under the constraint that $\Theta(T)$ is constant. The associated eigenvalue problem is $BT = kT$. It takes the form

$$\mu\left\{\Delta u + \left(\omega - \frac{1}{k}(3\omega - 1)\alpha\right) \operatorname{grad} \operatorname{div} u\right\} = 0,$$

with appropriate boundary conditions.

It follows from Cosserat spectrum theory that the spectrum of the operator B consists of a sequence $\{k_n\}$ of eigenvalues computed by means of the formula

$$\omega - \frac{1}{k_n}(3\omega - 1)\alpha = \omega_n,$$

so that

$$k_n = \frac{(3\omega - 1)\alpha}{\omega - \omega_n},$$

where ω_n is any of the eigenvalues of the Cosserat problem. The corresponding eigenvectors are

$$T_n = \frac{1}{k_n} \operatorname{div} u_n, \tag{9}$$

where u_n is the Cosserat eigenvector associated with the eigenvalue ω_n. There are some important engineering problems (see [1]) where the corresponding functional (8) is stationary under the constraint that $\Theta(T)$ is constant.

5. The diffusion problem in elastic media

Consider the coupled equations of the equilibrium elastodiffusion problem

$$\Delta C - k \operatorname{div} (C \operatorname{grad}(\operatorname{div} u - 3\alpha C)) + G = 0, \tag{10}$$

$$\Delta u + \omega \operatorname{grad} \operatorname{div} u = (3\omega - 1)\alpha \operatorname{grad} C, \tag{11}$$

$$C|_\Gamma = C_0, \quad u|_\Gamma = u_0 \quad \text{or} \quad \sigma_{ij}(u)n_j|_\Gamma = t_0, \tag{12}$$

where C is the admixture concentration, α is the expansion coefficient, u is the elastic displacement vector and G is the speed of admixture generation by the external sources.

This model is used widely in the theory of semi-conductors and thin films [5], in the interaction of bodies with powerful laser radiation [6] and in the valuation of strength of power-generating equipment [7]. The Cosserat spectrum theory provides a new method to obtain the analytic solutions of a number of important problems of this type. We state briefly the main idea of this approach.

First, the pair of functions (cf. (9))

$$C = \frac{1}{k_n} \operatorname{div} u_n, \quad u = u_n,$$

where

$$k_n = \frac{\alpha(3\omega - 1)}{\omega - \omega_n}, \quad \omega_n \neq -1,$$

satisfies the system (10), (11) with

$$G = -k\alpha \frac{1 - 3\omega_n}{\omega - \omega_n} |\operatorname{grad} C|^2.$$

In particular, for $G = 0$ the system (10), (11) admits solutions of the form

$$C = (a, x), \quad u = (u_j), \quad a = (a_j),$$

$$u_j = \alpha a_j(x_j^2 - x_{j+1}^2 - x_{j+2}^2)/2 + \alpha a_{j+1} x_j x_{j+1} + \alpha a_{j+2} x_j x_{j+2},$$

$$a_4 = a_1, \quad a_5 = a_2, \quad x_4 = x_1, \quad x_5 = x_2, \quad a_j = \text{const}.$$

This is a consequence of the fact that $\omega = 1/3$ is a point of the Cosserat spectrum of the second boundary value problem (7) for an arbitrary bounded domain Ω [8].

Next, we transform equation (10) to the form

$$-\frac{p_2}{2k}\Delta\left(1-\frac{k}{p_2}C\right)^2+\frac{k}{p_1}\nabla C\,\nabla(C-p_1\operatorname{div}u)+G=0,$$

where

$$\frac{1}{p_1}=\frac{\alpha(3\omega-1)}{\omega+1},\qquad \frac{1}{p_2}=-\frac{4\alpha}{\omega+1}.$$

We now consider the class of diffusion-induced displacements satisfying the condition

$$\nabla C = p_1\nabla\operatorname{div}u. \tag{13}$$

Under this condition, equation (10) is reduced to a linear one, and together with the first boundary condition in (12) permits us to determine the concentration C. Thus, the influence of the elastic medium on the diffusion process has an effect only on the dependence of the concentration C on the constant k/p_2.

If condition (13) is valid, then the displacement vector is defined from the equations

$$\operatorname{curl}\operatorname{curl}u=0,\qquad \operatorname{div}u=\frac{1}{p_1}C+\text{const}$$

and the second boundary condition in (12).

For example, we can find the spherically symmetric solution of (10), (11) for $G = 0$:

$$C(r)=\frac{p_2}{k}\left(1\pm\sqrt{A+\frac{B}{r}}\right),$$

$$u(r)=\frac{1}{p_1r^2}\int_{r_0}^{r}\tau^2C(\tau)\,d\tau+C_1r+C_2\frac{1}{r^2},\qquad r=|x|,\quad x\in\mathbb{R}^3,$$

where A, B, C_1 and C_2 are constants determined from the boundary conditions (12) and the sign in this formula depends on the sign of αk.

6. The Helmholtz decomposition problem

The splitting of a vector field into the sum of a gradient and the curl of a solenoidal vector field is often called the Helmholtz decomposition. In paper [9] this decomposition is considered when the domain Ω is a bounded subset of $\mathbb{R}^3$:

$$v(x)=\nabla\varphi(x)+\operatorname{curl}A(x),\qquad x\in\Omega,$$

where

$$\varphi(x)=-\operatorname{div}B(x),\qquad A(x)=\operatorname{curl}B(x)$$

and the vector $B = (B_1, B_2, B_3)$ is the solution of the problem

$$-\Delta B = v \text{ in } \Omega, \quad B|_{\partial\Omega} = 0 \text{ on } \partial\Omega, \quad i = 1, 2, 3. \tag{14}$$

Linear maps P_1, P_2 from $H_0 = L_2(\Omega; \mathbb{R}^3)$ into itself are then defined by

$$P_1 v = \nabla\varphi, \quad P_2 v = \operatorname{curl} A,$$

which raises the question: would it be of interest to know whether P_1 and P_2 are orthogonal projections on H_0 for general domains Ω? This question was still open in [10].

Here is a counterexample to an affirmative answer to this question. Let u_n be a Cosserat eigenvector of the first boundary value problem, that is,

$$(1 + \omega_n) \operatorname{grad} \operatorname{div} u_n - \operatorname{curl} \operatorname{curl} u_n = 0, \quad u_n|_{\partial\Omega} = 0;$$

alternatively, using the above notation, for $v_n = \omega_n \operatorname{grad} \operatorname{div} u_n$, $B_n = u_n$ we have

$$P_1 v_n = -\operatorname{grad} \operatorname{div} u_n = -\frac{1}{\omega_n} v_n,$$

$$P_2 v = \operatorname{curl} \operatorname{curl} u_n = (1 + \omega_n) \operatorname{grad} \operatorname{div} u_n = \frac{1 + \omega_n}{\omega_n} v_n.$$

Then the inner product of $P_1 v_n$ and $P_2 v_n$ is

$$(P_1 v_n, P_2 v_n) = -\frac{1 + \omega_n}{\omega_n^2} (v_n, v_n) \neq 0 \quad \text{if} \quad \omega_n \neq -1, \infty.$$

Therefore, P_1 and P_2 are neither idempotent nor orthogonal.

References

1. X. Markenscoff and M.V. Paukshto, The Cosserat spectrum in the theory of elasticity and applications, *Proc. R. Soc. Lond. A* (submitted for publication).
2. S.G. Mikhlin, N.F. Morozov and M.V. Paukshto, *The integral equations of the theory of elasticity*, Teubner, Stuttgart-Leipzig, 1995.
3. A.N. Kozhevnikov, On the second and third elasticity problems, *Dokl. Akad. Nauk SSSR* **302** (1988), 1308–1312.
4. R.V. Goldstein, On the three-dimensional elasticity problem for bodies with plane cracks of arbitrary discontinuities, *Izv. Russ. Acad. Sci. MTT* **3** (1979), 111–126 (Russian).
5. I.V. Belova and G.E. Murch, Thermal and diffusion-induced stresses in crystalline solids, *J. Appl. Phys.* **77** (1995), 127–134.

6. V.I. Emel'anov, Yu.G. Shlikov, Non-linear multi-mode theory of surface defect structures generating under the action of powerful laser radiation, *Izv. Russ. Acad. Sci. Phys.* **57** (1993), no. 12 18–38 (Russian).
7. I.J. Change, K.I. Kagawa, J.R. Rice and L.B. Sills, Non-equilibrium models for diffusive cavitation of grain interfaces, *Acta Met.* **27** (1979), 265–284.
8. S.G. Mikhlin, On the Cosserat functions, in *Problems of mathematical analysis*, Leningrad State University, 1966, 59–69 (Russian).
9. G. Auchmuty, Orthogonal decompositions and bases for three-dimensional vector fields, *Numer. Funct. Anal. Optim.* **15** (1994), 455–488.
10. G. Auchmuty, Potential representations of incompressible vector fields, Research Report UH/MD-210, University of Houston, 1996, 1–7.

Institute for Problems of Mechanical Engineering, Laboratory of Mathematical Methods for Mechanics and Materials, Russian Academy of Sciences, Bol'shoy 61, V.O., 199178 St. Petersburg, Russia

Contributed papers

R. AHASAN, S. VÄYRYNEN and H. VIROKANNAS

A mathematical model for the estimation of heat stress and sweat loss

1. Introduction

Heat stress contributes to metabolic heat production and sweating that may lead to discomfort, heat exhaustion, heat illness and loss of workers' performance. Sweating is closely related to heat stress because core (T_{core}) and skin (T_{sk}) temperatures are the main inputs to the thermoregulatory process in which sweat increases with body heat production (see [1] and [2]). Thus, a quantitative composite measurement of sweat loss has been developed by integrating climatic, physical and personal factors (Fig. 1) based on heat balance equations (see [3] and [4]). It has also been assumed that sweat production should correlate with the E_{req} and E_{max} in order to keep the body heat balanced [5]. Based on this, a comprehensive model, for example, $M_{\mathrm{SL}} = 27.9E_{\mathrm{req}}(E_{\mathrm{max}})^{-0.455}$ ($\mathrm{g/m^2h}$), has been used as a function of climatic condition, metabolic rate and work clothing [6]. This equation is derived from experimental exposures under controlled laboratory conditions, and is valid within a certain range ($E_{\mathrm{req}} = 50\ldots360$ $\mathrm{W/m^2}$ and $E_{\mathrm{max}} = 20\ldots525$ $\mathrm{W/m^2}$). In order to be applicable in an outdoor environment, this mathematical equation has been adjusted to show a high correlation between the measured and expected values. Using the new model as a predictive heat balance equation, this study followed the estimation of sweat loss, which also indicates heat stress. Thus, the aim of this paper is to identify heat stress and verify it by adjusting the measured and predictive values of sweat loss in hot working conditions (HWC). The results reveal that work tasks in HWC seem to be stressful in tropical developing countries because of excessive heat load. In this connection, occupational safety and health (OSH) measures are suggested which are yet to be implemented in many tropical developing countries.

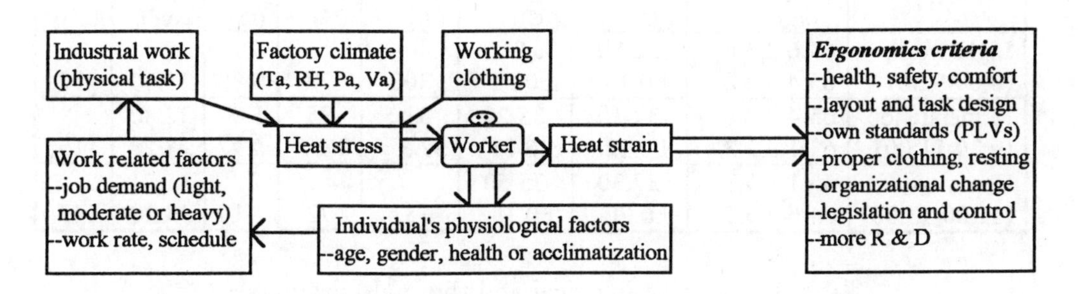

Fig. 1. Factors affecting the heat load on the workers in HWC.

2. Materials and procedure

Sixteen adult men (age 31, SD = ±2.3 years) participated in the experiment, who were involved with metal-handling tasks in Dhaka, Bangladesh. The subjects were in good health and acceptable physical condition, and considered to be acclimatised to strenuous work. The physiological parameters were measured at various work intensities under the existing climatic conditions (Table 1). These parameters were the metabolic rate ($M = 112$, 250 and 300 watts), the clothing coefficient ($I_T = 0.99$) and the body surface area ($A_D = 1.73$, SD $= \pm 0.2$ m^2) (see [7]–[9]). The subjects were advised to wear similar clothing (cotton shirt and khaki full pants), and they did not take any food and drinks, nor urinate during the tests. All clothed subjects were weighed on an electronic precision scale. Nude weight (in kg) was calculated by subtracting their clothing weight. Each exposure lasted for 1.5–2 hours, and the data were recorded every 30 minutes. The measured sweat losses (M_{SL}, g/m^{2}h) were calculated from the weight differences. The T_{oral} °C was recorded by a digital thermometer and considered equivalent to T_{core} °C [10]. The T_{sk} °C was measured from the chest, armpit and thigh by a skin thermistor. The air velocity (V_a, m/s), relative humidity (RH %) and ambient air temperature (T_a °C) were also measured at approximately 2 metres from ground level by a thermo-anemometer (Anlor, GGA-65) and a hygrometer (Humicor, 86-22492). The water vapour pressure (P_a, mmHg) and solar radiation ($SR_{\text{sun}} = 900$; SD $= \pm 13$, and $SR_{\text{shade}} = 130$; SD $= \pm 3$W/m^2) were obtained from the local meteorological department.

3. Calculations

In order to keep the body heat balanced (see [4] and [11]), it is necessary to have

$$E_{\text{req}} = (M - W_e) + H_c + H_r + H_L, \tag{1}$$

where M is the metabolic rate, W_e the external work, $H_c = 6.45 A_D (T_a - T_{\text{sk}})/I_T$ the convective heat transfer, $H_r = 1.5 A_D SR^{0.6}/I_T$ the radiative heat transfer, and $H_L = 0.047 A_D M_{\text{e.th}}/I_T$ (where $M_{\text{e.th}} = 5.67 \times 10^{-8}$ W/m^2K^4 is the Stephan-Boltzmann constant) is the long wave heat emission from the body into the atmosphere.

Work phase	Weight (kg)	T_{oral} (°C)	T_{sk} (°C)	T_a (°C)	RH (%)	V_a (m/s)	P_a mmHg	MS (g/h)
Light work (8:00-9:30)	65.6 ±3.6 65.4 ±2.8	36.80 ±0.15	34.50 ±0.20	30	77	0.80	33.60	77
Moderate job (9:30-11:00)	65.4 ±2.8 65.1 ±3.3	37.10 ±0.12	35.20 ±0.15	31 ***31.3***	75 ***74.7***	1.00 ***0.97***	32.80 ***33.2***	116
Heavy work (11:00-13:00)	65.1 ±3.3 64.6 ±5.2	37.30 ±0.20	35.50 ±0.18	33	72	1.10	33.20	147

Table 1. The physiological and thermal parameters.

Taking $W_e = 0$, neglecting respiratory and metabolic weight loss and using the required values from Table 1 led to an equality (1) of the form

$$E_{\text{req}} = M + [6.45 A_D (T_a - T_{\text{sk}})/I_T] + [1.5 A_D SR^{0.6}/I_T] + [0.047 A_D M_{\text{e.th}}/I_T]. \tag{2}$$

However, the original equation was based on $E_{\max}$ under the assumption that outdoor conditions can cause a discrepancy, exhibited by the correlation between the measured and predicted values of T_{rectal}. Consequently, the predictive equation was adjusted through the separate evaluation of the radiative heat exchange (RHE) for short wave ($H_s = 1.5SR^{0.6}/I_T$) and long wave (H_L), and their insertion in E_{req}. Again, considering radiative and convective heat transfer (H_{r+c}) separately, $E_{\max}$ was found from (2) for different values of $T_{\text{oral}} \approx T_{\text{rectal}}$, namely

$$T_{\text{rectal}} = 36.75 + 0.004(M - W_e) + 0.0011(H_{r+c}) + 0.8e^{0.0047(E_{\text{req}} - E_{\max})}, \quad (3)$$

$$T_{\text{rectal}} = 36.75 + 0.004(M - W_e) + 0.0011H_c + 0.0025H_r + 0.8e^{0.0047(E_{\text{req}} - E_{\max})}. \quad (4)$$

4. Results

The M_{SL} are calculated from (3) and (4) by considering $T_{\text{rectal}} = T_{\text{oral}}$ for different work rates at 36.8°C, 37.1°C and 37.3°C. The results for thermal sensation taken from subjective ratings (questionnaire) are summarised in Table 2.

Body parts	Comfortable (%)	Uncomfortable (%)	Body parts	Comfortable (%)	Uncomfortable (%)
Whole body	43.8	56.2	Hand/arms	62.5	37.5
Face	37.5	62.5	Thigh/legs	68.8	31.2
Chest	50.0	50.0	Toe/feet	43.8	57.2
Back	56.3	43.7	Clothing	37.5	62.5

Table 2. The workers' thermal sensation ($N = 16$).

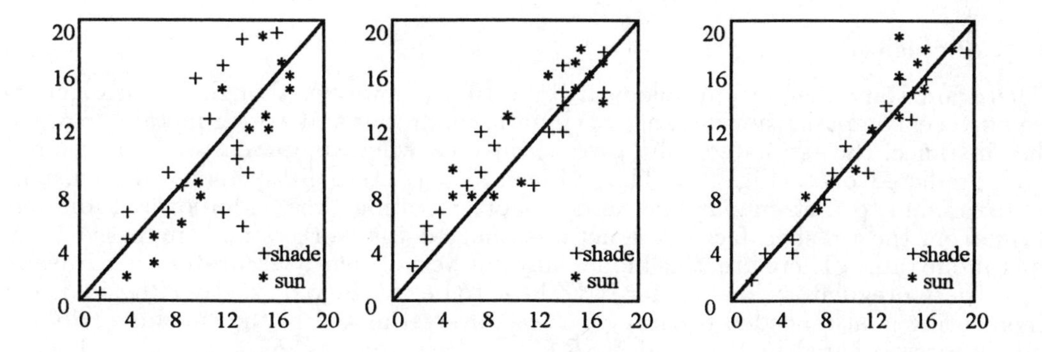

Fig. 2. The correlation between measured and predicted sweat loss. The horizontal axis denotes the measured sweat rate and the vertical axis indicates the predicted sweat rate.

The measured sweat losses were compared with the predicted rates. The correlation between them was statistically significant ($p < 0.0001$), but the sum square error was still high (Fig. 2). It was assumed that a component which expresses a fraction of heat dissipation from the body to the surrounding environment was missing [6], and this factor was supposed to be connected with H_L in (1) and (2), giving a linear correlation ($r = 0.97$, $p < 0.001$) between the measured and predicted values. However, the predicted values were overestimated compared to the actual and measured values of radiative heat exchange (RHE) in (3) and (4). This adjustment for H_r also revealed a higher correlation between the measured and predicted values ($r = 0.99$, $p < 0.0001$). Here the sum square error was supposed to be low, but it was assumed that the value of the heat from radiation and convection should be modified.

5. OSH measures

The workers' heat adaptation capacity depends on their acclimatisation level and physical fitness, and thus a reduction in their physical load through adequate work rests is important. For instance, group work and better job rotation or a change of work tasks are beneficial in HWC. Participatory training, which can improve work techniques, and self-pacing are also effective. The work place design plays an important role in that it includes a better arrangement of the tasks, schedule and changes in the layout. In excessive heat, short exposure times are recommended for those with comparatively lower metabolic rates. Simple and practical measures (natural and cross ventilation, heat barriers) should be implemented. Lightweight, washable, fire-resistant and vapour-permeable fabrics are recommended for general use, but aluminium clothing is more efficient against radiant heat (see [15] and [16]). To provide thermal comfort, T_{sk} has to be within permissible limit values (PLVs) and evaporative heat loss must be lowered. The NIOSH [17] and ISO's [18] recommendations are helpful in reducing environmental heat load. Still, certain important standards, such as the index of heat stress and local legislation, have yet to be implemented in many tropical developing countries. Owing to their poverty and socio-cultural traditions, extensive changes may not be possible, but specific and simple modifications can be highly beneficial.

6. Conclusion

There are some visible physiological limitations, however, though the predicting equation encompasses a wide range of climatic conditions and worker metabolic rates. For instance, the estimated values were scattered when compared to the measured and predicted ones (Fig. 2). The subjects' T_{oral} was a little higher than usual, perhaps due to a strenuous task and a shorter resting time. Moreover, radiation heat from the tin-shed factory structures affected the workers and increased their metabolic rates. Therefore, this model may not be sufficiently accurate, which means that thermoregulatory barriers for RHE have to be considered exactly. Physiological monitoring is also needed because M_{SL} was overestimated in the sunshine (SR_{sun}) and underestimated in the shade (SR_{shade}). However, the E_{req} and E_{max} lied in the predictive equation, which was a great advantage. In addition, possible errors might have arisen since few parameters were collected from the local meteorological department, 5 km away from the work sites; moreover, I_T was taken from ISO's table and A_D from DuBois and DuBois [9]. The RH inside the factory was lower than outside, perhaps because metal scraps were melted inside the factory, which reduced the humidity content. During this study, the weather was sunny but mild. Still, heat stress remains one of the most frequent complaints (as also learned from the questionnaire). The authors believe that the occurrence of heat disorders and the frequency of occurrence based on the effects of heat is not easy to assess correctly because of compounding factors, for example, age, gender, acclimatisation level, nutrition, and tropical diseases. Based on these considerations, more research studies are recommended in the assessment strategy for the real development of HWC.

References

1. A.P. Gagge and Y.N. Nishi, *Heat exchange between human skin and thermal environment: handbook of physiology, reactions to environmental agents*, American Physiological Society, 1977, 69–91.
2. R.W. Bullard, M.R. Banerjee and B.A. MacIntyre, The role of skin in negative feedback regulation of eccrine sweating, *Internat. J. Biometeorology* **11** (1967), 93–104.
3. ISO-7933, *Hot environments—analytical determination and interpretation of thermal stress using calculation of required sweat rate*, Internat. Standard Org., 1989.
4. B. Givoni and R.F. Goldman, Predicting rectal temperature response to work, environment and clothing, *J. Appl. Physiology* **32** (1972), 812–822.
5. K. Lustinec, *Sweat rate, its prediction and interpretation*, Archives of Science Physiology, 1973, 127–136.
6. Y. Shapiro, K.B. Pandolf and R.F. Goldman, Predicting sweat loss response to exercise, environment and clothing, *European J. Appl. Physiology* **48** (1982), 83–96.
7. M.R. Ahasan, S. Väyrynen and H. Virokannas, Thermo-regulatory response and muscle fatigue for steel mill workers, *Ergonomics* (in press).
8. ISO-9920, *Estimation of thermal characteristics of a clothing ensemble*, Internat. Org. Standardisation, Geneva, 1993.
9. D. DuBois and E.F. DuBois, Clinical calorimetry X: a formula to estimate approximate surface area if height and weight are known, *Arch. Internal Medicine* **17** (1961), 863–871.
10. P. Mairiaux, J.C. Sagot and V. Candas, Oral temperature as an index of core temperature during heat transients, *European J. Appl. Physiology* **50** (1983), 331–341.
11. A.P. Gagge, A new phyiological variable associated with sensible and insensible perspiration, *Amer. J. Physiology* **120** (1973), 277–287.
12. M. Millican, R.C. Baker and G.T. Conk, Controlling heat stress-administrative vs physical control, *Amer. Industrial Hygiene Assoc. J.* **42** (1981), 411–416.
13. ILO Geneva, *Proc. on the improvement of working conditions and productivity in small and medium sized enterprises: intensive training course, India, 1985.*
14. P.O. Fanger, *Thermal comfort*, McGraw-Hill, New York, 1970.

15. M.R. Ahasan, S. Väyrynen and H. Virokannas, Low cost protective clothing for a tropical climate, in *Proc. of the Virtual Conf. CybErg96, Curtin, 1996*, http://www.curtin.edu.au/conference/cyberg.

16. G.W. Crockford, R.N. Sen and J. Spencer, The design of work-wear and protective clothing for use in tropics, *Indian J. Physiology Allied Sci.* **38** (1984), 125–133.

17. NIOSH, *Criterion for recommended standards on occupational exposure to a hot environment*, US National Institute for Occupational Safety and Health, DHHS, 1986, 86–113.

18. ISO–7730, *Moderate thermal environment: determination of PMV and PPD indices, specification of conditions of thermal comfort*, Internat. Standard Org., 1994.

R.A., S.V.: Division of Work Science, University of Oulu, 90570 Oulu, Finland

H.V.: Department of Public Health Sciences and General Practice, University of Oulu, 90570 Oulu, Finland

R. AHASAN, S. VÄYRYNEN and H. VIROKANNAS

Prediction of permissible workload using fuzzy mathematics

1. Introduction

Since work-stress has a strong correlation to human perception, subjective ratings are supposed to be validated in workload calculations (see [1] and [2]). The job-demand (that is, stressors) generally arises from physical and mental responses which have a significant influence on subjective ratings, and thus by measuring human attitudes, the existence and level of workload can be predicted using mathematical structures (that is, fuzzy concepts). In this case, the use of linguistic variables (for example, "heavy", "uncomfortable", "permissible" or "not permissible") in which fuzzy methods can be defined, is likely to be useful to evaluate physical stress. Here, human perception of workload due to fuzziness rather than randomness is an important criterion to be considered. Fuzzy set theory was actually introduced in [3] for the representation of uncertainty in expression, and allows information to be collected from ill-known data. Therefore, the concept of fuzziness has been used in ergonomic task analysis as it deals with human-based uncertainty [4].

To quantify subjective categories of load heaviness, the level of perceived stresses as linguistic values needs to be evaluated for a certain limit of workloads that are physically appreciable for a certain population. In this aspect, there are several ways that fuzzy methods can be applied; for example, the use of linguistic variables from strain parameters by suitable subjective ratings, and fuzzy regression analysis from the MFs (membership functions) in a predictive equation (Fig. 1). Here, MFs are a quantitative form of a subjective degree of belief that can give expression to a fuzzy set [5]. Thus, in this paper, the potential application of fuzzy concepts is illustrated through the investigation of "permissible" physical workload, from a case study in manual lifting tasks.

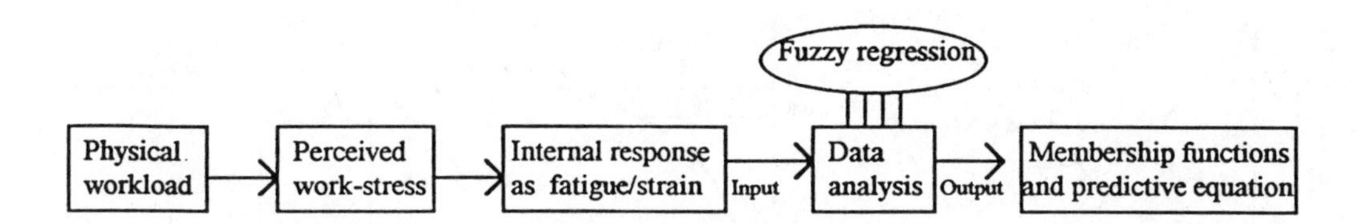

Fig. 1. The concept of fuzzy regression analysis in workload calculations.

2. Workload and fuzziness

To quantify subjective categories of workload heaviness from physical tasks, levels of perceived stresses as linguistic values are needed. Physical workload generally arises from musculoskeletal systems, for example, lifting, carrying, pushing, pulling, loading or unloading tasks. Moreover, the handling frequency, duration of activity and moving distances are also directly proportional to the weight of load and pace of tasks. To define these stress parameters by means of linguistic values and MFs, we consider the load weight S_1, the load handling frequency S_2, the task duration S_3 and the moving distance S_4.

Now to define a fuzzy subset A, a membership value between 0 and 1 must be assigned to each element, which represents its grade of membership (GM) in A. In this case, a reference value should be defined in order to determine whether suitable ratings can be assigned to an individual's perception. In lifting tasks, the reference value indicates the weight of the load that a person most likely can handle safely; however, the characteristic function can allow various values of GM from the subjective ratings to the elements of a given fuzzy set [6]. We consider S_1, S_2, S_3, S_4 defined on a five-point subjective scale: $S_1 \rightarrow$ "very light", "light", "medium", "heavy" and "very heavy"; $S_2 \rightarrow$ "very low", "low", "medium", "high" and "very high"; $S_3 \rightarrow$ "very short", "short", "medium", "long" and "very long"; and $S_4 \rightarrow$ "very close", "close", "medium", "far" and "too far". We consider the MFs that express fuzzy sets for $S_1 = (x_1, x_2, x_3, \ldots, x_n)$, where $x_1, x_2, x_3, \ldots, x_n$ are base variables. The perceived workload for S_1 by a fuzzy subset can now be expressed, for example, as follows:

$$
\begin{aligned}
A &= \sum \frac{U_A}{x_i} \\
&= [0/1, 0.2/7, 0.4/10, 0.6/20, 0.7/25, 1/35, 0.8/38, 0.7/40, 0.2/42, 0.1/45, 0/50]. \quad (1)
\end{aligned}
$$

Equation (1) expresses the grade of MFs (GMFs) of stresses from lifting different weights of loads. Here, U_A denotes the universe of discourse or base variables (that is, GMFs), and x_i denotes the load weight. The stress level is higher when the load weight makes the worker exceed his usual lifting capacity. For example, up to a certain vertical distance, the perceived workload for a certain worker is 0.7 (which indicates "medium") to lift 25 kg in a specific circumstance (Fig. 2a), and may be "1" for 35 kg (which indicates "very heavy"). Beyond this, the lifting capacity expresses a negative attitude. It means that at 0.7 the lifting load is permissible; after that limit, it will be heavy or not permissible.

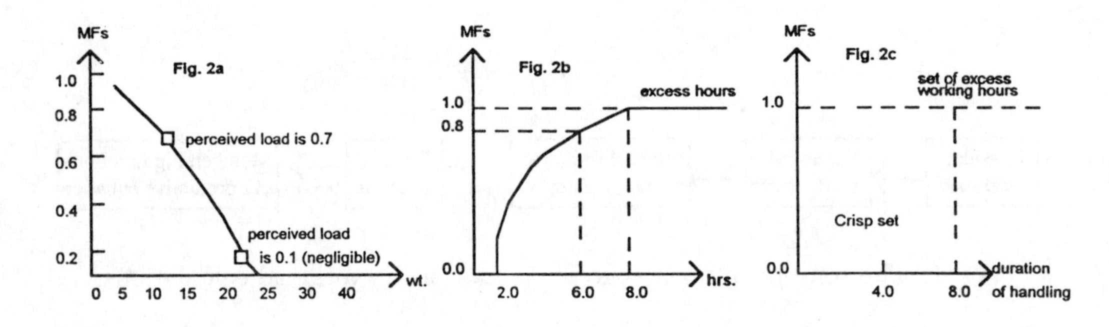

Fig. 2. Representation of MFs in lifting load, duration of work and frequency of handling.

Similarly, S_2, S_3 and S_4 can be examined for permissible limit values from the subjective scales. Equation (1) can also be expressed in terms of the GMFs from other stressors, such as duration of work (Fig. 2b) and frequency of work tasks (Fig. 2c). The vertical axis defines the quantification of the degree of ambiguity, that is, $0 \leq U \leq 1$, where U is quantified over a specific range of intervals (0 to 1) that represent stress levels for each linguistic variable, and then the MFs can be calculated as an expression of a fuzzy subset (1). The intervals represent the GMs for each linguistic variable in the set.

Environmental stressors can also be assessed within the range from 0 to 1. Here the grade is binary, either 0 or 1, and the crisp set (Fig. 2c) has only two grades. The MFs with the corresponding support set will be "1" for uncomfortable and "0" for a comfortable feeling. The same method could be applied to postural workload. If the subject has any postural discomfort from frequent standing, bending or twisting, then the MFs with the corresponding support set are "1" and "0" when the answer is "no".

3. A case study for metal handling workers

Twenty six healthy male subjects (age 24–41 years) who had no spinal problems (known from interview) were asked to lift metal bars with weights ranging from 2 kg to 20 kg, with increments of 5 kg, from the ground to shoulder level. The order of the (identical) metal bars for each subject was randomised. The subjects were also advised to do pre-lifting practice in their spare time, bearing in mind that their tasks should continue for an 8 hour work-day. They lifted the bars repeatedly with a frequency of two lifts per minute, until they could make a decision about the heaviness of the weight of the loads. Since the universe of discourse (base or subjective variables) was divided over the region (0, 1), the stress level could be the support set corresponding to the highest GMFs. The corresponding GMFs were then found by means of the fuzzy subset A, where X is precisely defined by the association with each factor x. According to [7], let $X = f\{x\}$ be a collection of contributing factors x (numbers between zero and one) which represent the GM in A. The psycho-physical technique was applied to quantify the subjective scale of load heaviness. The subjects were also asked to judge the degree of compatibility of lifting with linguistic categories of load acceptability as a maximum acceptable load (MAL) and a maximum comfortable load (MCL).

4. Results

The subjective heaviness in lifting tasks is represented in Fig. 3. A statistical analysis [8] of the differences between compatibility values of load selection for MAL and MCL showed significant effects ($p < 0.05$) for a load of 15 kg lifted at shoulder level.

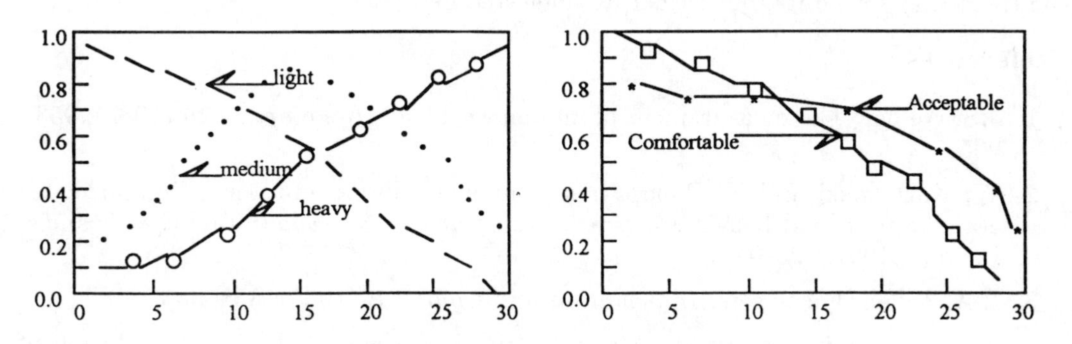

Fig. 3. Linguistic values of load heaviness and fuzzy compatibility functions for MAL and MCL.

The results indicate which subjects should not lift more than 15 kg, which exceeds the MCL, in this case.

5. Conclusion

Physical workload is an important factor that determines an individual's ability to accomplish tasks. Due to variations in individuals' performance and subjective ratings, a considerable problem can occur in the workload calculation with reasonable accuracy. For instance, perception of physical workload should be considered not only at the beginning, but also at the end of work. The workload can also be increased in the course of the work-shifts. A weight assessed as "very heavy" may mean either that the subject could just handle it, or that he could not handle it at all. In this case, the "very heavy" estimation should consider MAL for a single lift. Therefore, a revision of psycho-physical approaches are needed so that the subjects should select MCL rather than MAL in order to prevent over-exertion injury. Moreover, all measuring techniques are sensitive to different problems, and that is why the results obtained may not be sufficiently accurate.

For example, a certain weight cannot be assigned with absolute certainty to one class or another, and likewise, a weight belongs to several or even all classes with different MFs. Also, human decision can create differences in the MFs' rule (see [9] and [10]). By contrast, one's certain subjective heaviness or "set" has no absolute and certain boundary. This means that the change from the membership to non-membership function develops with the weight of load gradually, not abruptly (see [11] and [12]). However, a fuzzy concept can often objectively identify points of potentially high workload and diagnose their causes. It may augment conventional statistical techniques in the analysis of human data and offer useful supplements for reliability analysis and regression. Moreover, it gives an idea for comparing human concepts to some degree, and indicates the power of each objective with respect to the next higher level of decision (see [13] and [14]). As a research language, fuzzy logic will perhaps not be accurate in behavioural sciences like human factor engineering or ergonomics, because of vagueness in human perception. However, there it is still possible to assess the exact level of stress (that is, MCL and/or MAL as permissible workload) using fuzzy techniques (see [14] and [15]). In this regard, an understanding of the meaning of workload heaviness is to be developed by effective strategies. To develop a mathematical model that is sufficiently accurate for the prediction of permissible workloads, empirical assessment must also be eliminated. Moreover, the prediction of permissible workloads using fuzzy concepts should not be treated as set limits, but rather as suggested guidance.

References

1. J.S. Weiner, The measurement of human workload, *Ergonomics* **25** (1982), 953–965.

2. W. Karwowski and N. Pongpatana, Linguistic interpretation in human categorisation of load heaviness, *Proc. Designing for Everyone*, Taylor & Francis, London, 1991, 425–427.

3. L.A. Zadeh, Fuzzy sets, *Information and Control* **8** (1965), 338–353.

4. E.S. Jung and A. Freivalds, Multiple criteria decision making for resolution of conflicting ergonomic knowledge in manual materials handling, *Ergonomics* **34** (1991), 1351–1356.

5. I.B. Turksen, Measurement of membership functions, *Proc. Appl. Fuzzy Set Theory in Human Factors*, Elsevier, 1986, 55–67.
6. H. Luczak and S. Gee, Fuzzy modelling of relations between physical weight and perceived heaviness: effects of size-weight, illusion in industrial lifting, *Ergonomics* **32** (1989), 823–837.
7. W. Karwowski and A. Mital, Development of safety index for manual lifting tasks, *Applied Ergonomics* **17** (1986), 58–64.
8. W. Karwowski and N. Pongpatana, Fuzzy modelling of load heaviness: perspective of acceptability and comfortability criteria in manual lifting, *Proc. Ergonomics in Materials Handlings and Information Processing Work*, Poland, 1993, 103–107.
9. R.R. Yager, Multiple objective decision-making using fuzzy sets, *Int. J. Man-Machine Studies* **9** (1977), 375–382.
10. S.H. Snook, Psychophysical considerations in the permissible load, *Ergonomics* **28** (1985), 327–330.
11. T. Terano, K. Asai and M. Sugeno, *Fuzzy systems, theory and applications*, Academic Press, New York, 1992.
12. B.C. Jiang and J.L. Smith, Fuzzy modelling of combined manual materials handling capacity, *Proc. Appl. Fuzzy Sets in Human Factors*, Elsevier, 1986, 349–360.
13. N. Moray, B. King, B. Turksen and K. Waterton, A closed-loop causal model of workload based on comparison of fuzzy and crisp measurement techniques, *Human Factors* **29** (1987), 339–348.
14. J.G. Chen, H.S. Jung and B.J. Peacock, A fuzzy sets modelling approach for ergonomic workload stress analysis, *Int. J. Industrial Ergonomics* **13** (1994), 189–216.
15. S.S. Asfour, T.M. Khalil, A.M. Genaidy and S.M. Waly, Evaluation of lifting capacity determination methodologies: a fuzzy set approach, *Proc. Appl. Fuzzy Sets in Human Factors*, Elsevier, 1986, 339–347.

R.A., S.V.: Division of Work Science, University of Oulu, 90570 Oulu, Finland

H.V.: Department of Public Health Sciences and General Practice, University of Oulu, 90570 Oulu, Finland

Yu.A. BALOSHIN and A.V. KOSTIN

An approach to the exterior Dirichlet problem in $\mathbb{R}^2$ with applications to wave scattering

The study of phenomena arising from the scattering of waves from rough surfaces is of interest in many fields of physics and engineering. For instance, one of the most promising application in optics seems to be the employment of a laser beam scattering for non-destructive inspection of topography and optical quality of surfaces.

In dealing with scattering from rough surfaces, approximate methods, such as perturbation techniques [1] or the Kirchhoff approximation [2], have gained ground. Unfortunately, the validity of these methods proves insufficient to handle interaction with resonance roughnesses, as well as grazing incidence geometries. For this reason a large amount of research, reported in the last 15 years, has been devoted to rigorous methods based, in particular, on boundary integral equations. (A review may be found in [3].)

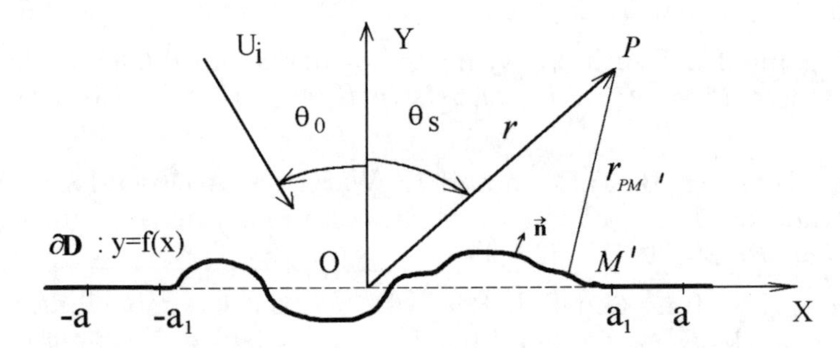

Fig. 1. Geometry of the problem.

In this paper we investigate the external 2D Dirichlet boundary value problem (BVP) for the Helmholtz equation:

$$\Delta U(P) + k^2 U(P) = 0, \quad U|_{\partial D} = 0, \tag{1}$$

where U is the total field, $P = (x, y) \in \mathbb{R}^2$, $k = 2\pi/\lambda \in \mathbb{R}$ is the wave number, λ is the wavelength, $\partial D = \{(x, f(x)) : x \in \mathbb{R}\}$ with $f \in \mathbb{C}^2$ and $\operatorname{supp} f \subseteq [-a_1, a_1]$ (see Fig. 1). This BVP determines the reflection of an electromagnetic harmonic TE-polarized wave from a 1D perfectly conducting rough surface with boundary ∂D, as well as the scattering by a sound-soft wall in acoustics.

It is convenient to represent (1) in the equivalent form

$$\Delta U_s(P) + k^2 U_s(P) = 0, \quad U_s|_{\partial D} = -(U_i + U_m)|_{\partial D}, \tag{2}$$

The authors are indebted to V.V. Zalipaev for stimulating discussions and help with various aspects of the work.

where $U_s = U - U_i - U_m$ is the scattered field, $U_i = \exp[i(\alpha_0 x - \beta_0 y)]$ is the incident plane wave, $U_m = -\exp[i(\alpha_0 x + \beta_0 y)]$ is the mirror-reflected wave with respect to the plane $y = 0$, $\alpha_0 = k \sin\theta_0$, $\beta_0 = k\cos\theta_0$ and θ_0 is the angle of incidence. If the radiation condition on U_s is imposed, then BVP (2) has a unique solution [4].

An attempt to solve BVP (2) was made in [5], where an integral equation of the first kind on an unbounded domain with a weakly singular kernel for an unknown distribution of surface current density was obtained. In order to solve that equation, it was suggested that the integration domain should be truncated, which seems untenable from the mathematical and physical points of view. We introduce a new rigorous formalism, which reduces the problem to Fredholm integral equations of the second kind and ensures a high accuracy and stability of computations.

We seek the solution of (2) as a double layer potential

$$U_s(P) = \int_{\partial D} \mu(M') \frac{\partial H_0^{(1)}(kr_{PM'})}{\partial n'}\, ds', \tag{3}$$

where $\partial H_0^{(1)}(kr_{PM'})/\partial n'$ is the normal derivative of the Hankel function of the first kind and order zero, $r_{PM'}$ is the distance between the observation point P and $M' \in \partial D$, $\mu(M')$ is the density of the scattered field sources to be found and ds' is the element of arc along ∂D.

The integral equation for $\mu(x)$ can be obtained from (3) by passing to the limit as $P \to M \in \partial D$:

$$\mu(x) + \int_{-\infty}^{\infty} K(x, x')\, \mu(x')\, dx' = g(x), \tag{4}$$

where the kernel $K(x, x') = \dfrac{ik}{2} \dfrac{H_1^{(1)}(kr_{MM'})}{r_{MM'}} [f(x) - f(x') - f'(x')(x - x')]$ is continuous, and $g(x) = -\exp(i\alpha_0 x) \sin[\beta_0 f(x)]$. We reduce (4) to a Fredholm integral equation of the second kind. For this purpose we split $\mu(x)$, writing $\mu(x) = \mu_I(x) + \mu_E(x)$, where

$$\mu_I(x) \equiv \begin{cases} \mu(x), & |x| < a, \\ 0, & |x| > a, \end{cases} \qquad \mu_E(x) \equiv \begin{cases} 0, & |x| < a, \\ \mu(x), & |x| > a, \end{cases} \qquad a > a_1. \tag{5}$$

Further, noting that $K(x, x') = 0$ for any $|x|, |x'| > a_1$, we can transform (4) into an equation for the internal density μ_I:

$$\mu_I(x) + \int_{|x'|<a} [K(x, x') + \tilde{K}(x, x')]\, \mu_I(x')\, dx' = g(x), \quad |x| < a, \tag{6}$$

where

$$\tilde{K}(x, x') = -\int_{|x''|>a} K(x, x'') K(x'', x')\, dx'' \tag{7}$$

and the external density μ_E is expressed in terms of the internal one:

$$\mu_E(x) = -\int_{|x'|<a_1} K(x, x') \mu_I(x')\, dx'. \tag{8}$$

It is worth noting that setting the width $a - a_1$ of the transient zone large enough, we can use integration by parts to obtain an asymptotic estimate for $\tilde{K}(x, x')$. Thus, solving (6), for example, by a moment method and using (8), the density $\mu(x)$ can be calculated for any x.

Once the density μ has been obtained, the scattered field U_s can be deduced from (3). We now write (3) in the polar coordinate system (r, θ_s) as

$$U_s(r, \theta_s) = [2\pi/(kr)]^{1/2} \exp[i(kr - \pi/4)]A(\theta_s, \theta_0) + O[(kr)^{-3/2}],$$

where

$$A(\theta_s, \theta_0) = \frac{i}{\pi} \int_{-\infty}^{\infty} [\alpha_s f'(x) - \beta_s] \exp\{-i[\alpha_s x + \beta_s f(x)]\}\mu(x)\, dx \tag{9}$$

is the scattering amplitude, $\alpha_s = k \sin\theta_s$ and $\beta_s = k\cos\theta_s$. Substituting (5) and (7) into (9), we represent $A(\theta_s, \theta_0)$ as $A(\theta_s, \theta_0) = A_I(\theta_s, \theta_0) + A_E(\theta_s, \theta_0)$, where $A_I(\theta_s, \theta_0)$ is given by (9) with the truncated integration domain $[-a, a]$. As for A_E, after cumbersome calculations we find that, in terms of μ_I,

$$\begin{aligned}
A_E(\theta_s, \theta_0) = &-\frac{k}{\pi(2\pi i)^{1/2}} \int_{|x|<a_1} \mu_I(x) \exp(-i\alpha_s x) \\
&\times\Bigg\{\beta_s\left(if(x) + \frac{\varepsilon f'(x)}{8k}\right)\left[\frac{SI_1(\Lambda^{(+)}, b^{(+)}, \varepsilon)}{(\Lambda^{(+)})^{1/2}} + \frac{SI_1(\Lambda^{(-)}, b^{(-)}, -\varepsilon)}{(\Lambda^{(-)})^{1/2}}\right] \\
&+ \frac{\beta_s f'(x)}{k}\left[\frac{1 - i[8\Lambda^{(+)}(1 + [b^{(+)}]^2)^{1/2}]^{-1}}{[\Lambda^{(+)}(1 + [b^{(+)}]^2)^{1/2}]^{1/2}} \exp\{i\Lambda^{(+)}[(1 + [b^{(+)}]^2)^{1/2} + \varepsilon]\}\right. \\
&\left. - \frac{1 - i[8\Lambda^{(-)}(1 + [b^{(-)}]^2)^{1/2}]^{-1}}{[\Lambda^{(-)}(1 + [b^{(-)}]^2)^{1/2}]^{1/2}} \exp\{i\Lambda^{(-)}[(1 + [b^{(-)}]^2)^{1/2} - \varepsilon]\}\right] \\
&+ i\varepsilon f'(x)\{[(1+\varepsilon)\Lambda^{(+)}]^{1/2} SI_0(\Lambda^{(+)}, b^{(+)}, \varepsilon) + [(1-\varepsilon)\Lambda^{(-)}]^{1/2} SI_0(\Lambda^{(-)}, b^{(-)}, -\varepsilon)\}\Bigg\}\, dx \\
&+ o([k(a - a_1)]^{-3/2}),
\end{aligned} \tag{10}$$

where $\varepsilon = \sin\theta_s$, $\Lambda^{(\pm)} = k(a \pm x)$, $b^{(\pm)} = f(x)/(a \pm x)$, and

$$\begin{aligned}
SI_0(\Lambda, b, \varepsilon) &= (1+\varepsilon)^{1/2} \int_1^{\infty} \frac{\exp\{i\Lambda[(t^2 + b^2)^{1/2} + \varepsilon t]\}}{(t^2 + b^2)^{1/4}}\, dt, \\
SI_1(\Lambda, b, \varepsilon) &= \int_1^{\infty} \frac{\exp\{i\Lambda[(t^2 + b^2)^{1/2} + \varepsilon t]\}}{(t^2 + b^2)^{3/4}}\, dt, \quad \Lambda \gg 0, \quad b \in \mathbb{R}, \quad |\varepsilon| \le 1.
\end{aligned} \tag{11}$$

The integrals (11) can be evaluated by means of the uniform asymptotic technique [6].

The calculations of $A(\theta_s, \theta_0)$ have been tested against the energy balance criterion (EBC) [5] $\Delta\mathcal{E} = \left|1 - [2\,\mathrm{Re}\, A(\theta_0, \theta_0)]^{-1} \int_{-\pi/2}^{\pi/2} |A(\theta_s, \theta_0)|^2\, d\theta_s\right| \equiv 0$, closely connected with the relative computational error. We note that $A(\pm\pi/2, \theta_0) \equiv 0$.

To illustrate the results obtained, we consider the scattering from a finite deep resonance grating

$$f(x) = h\eta(x)\cos(2\pi x/d), \tag{12}$$

where $\eta(x)$ is the cutting function [7], $h = d = \lambda$, $a = 25\lambda$ and $a_1 = 20\lambda$ (see Fig. 2). The number of matching points for solving (6) was 1000. In Fig. 3 the energy defect $\Delta\mathcal{E}$ (in percentage) versus the angle of incidence for the integral equation method discussed (solid line) and offered in [5] (broken line) is plotted. It is obvious that the scattering amplitude calculated following [5] is incompatible with EBC and, thus, should be rejected. On the other hand, the proposed method yields very accurate data (max $\Delta\mathcal{E} < 0.3\%$).

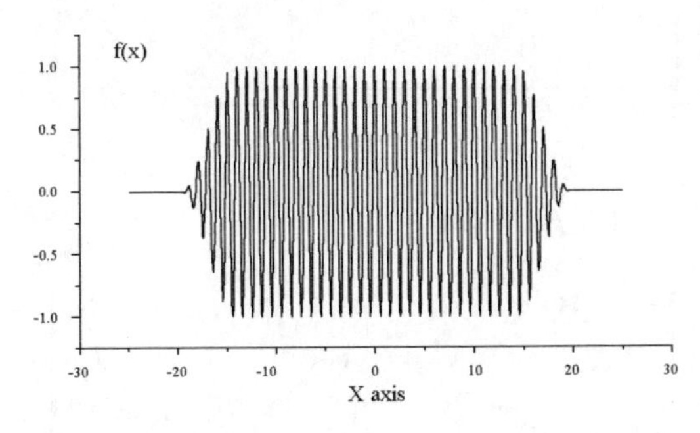

Fig. 2. The profile $f(x)$ of the finite resonance grating (12). The X- and Y-axes are plotted per wavelength.

Fig. 3. The energy defect $\Delta\mathcal{E}$ vs the angle of incidence θ_0. Solid line: the new approach; broken line: as suggested in [5].

Similar conclusions are valid for large-scale, steep, smooth or resonance modulated asperities, including the grazing incidence geometry. Figs. 4–6 show the modulus of the density $\mu(x)$ and the corresponding scattered field intensity at infinity $|A(\theta_s, \theta_0)|^2$ for angles of incidence $\theta_0 = 0°$ (Fig. 4), $\theta_0 = 30°$ (Fig. 5) and $\theta_0 = 89°$ (Fig. 6).

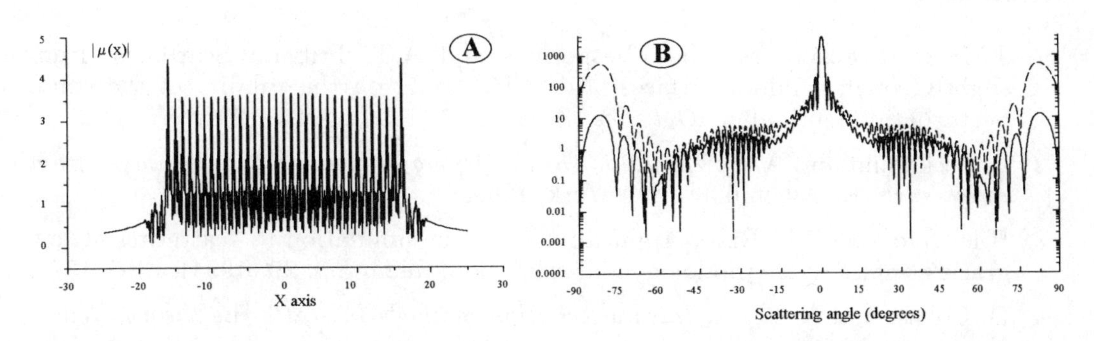

Fig. 4. (**A**): the modulus of the density $\mu(x)$. (**B**): the scattered intensity at infinity $|A(\theta_s, \theta_0)|^2$ vs the scattering angle θ_s. Solid line: the new approach; broken line: as suggested in [5]. Angle of incidence: $\theta_0 = 0°$.

We note that for a sufficiently shallow roughness (for example, for (12) with $h < 0.2\lambda$) and quasinormal incidence both approaches give qualitatively resembling results, though the error of our calculations is, as a rule, smaller by two orders of magnitude.

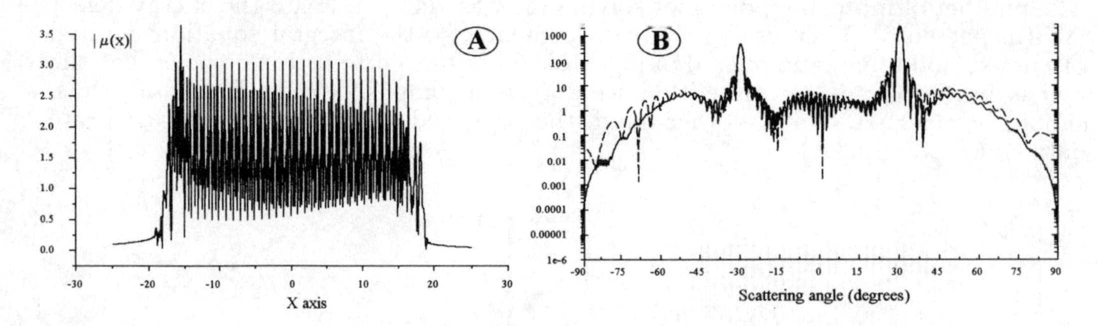

Fig. 5. Same as Fig. 4 for $\theta_0 = 30°$.

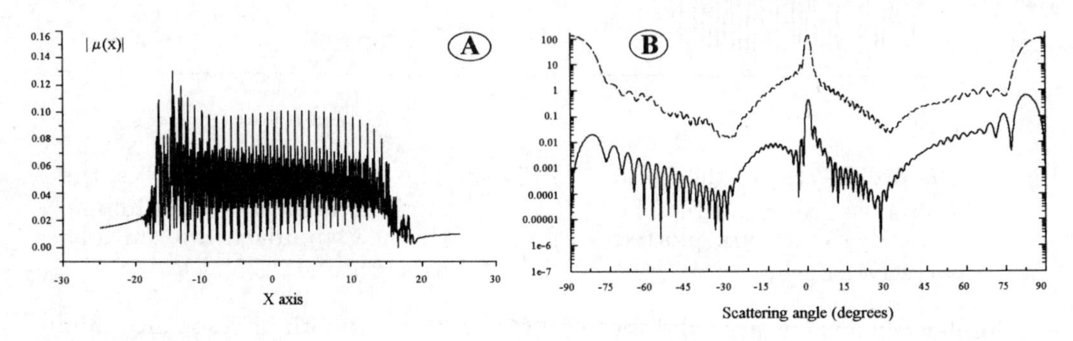

Fig. 6. Same as Fig. 4 for $\theta_0 = 89°$.

References

1. J.M. Soto-Crespo, M. Nieto-Vesperinas and A.T. Friberg, Scattering from slightly rough random surfaces: a detailed study on the validity of the small perturbation method, *J. Opt. Soc. Amer. A* **7** (1990), 1185–1201.
2. P. Beckmann and A. Spizzichino, *The scattering of electromagnetic waves from rough surfaces*, Macmillan, New York, 1963.
3. H.D. Ngo and C.L. Rino, Application of beam simulation to scattering at low grazing angles. 1. Methodology and validation, *Radio Sci.* **29** (1994), 1365–1379.
4. D. Colton and R. Kress, *Integral equation methods in scattering theory*, Wiley, New York, 1983.
5. D. Maystre, Electromagnetic scattering from perfectly conducting rough surfaces in the resonance region, *IEEE Trans. Antennas and Propagation* **31** (1983), 885–895.

6. V.A. Borovikov, *Uniform stationary phase method*, Peter Peregrinus, London, 1995.

7. V.S. Vladimirov, *Equations of mathematical physics*, Mir, Moscow, 1984.

Department of Theoretical Physics, St. Petersburg Institute of Fine Mechanics and Optics (Technical University), Sablinskaya 14, 197101 St. Petersburg, Russia

I.V. BOIKOV

Iterative methods of solution for integral equations with convolution

In this paper we construct iterative methods for the solution of integral equations with convolution of the form

$$\lambda x(t_1,\ldots,x_l)+\int_a^b\cdots\cdot\int_a^b h(t_1-\tau_1,\ldots,t_l-\tau_l)x(\tau_1,\ldots,\tau_l)d\tau_1\ldots d\tau_l$$
$$=g(t_1,\ldots,t_l)\quad(\lambda\geq 0).\quad (1)$$

Let $H(\omega_1,\ldots,\omega_l)=F(h(t_1,\ldots,t_l))$, $X(\omega_1,\ldots,\omega_l)=F(x(t_1,\ldots,t_l))$ and $G(\omega_1,\ldots,\omega_l)=F(g(t_1,\ldots,t_l))$. We denote by F and F^{-1} the operators of the Fourier transformation and inverse Fourier transformation. We show that for any kernel $h(t_1,\ldots,h_l)$ with piecewise continuous Fourier transform $H(\omega_1,\ldots,\omega_l)$ there are iterative methods for the solution of (1).

1. We consider the equation

$$\lambda x(t)+\frac{1}{\sqrt{2\pi}}\int_a^b h(t-\tau)x(\tau)d\tau=g(t),\quad (2)$$

where $h(t),\ g(t)\in L_2[a,b]$, $\lambda\geq 0$ and $-\infty\leq t\leq\infty$.

Let P_{ab} be the projection defined by $P_{ab}f(t)=f(t)$ if $t\in[a,b]$ and $P_{ab}f(t)=0$ if $t\notin[a,b]$.

Using the operator P_{ab}, we can rewrite (2) as

$$\lambda x(t)+P_{ab}\Big(\frac{1}{\sqrt{2\pi}}\int_{-\infty}^{\infty} h(t-\tau)P_{ab}(x(\tau))d\tau\Big)=g^*(t),\quad -\infty\leq t\leq\infty,\quad (3)$$

where $g^*(t)=f(t)$ for $t\in[a,b]$ and $g^*(t)=0$ for $t\notin[a,b]$.

We extend the function $h(t)$ from the segment $[a-b,b-a]$ to the axis $(-\infty,\infty)$ and denote the new function by $h^*(t)$. Let $H^*(\omega)=F(h^*(t))$ and $G^*(\omega)=F(g^*(t))$.

We say that the function $\lambda+H^*(\omega)$ satisfies the condition A if the increment of its argument is not greater than π. In this case there is a constant γ such that $q=\sup\limits_{-\infty<\omega<\infty}\mid 1-\gamma(\lambda+H^*(\omega))\mid\leq 1$.

We propose the following iterative methods for the solution of (2). For $q<1$ we use the iterative process

$$x_{n+1}(t)=x_n(t)-\gamma\Big(\lambda x_n+\frac{1}{\sqrt{2\pi}}\int_a^b h(t-\tau)x_n(\tau)d\tau-g(t)\Big),\quad a\leq t\leq b,\quad (4)$$

$n = 0, 1, \ldots$ For $q = 1$ we use the iterative process

$$x_{n+1}(t) = \alpha_n x_n(t) + (1 - \alpha_n)(x_n(t)$$
$$- \gamma\left(\lambda x_n(t) + \frac{1}{\sqrt{2\pi}}\int_a^b h(t-\tau)x_n(\tau)d\tau - g(t)\right), \quad a \le t \le b, \tag{5}$$

$n = 0, 1, \ldots$, where $0 < \alpha \le \alpha_n \le \beta < 1$.

We study these iterative processes in the space $L_2[-1, 1]$.

Theorem 1. *If the function $\lambda + H^*(\omega)$ satisfies the condition A with $q < 1$, then the iterative processes* (4) *with any starting point converge to the solution $x^*(t)$ of* (2) *with convergence rate Bq^n, $B =$ const.*

Proof. We must prove that the operator

$$Kx \equiv x(t) - \gamma\left(\lambda x(t) + \frac{1}{\sqrt{2\pi}}\int_a^b h(t-\tau)x(\tau)d\tau\right)$$

is a contraction in the space $L_2[a, b]$. Let $\hat{x}(t) = x(t)$ for $t \in [a, b]$ and $\hat{x}(t) = 0$ for $t \notin [a, b]$. Also, let $\hat{X}_{ab}(\omega) = F(\hat{x}(t))$. It is easy to see that $\|P_{ab}\|_{L_2[a,b]} = 1$. Since, as is well known, $\|F\|_{L_2(-\infty,\infty)} = \|F^{-1}\|_{L_2(-\infty,\infty)} = 1$, it follows that

$$\|Kx\|_{L_2[a,b]} = \left\|P_{ab}(\hat{x}(t) - \gamma(\lambda\hat{x}(t) + \frac{1}{\sqrt{2\pi}}\int_{-\infty}^{\infty} h^*(t-\tau)P_{ab}\hat{x}(\tau)d\tau))\right\|_{L_2(-\infty,\infty)}$$
$$\le \left\|P_{ab}\hat{x}(t) - \gamma(\lambda P_{ab}\hat{x}(t)) - \frac{1}{\sqrt{2\pi}}\int_{-\infty}^{\infty} h^*(t-\tau)P_{ab}\hat{x}(\tau)d\tau))\right\|_{L_2(-\infty,\infty)}$$
$$= \|\hat{X}_{ab}(\omega) - \gamma(\lambda\hat{X}_{ab}(\omega) - H^*(\omega)\hat{X}_{ab}(\omega)\|_{L_2(-\infty,\infty)}$$
$$\le \sup_{\omega}|1 - \gamma(\lambda - H^*(\omega))|\|\hat{X}_{ab}(\omega)\ \|_{L_2(-\infty,\infty)} \le q\|x(t)\|_{L_2[a,b]}.$$

We now finish the proof by applying Banach's fixed point theorem [1].

Theorem 2. *If* (2) *has a unique solution $x^*(t)$, and the function $\lambda + H^*(\omega)$ satisfies condition A, then the iterative process* (5) *converges to the solution $x^*(t)$ in the metric of the space $L_2[a, b]$.*

Proof. Repeating the argument in the proof of Theorem 1, we see that

$$\|Kx\|_{L_2[a,b]} \le \sup_{\omega}|1 - \gamma(\lambda + H^*(\omega))|\|x\|_{L_2[a,b]} \le \|x\|_{L_2[a,b]}.$$

From this it follows that $\|K\|_{L_2[a,b]} = 1$. Using this equality and Oblomskii's theorem [2], we arrive at the desired assertion.

2. We consider the equation

$$\lambda x(t) + \frac{1}{\sqrt{2\pi}} \int_{-\infty}^{\infty} h(t-\tau)x(\tau)d\tau = g(t), \quad \lambda \geq 0. \tag{6}$$

Let $H(\omega)$ be a piecewise-continuous function, and suppose that condition A is not satisfied. For any $\epsilon > 0$ there are numbers T_0 and T_N such that

$$\left[\int_{-\infty}^{T_0} |X(\omega)|^2 d\omega + \int_{T_N}^{\infty} |X(\omega)|^2 d\omega\right]^{1/2} \leq \epsilon.$$

Now we select points $T_0, \ldots, T_N$, $T_0 < T_1 < \ldots < T_N$, such that the increment of the argument of the function $\lambda + H(\omega)$ is less than π for $\omega \in \Lambda_k$, $\Lambda_k = [T_k, T_{k+1}]$, $k = 0, 1, \ldots, N-1$. In this case we can choose the constants γ_k so that

$$\sup_{\omega} |1 - \gamma_k(\lambda + H(\omega))| = q_k < 1 \text{ for } \omega \in \Lambda_k, \; k = 0, 1, \ldots, N-1.$$

Let $q = \max_{0 \leq k \leq N-1} q_k$, and let $E_k(\omega)$ be the characteristic function of Λ_k, $k = 0, 1, \ldots, N-1$; that is, $E_k(\omega) = 1$ if $\omega \in \Lambda_k$ and $E_k(\omega) = 0$ if $\omega \notin \Lambda_k$. Let $x_j^*(t) = F^{-1}(E_j(\omega)X^*(\omega))$, $x_{0j}(t) = F^{-1}(E_j(\omega)X_0(\omega))$ and $g_j(t) = F^{-1}(E_j(\omega)G(\omega))$.

For the solution of (6) we use the iterative process

$$x_{n+1,j}(t) = x_{n,j}(t) - \gamma_j\left(\lambda x_{n,j}(t) + \frac{1}{\sqrt{2\pi}} \int_{-\infty}^{\infty} h(t-\tau)x_{n,j}(\tau)d\tau - g_j(t)\right), \tag{7}$$

$j = 0, \ldots, N-1$, $n = 0, 1, \ldots$, if $q < 1$, or the iterative process

$$\begin{aligned} x_{n+1,j} = \alpha_{n,j}x_{n,j}(t) - (1-\alpha_{n,j})(x_{n,j}(t) \\ - \gamma_j\left(\lambda x_{n,j} + \frac{1}{\sqrt{2\pi}} \int_{-\infty}^{\infty} h(t-\tau)x_{n,j}(\tau)d\tau - g_j(t))\right), \end{aligned} \tag{8}$$

$j = 0, \ldots, N-1$, $n = 0, 1, \ldots$

The approximate solution $x_{n+1}(t)$ of (6) is defined by

$$x_{n+1}(t) = \sum_{j=0}^{N-1} x_{n+1,j}(t). \tag{9}$$

Theorem 3. *If (6) has a unique solution $x^*(t)$, then the sequence x_{n+1} converges to x^* and $\|x^*(t) - x_{n+1}(t)\|_{L_2} \leq A(q^n) + \epsilon$.*

Proof. We prove the convergence of the iterative processes (7), (9) to the solution $x^*(t)$ of (6). The function $x^*(t)$ belongs to $L_2(-\infty, \infty)$. For any $\epsilon > 0$ there is a constant T such that

$$\int_{-\infty}^{-T} |X^*(\omega)|^2 d\omega + \int_{T}^{\infty} |X^*(\omega)|^2 d\omega < \epsilon^2.$$

As the starting point, we take $x_0(t)$ to be a function with compact support on $[-T, T]$.

Repeating the arguments in the proof of Theorem 1, we show that the sequences $x_{n+1,j}(t)$ converge to $x_j^*(t)$ as $n \to \infty$ and $T \to \infty$, and estimate the error of the iterative processes (7), (9).

Theorem 4. *If* (6) *has a unique solution* $x^*(t)$ *and the function* $\lambda + H(\omega)$ *is equal to zero at some points in* $(-\infty, \infty)$, *then the iterative processes* (8), (9) *converge to* $x^*(t)$.

This assertion is proved by means of the arguments used in Theorems 2 and 3.

We consider the following method of approximate solution for (6). Applying the Fourier transformation to (6), we arrive at the equation $(\lambda + H(\omega))X(\omega) = G(\omega)$. Let D be a real number. We introduce the set of points $\omega_k = -D + kD/N$, $k = 0, 1, \ldots, 2N$. Let $\lambda + H(\omega_k) \neq 0$, $k = 0, 1, \ldots, 2N$. We consider the iterative process

$$X_{n+1}(\omega_k) = X_n(\omega_k) - \gamma_k((\lambda + H(\omega_k))X_n(\omega_k) - G(\omega_k)),$$

$k = 0, 1, \ldots, 2N$, $n = 0, 1, \ldots,$ where the constant γ_k is chosen so that

$$q_k = |1 - \gamma_k(\lambda + H(\omega_k))| \leq 1/2, \quad k = 0, 1, \ldots, 2N.$$

It is easy to see that for any $k = 0, 1, \ldots, 2N$ the iterations $X_{n+1}(\omega_k)$ converge to $\bar{X}(\omega_k)$ at a rate of at least Aq_k^n.

Using the inverse Fourier transformation, we calculate the approximate values of the solution $x^*(t)$ of (6).

3. So far we have investigated one-dimensional integral equations. All the above results can be extended to many-dimensional integral equations with convolution. As an example, we consider Theorem 3 for (1) with $l = 2$. Let $h(t_1, t_2)$, $g(t_1, t_2)$ $\in L_2[-\infty, \infty]$.

We say that the function $\lambda + H(\omega_1, \omega_2)$ satisfies condition A if the increment of the argument of $\lambda + H(\omega_1, \omega_2)$ does not exceed π for $-\infty \leq \omega_1, \omega_2 \leq \infty$.

Suppose that $H(\omega_1, \omega_2)$ is piecewise continuous and that $\lambda + H(\omega_1, \omega_2) \neq 0$ for any $(\omega_1, \omega_2), -\infty < \omega_1, \omega_2 < \infty$. Suppose also that condition A is not satisfied. Let $x^*(t_1, t_2) \in L_2(-\infty, \infty)$ be the unique solution of (1). For any $\epsilon > 0$, there are numbers T_0 and T_N such that

$$\left[\left(\int_{-\infty}^{T_0}\int_{-\infty}^{\infty} + \int_{T_N}^{\infty}\int_{-\infty}^{\infty} + \int_{T_0}^{T_N}\int_{-\infty}^{T_0} + \int_{T_0}^{T_N}\int_{T_N}^{\infty}\right)|X^*(\omega_1, \omega_2)|^2 d\omega_1 d\omega_2\right]^{1/2} \leq \epsilon.$$

Now we select points $T_0, \ldots, T_N, T_0 < T_1 < \ldots < T_N$ such that the increment of the argument of the function $\lambda + H(\omega_1, \omega_2)$ is less than π for $\omega \in \Lambda_{kl}$, where $\Lambda_{kl} = [T_k, T_{k+1}; T_l, T_{l+1}]$, $k, l = 0, 1, \ldots, N - 1$.

In this case, for any parallelepiped Λ_{kl}, $k,l = 0,1,\ldots,N-1$, we can select the constant γ_{kl} so that $\sup\limits_{(\omega_1,\omega_2)\in\Lambda_{kl}} |1-\gamma_{kl}(\lambda + H(\omega_1,\omega_2))| \leq q_{kl} < 1$. Let $q = \max\limits_{kl} q_{kl}$, let $E_{kl}(\omega_1,\omega_2)$ be the characteristic function of the parallelepiped Λ_{kl}, $k,l = 0,1,\ldots,N-1$, and let $x_{kl}(t_1,t_2) = F^{-1}(E_{kl}(\omega_1,\omega_2)X(\omega_1,\omega_2))$, $x_{0kl}(t_1,t_2) = F^{-1}(E_{kl}(\omega_1,\omega_2)X_0(\omega_1,\omega_2))$, and $g_{kl} = F^{-1}(E_{kl}(\omega_1,\omega_2)G(\omega_1,\omega_2))$.

For the solution of (1) we use the iterative process

$$x_{kl}^{n+1}(t_1,t_2) = x_{kl}^{n}(t_1,t_2) - \gamma_{kl}\Bigg(\lambda x_{kl}^{n}(t_1,t_2) + \frac{1}{2\pi}\int_{-\infty}^{\infty}\int_{-\infty}^{\infty} h(t_1-\tau_1,t_2-\tau_2)x_{kl}^{n}(\tau_1,\tau_2)d\tau_1,d\tau_2 - g_{kl}(t_1,t_2)\Bigg),$$

$k,l = 0,1,\ldots,N-1$, $n = 0,1,\ldots$,

$$x^{n+1}(t_1,t_2) = \sum_{k=0}^{N-1}\sum_{l=0}^{N-1} x_{kl}^{n+1}(t_1,t_2).$$

Theorem 5. *If* (1) *has a unique solution* $x^*(t_1,t_2)$, *then the sequence* $x^{n+1}(t_1,t_2)$ *converges to* $x^*(t_1,t_2)$ *and* $\|x^*(t_1,t_2) - x^{n+1}(t_1,t_2)\| \leq A(q^n + \epsilon)$.

References

1. L.A. Lyusternik and V.I. Sobolev, *Elements of functional analysis,* Nauka, Moscow, 1965 (Russian).
2. L.A. Oblomskii, Iterative methods for linear equations in Banach spaces, *Zh. Vychisl. Mat. i Mat. Fiz.* **8** (1968), 417–426.

Penza Technical State University, Penza, Russia

E-mail: cnit@diamond.stup.ac.ru

L.P. CASTRO

Integral equations of convolution type on unions of intervals

1. Introduction

The main purpose of this note is to study, on the union of one finite and one infinite intervals, systems of convolution equations of the form

$$\int_\Omega \mathcal{K}(\xi - x)\varphi(x)dx = \psi(\xi), \quad \xi \in \Omega, \tag{1}$$

from the point of view of their generalised invertibility.

Here we would like to extend the theory in the following way: (i) to construct equivalence relations between the operator associated with equation (1) and a matrix Wiener-Hopf operator on the half-line, based on the method of decomposition of higher order Wiener-Hopf operators presented in [1]; (ii) to represent generalised inverses explicitly when they exist.

2. Definitions and notation

For functions φ in the space $\mathcal{S}(\mathbb{R})$ of rapidly decreasing functions on $\mathbb{R}$, the Fourier transform is defined by

$$\mathcal{F}\varphi(\xi) = \int_\mathbb{R} e^{i\xi x}\varphi(x)\,dx.$$

In $\mathcal{S}'(\mathbb{R})$, the space of generalised functions of slow growth on $\mathbb{R}$, the Fourier transformation is defined by duality.

Let $1 < p < \infty$. As usual, $L^p(\mathbb{R})$ denotes the Banach space of Lebesgue measurable functions φ on $\mathbb{R}$ such that $|\varphi|^p$ is integrable.

For a given domain $\Gamma \subseteq \mathbb{R}_+$, we denote by $L^p_\Gamma(\mathbb{R})$ the closed subspace of $L^p(\mathbb{R})$ whose elements have support in $\overline{\Gamma}$. The space $L^p_\Gamma(\mathbb{R})$ is endowed with the subspace topology. For the sake of simplicity, we write $L^p_+(\mathbb{R})$ for $L^p_{\mathbb{R}_+}(\mathbb{R})$. We use the projection operators P_Γ and Q_Γ that map $\left[L^p_+(\mathbb{R})\right]^n$ on to $[L^p_\Gamma(\mathbb{R})]^n$ and $\left[L^p_{\mathbb{R}_+\setminus\bar{\Gamma}}(\mathbb{R})\right]^n$, respectively.

Let

$$\Omega =]0,a[\cup]b,+\infty[,$$

where $0 < a < b < +\infty$. We focus our interest on the study of the operator associated with equation (1) in the context of L^p spaces. Thus, we investigate operators of the form

$$W_{K,\Omega} : [L^p_\Omega(\mathbb{R})]^n \to [L^p_\Omega(\mathbb{R})]^n, \quad W_{K,\Omega}\varphi = P_\Omega(\mathcal{K} * \varphi), \tag{2}$$

where $*$ denotes the convolution operation and the kernel $\mathcal{K}$ is a given $n \times n$ matrix-valued function with components in $\mathcal{S}'(\mathbb{R})$. We assume that the Fourier transform $K = \mathcal{F}\mathcal{K}$ of $\mathcal{K}$, is an $n \times n$ element of the L^p-multiplier algebra [2]. Under these assumptions, the linear operator $W_{K,\Omega}$ is well defined and bounded. These operators are sometimes regarded as pseudodifferential or Wiener-Hopf operators (see [2]–[4]).

3. An extension method

First we relate the convolution operator $W_{K,\Omega}$ to a Wiener-Hopf operator of higher order that acts on the half-line.

Theorem 1. *The convolution operator $W_{K,\Omega}$ defined in (2) is equivalent after extension to the operator $W : \left[L^p_+(\mathbb{R})\right]^n \to \left[L^p_+(\mathbb{R})\right]^n$ defined by*

$$W = W_{K,\mathbb{R}_+} - P_{[a,+\infty[}(W_{K,\mathbb{R}_+} - I) + P_{[b,+\infty[}(W_{K,\mathbb{R}_+} - I), \tag{3}$$

which has the form of a Wiener-Hopf operator with oscillating Fourier symbol.

Proof. Attending to the direct sum

$$\begin{aligned}\left[L^p_+(\mathbb{R})\right]^n &= P_\Omega\left[L^p_+(\mathbb{R})\right]^n \oplus Q_\Omega\left[L^p_+(\mathbb{R})\right]^n \\ &= \left[L^p_\Omega(\mathbb{R})\right]^n \oplus \left[L^p_{\mathbb{R}_+\setminus\bar{\Omega}}(\mathbb{R})\right]^n,\end{aligned}$$

we write the operator $\widetilde{W}_{K,\Omega} = P_\Omega W_{K,\mathbb{R}_+} P_\Omega + Q_\Omega : \left[L^p_+(\mathbb{R})\right]^n \to \left[L^p_+(\mathbb{R})\right]^n$ in the form

$$\widetilde{W}_{K,\Omega} = \begin{bmatrix} W_{K,\Omega} & 0 \\ 0 & I_{|Q_\Omega[L^p_+(\mathbb{R})]^n} \end{bmatrix}.$$

In this sense we have that $W_{K,\Omega}$ is equivalent after extension to the paired Wiener-Hopf operator $\widetilde{W}_{K,\Omega}$. We now prove that $\widetilde{W}_{K,\Omega}$ and W are equivalent operators. First we note that

$$P_\Omega = I - P_{[a,+\infty[} + P_{[b,+\infty[}\,, \quad Q_\Omega = P_{[a,+\infty[} - P_{[b,+\infty[}$$

(where we identify $P_{[h,+\infty[} = P_{\mathbb{R}_+}\mathcal{F}^{-1}e^{i\xi h}\mathcal{F}P_{\mathbb{R}_+}\mathcal{F}^{-1}e^{-i\xi h}\mathcal{F}$, $h = a, b$). Thus, it follows that

$$\begin{aligned}\widetilde{W}_{K,\Omega} &= (I - P_\Omega W_{K,\mathbb{R}_+} Q_\Omega)(P_\Omega W_{K,\mathbb{R}_+} + Q_\Omega) \\ &= (I - P_\Omega W_{K,\mathbb{R}_+} Q_\Omega)W,\end{aligned} \tag{4}$$

in which $I - P_\Omega W_{K,\mathbb{R}_+} Q_\Omega : \left[L^p_+(\mathbb{R})\right]^n \to \left[L^p_+(\mathbb{R})\right]^n$ is a bounded linear operator invertible by $I + P_\Omega W_{K,\mathbb{R}_+} Q_\Omega$.

Theorem 2. *If W^- is a generalised inverse of W, then*

$$W^-_{K,\Omega} : \left[L^p_\Omega(\mathbb{R})\right]^n \to \left[L^p_\Omega(\mathbb{R})\right]^n, \quad W^-_{K,\Omega} = P_\Omega W^-_{|P_\Omega[L^p_+(\mathbb{R})]^n}$$

is a generalised inverse of $W_{K,\Omega}$.

Proof. Suppose that W^- is a generalised inverse of W. From the equivalence relation (4) we immediately obtain a generalised inverse of $\widetilde{W}_{K,\Omega}$, namely

$$\widetilde{W}^-_{K,\Omega} = W^-(I + P_\Omega W_{K,\mathbb{R}_+} Q_\Omega). \tag{5}$$

On the other hand, writing $\widetilde{W}_{K,\Omega}\widetilde{W}^-_{K,\Omega}\widetilde{W}_{K,\Omega} = \widetilde{W}_{K,\Omega}$ in matrix form, we have

$$\begin{bmatrix} W_{K,\Omega} & 0 \\ 0 & I_{|Q_\Omega[L^p_+(\mathbb{R})]^n} \end{bmatrix} \begin{bmatrix} P_\Omega\left(\widetilde{W}^-_{K,\Omega}\right)_{|P_\Omega[L^p_+(\mathbb{R})]^n} & P_\Omega\left(\widetilde{W}^-_{K,\Omega}\right)_{|Q_\Omega[L^p_+(\mathbb{R})]^n} \\ Q_\Omega\left(\widetilde{W}^-_{K,\Omega}\right)_{|P_\Omega[L^p_+(\mathbb{R})]^n} & Q_\Omega\left(\widetilde{W}^-_{K,\Omega}\right)_{|Q_\Omega[L^p_+(\mathbb{R})]^n} \end{bmatrix}$$
$$\times \begin{bmatrix} W_{K,\Omega} & 0 \\ 0 & I_{|Q_\Omega[L^p_+(\mathbb{R})]^n} \end{bmatrix} = \begin{bmatrix} W_{K,\Omega} & 0 \\ 0 & I_{|Q_\Omega[L^p_+(\mathbb{R})]^n} \end{bmatrix}.$$

Consequently, we obtain

$$W_{K,\Omega}\left[P_\Omega\left(\widetilde{W}^-_{K,\Omega}\right)_{|P_\Omega[L^p_+(\mathbb{R})]^n}\right] W_{K,\Omega} = W_{K,\Omega}. \tag{6}$$

Thus, using (5) in the identity (6), we find the following generalised inverse of $W_{K,\Omega}$:

$$W^-_{K,\Omega} : [L^p_\Omega(\mathbb{R})]^n \to [L^p_\Omega(\mathbb{R})]^n,$$
$$W^-_{K,\Omega} = P_\Omega\left[W^-(I + P_\Omega W_{K,\mathbb{R}_+} Q_\Omega)\right]_{|P_\Omega[L^p_+(\mathbb{R})]^n} = P_\Omega W^-_{|P_\Omega[L^p_+(\mathbb{R})]^n}.$$

Theorem 3. *The operator W defined by* (3) *is equivalent after extension to*

$$\mathcal{W} : \left[L^p_+(\mathbb{R})\right]^{3n} \to \left[L^p_+(\mathbb{R})\right]^{3n}, \quad \mathcal{W} = \begin{bmatrix} W_{K,\mathbb{R}_+} & -P_{[b,+\infty[} & P_{[a,+\infty[} \\ W_{K,\mathbb{R}_+} - I & I & 0 \\ W_{K,\mathbb{R}_+} - I & 0 & I \end{bmatrix}. \tag{7}$$

Proof. If we apply to W the iteration method presented in [1] for the decomposition of higher order Wiener-Hopf operators, we find two new invertible and bounded linear operators E and F defined by

$$E : \left[L^p_+(\mathbb{R})\right]^{3n} \to \left[L^p_+(\mathbb{R})\right]^{3n}, \quad E = \begin{bmatrix} I & P_{[b,+\infty[} & -P_{[a,+\infty[} \\ 0 & I & 0 \\ 0 & 0 & I \end{bmatrix}, \tag{8}$$

$$F : \left[L^p_+(\mathbb{R})\right]^{3n} \to \left[L^p_+(\mathbb{R})\right]^{3n}, \quad F = \begin{bmatrix} I & 0 & 0 \\ I - W_{K,\mathbb{R}_+} & I & 0 \\ I - W_{K,\mathbb{R}_+} & 0 & I \end{bmatrix}. \tag{9}$$

These two operators allow us to write the equivalence relation after extension in the explicit form

$$W \oplus I_{\left[L^p_+(\mathbb{R})\right]^{2n}} = E\mathcal{W}F. \tag{10}$$

The following assertion is a direct consequence of Theorems 1 and 3.

Corollary. *The operator $W_{K,\Omega}$ is equivalent after extension to the system operator $\mathcal{W}$ defined in* (7).

We are now in a position to formulate the main result of this note. In the case where the operator $W_{K,\Omega}$ is generalised invertible, we are able to obtain generalised inverses of $W_{K,\Omega}$ in terms of the generalised inverses of an extended operator that acts on the half-line.

Let $(A)_{11} : \left[L^p_+(\mathbb{R})\right]^n \to \left[L^p_+(\mathbb{R})\right]^n$ denote the restriction to the first n component spaces of an operator $A : \left[L^p_+(\mathbb{R})\right]^{3n} \to \left[L^p_+(\mathbb{R})\right]^{3n}$.

Theorem 4. *If $\mathcal{W}^- : \left[L^p_+(\mathbb{R})\right]^{3n} \to \left[L^p_+(\mathbb{R})\right]^{3n}$ is a generalised inverse of $\mathcal{W}$, then*

$$W^-_{K,\Omega} : [L^p_\Omega(\mathbb{R})]^n \to [L^p_\Omega(\mathbb{R})]^n, \quad W^-_{K,\Omega} = P_\Omega\left[\left(F^{-1}\mathcal{W}^-E^{-1}\right)_{11}\right]_{|P_\Omega[L^p_+(\mathbb{R})]^n}$$

is a generalised inverse of the convolution operator $W_{K,\Omega}$ (with E^{-1} and F^{-1} being the inverses of E and F given in (8) *and* (9), *respectively).*

Proof. From Theorem 3 (see (10)) we know that

$$\mathcal{W} = E^{-1}\begin{bmatrix} W & 0 & 0 \\ 0 & I & 0 \\ 0 & 0 & I \end{bmatrix} F^{-1}. \tag{11}$$

Thus, if we use (11) in $\mathcal{W}\mathcal{W}^-\mathcal{W} = \mathcal{W}$, a straightforward calculation leads us to

$$\begin{bmatrix} W\left(F^{-1}\mathcal{W}^-E^{-1}\right)_{11}W & * & * \\ * & * & * \\ * & * & * \end{bmatrix} = \begin{bmatrix} W & 0 & 0 \\ 0 & I & 0 \\ 0 & 0 & I \end{bmatrix},$$

which yields a generalised inverse of W, $W^- = \left(F^{-1}\mathcal{W}^-E^{-1}\right)_{11}$. Now it only remains to apply Theorem 2 to obtain the statement.

Similarly, we can find generalised inverses of $\mathcal{W}$ in terms of the generalised inverses of $W_{K,\Omega}$. However, this is less important for applications and we omit it.

4. Related facts

We note that the existence of generalised inverses of the operators under consideration can be characterised by the so-called cross-factorisation of $\mathcal{F}^{-1}K\mathcal{F}$ with respect to $L^p(\mathbb{R})$ and an intermediate space (with corresponding projectors) (see [4]–[7]).

Under the additional assumption that $W_{K,\Omega}$ is a Fredholm operator, other equivalence after extension relations are now studied and can be used more efficiently.

Theorem 5. *If the operator $W_{K,\Omega}$ is a Fredholm operator, then $W_{K,\Omega}$ is equivalent after extension to*

$$\mathcal{T} : \left[L_+^p(\mathbb{R})\right]^{3n} \to \left[L_+^p(\mathbb{R})\right]^{3n},$$

$$\mathcal{T} = \begin{bmatrix} P_{\mathbb{R}_+}\mathcal{F}^{-1}e^{-i\xi a}\mathcal{F} & 0 & 0 \\ 0 & P_{\mathbb{R}_+}\mathcal{F}^{-1}e^{-i\xi(b-a)}\mathcal{F} & 0 \\ W_{K,\mathbb{R}_+} & -P_{\mathbb{R}_+}\mathcal{F}^{-1}e^{i\xi a}\mathcal{F} & P_{\mathbb{R}_+}\mathcal{F}^{-1}e^{i\xi b}\mathcal{F}W_{K,\mathbb{R}_+} \end{bmatrix}. \qquad (12)$$

The advantage of $\mathcal{T}$ (compared to $\mathcal{W}$) is its triangular matrix form. This simplifies considerably the factorisations of the symbols of this kind of operator (see [8]).

Unfortunately, the general case has not been solved yet.

Problem. *As far as the author knows, the question of equivalence after extension of the operators $W_{K,\Omega}$ and $\mathcal{T}$ (see (2) and (12)) is still open in the cases where $p \neq 2$ and the operator $W_{K,\Omega}$ is not Fredholm, and when $p = 2$ and $W_{K,\Omega}$ is not a generalised invertible operator.*

Acknowledgment. The author would like to thank Professor F.-O. Speck for many valuable discussions on the subject of this note.

References

1. L.P. Castro and F.-O. Speck, On the inversion of higher order Wiener-Hopf operators, *J. Integral Equations Appl.* **8** (1996), 269–285.
2. R. Duduchava, *Integral equations with fixed singularities*, Teubner, Leipzig, 1979.
3. G. Èskin, *Boundary value problems for elliptic pseudodifferential equations*, American Mathematical Society, Providence, R.I., 1981.
4. F.-O. Speck, *General Wiener-Hopf factorization methods*, Pitman Res. Notes Math. **119**, Longman, Harlow, 1985.
5. F.-O. Speck, On the generalized invertibility of Wiener-Hopf operators in Banach spaces, *Integral Equations Operator Theory* **6** (1983), 458–465.
6. M.C. Câmara, A.B. Lebre and F.-O. Speck, Meromorphic factorization partial index estimates and elastodynamic diffraction problems, *Math. Nachr.* **157** (1992), 291–317.
7. L.P. Castro and F.-O. Speck, On the characterization of the intermediate space in generalized factorizations, *Math. Nachr.* **176** (1995), 39–54.
8. Yu.I. Karlovich and I.M. Spitkovskii, Factorization problem for almost periodic matrix-functions and Fredholm theory of Toeplitz operators with semi-almost periodic matrix symbols, *Lecture Notes in Math.* **1043** (1984), 279–282.

Department of Mathematics, University of Aveiro, 3810 Aveiro, Portugal

E-mail: lcastro@mat.ua.pt

A. CHARAFI and D.R. MATRAVERS

On the use of radial basis functions in boundary integral equation methods

1. Introduction

Domain integrals occur in the numerical solution of inhomogeneous partial differential equations (PDE) by boundary integral equation methods (BIEM). Although they do not usually introduce new unknowns, they may be computationally expensive to evaluate. One approach to this problem is to transform the domain integrals into equivalent boundary integrals. This can be achieved by approximating the inhomogeneous term with a series of radial basis functions (RBF) for which the equation is analytically integrable, and so can provide an approximate particular solution of the inhomogeneous PDE.

In this paper we carry out numerical experiments on the Poisson equation to compare the performance of three types of RBF, namely the polynomial $1+r$, the multiquadric (MQ), and the thin plate spline (TPS). We study the relationship between the convergence of the radial basis approximation to the inhomogeneous term and the convergence of the numerical results to the solution of the Poisson equation.

2. Boundary integral equation methods

We consider the boundary value problem

$$\begin{aligned} \Delta u &= f \quad \text{in } \Omega, \\ u &= g_1 \quad \text{on } \Gamma_1, \\ \frac{\partial u}{\partial n} &= g_2 \quad \text{on } \Gamma_2, \end{aligned} \tag{1}$$

where $\Omega \subset \mathbb{R}^d$ and $\partial\Omega = \Gamma_1 \cup \Gamma_2$.

Equation (1) can be transformed into the integral equation

$$\theta(x)u(x) + \int_{\partial\Omega} u(\xi)\frac{\partial E_d}{\partial n}(x-\xi)d\xi = \int_{\partial\Omega} \frac{\partial u}{\partial n}(\xi)E_d(x-\xi)d\xi - \int_{\Omega} f(\xi)E_d(x-\xi)d\xi, \tag{2}$$

where $x \in \partial\Omega$, $\theta(x)$ is a geometric constant and E_d is a fundamental solution of the Laplace equation in $\mathbb{R}^d$.

If the function $f(x)$ is identically zero, then relation (2) describes a purely BIE, and to solve it numerically we need to discretise only the boundary $\partial\Omega$.

For inhomogeneous equations we need to evaluate the domain integral

$$\int_{\Omega} f(\xi) E_d(x-\xi) d\xi,$$

or to produce a method of avoiding it. We will discuss methods which eliminate the inhomogeneities via exact or approximate particular solutions.

Supose that we can find a particular solution $\hat{u}$ that satisfies the inhomogeneous equation

$$\Delta \hat{u} = f. \tag{3}$$

By taking $v = u - \hat{u}$, problem (1) becomes

$$\begin{aligned} \Delta v &= 0 \quad \text{in } \Omega, \\ v &= g_1 - \hat{u} \quad \text{on } \Gamma_1, \\ \frac{\partial v}{\partial n} &= g_2 - \frac{\partial \hat{u}}{\partial n} \quad \text{on } \Gamma_2. \end{aligned} \tag{4}$$

In the integral equation for v there is no domain integral, but the boundary integrals have new terms. In this method $\hat{u}$ has to be determined analytically, so it is important to determine for which functions (3) can be integrated.

For some radial functions f, a particular solution $\hat{u}$ can be determined and equation (2) can be reduced to a purely BIE.

In general, the inhomogeneous term $f(x)$ can be approximated by a function $s(x)$ of the form

$$s(x) = \sum_{k=1}^{m} \lambda_k \phi(\|x - x_k\|), \tag{5}$$

where ϕ is an RBF, $\{x_k\}$ are m different points of Ω and the coefficients λ_k are defined by the equations

$$s(x_j) = f(x_j), \quad j = 1, 2, \ldots, m.$$

If we denote by $\hat{u}_k$ a particular solution of $\Delta u = \phi(\cdot - x_k)$, we can then approximate the inhomogeneous BVP by

$$\begin{aligned} \Delta v &\simeq 0 \quad \text{in } \Omega, \\ v &= g_1 - \hat{u} \quad \text{on } \Gamma_1, \\ \frac{\partial v}{\partial n} &= g_2 - \frac{\partial \hat{u}}{\partial n} \quad \text{on } \Gamma_2, \end{aligned} \tag{7}$$

where $v = u - \hat{u}$ and $\hat{u} = \sum_{k=1}^{m} \lambda_k \hat{u}_k$.

Problem (7) can then be transformed into a purely BIE and solved without domain integration.

3. Radial basis function approximations

Definition 1. By a radial basis function we understand any continuous function $\phi : \mathbb{R}^+ \longrightarrow \mathbb{R}$ for which the interpolation problem is uniquely solvable, that is,

$$\forall \Omega \subset \mathbb{R}^d,\ \forall f \in \mathcal{C}(\Omega),\ \forall \{x_1, \dots, x_m\} \subset \Omega \ \ \exists!\, (\lambda_1, \dots, \lambda_m) \in \mathbb{R}^m$$

such that

$$f(x_i) = \sum_{j=1}^{m} \lambda_j \phi(\|x_i - x_j\|), \quad i = 1, \dots, m,$$

where $\|\cdot\|$ is the Euclidean norm on $\mathbb{R}^d$.

We remark that this definition differs from the one usually quoted in the literature, which relies on particular cases. The existence of such functions is guaranteed by Michelli's theorem [1] about the non-singularity of the interpolation matrix obtained with $\phi = \mathrm{id}_{\mathbb{R}^+}$.

Conjecture 1. *Let Ω be a compact subset of $\mathbb{R}^d$ and ϕ a radial basis function. Then $\mathcal{C}(\Omega)$ is the closed linear span of the set of functions $\{\phi(\|\cdot - y\|)/y \in \Omega\}$.*

In other words, we conjecture that continuous functions from Ω to $\mathbb{R}$ can be approximated uniformly by RBF. We know that the conjecture is true for some RBF (see [1]).

In recent years the theory of RBF [1] has undergone intensive research, and considerable success has been achieved in using them to interpolate multivariate data and functions. The choice of radial basis functions has been dictated by their use in solving physical and engineering problems, before one even thought of developing a theory. Here are the most popular choices of ϕ.

- $\phi(r) = 1 + r$, first studied by Brebbia and Nardini [2] in their original paper on the dual reciprocity method; an ad-hoc choice which is not intuitive but works well.
- $\phi(r) = r^2 \log(r) + p(x)$ (where $p(x)$ is a polynomial), called thin plate spline (TPS). This is a more "natural" choice. A theorem by Duchon [3] states that in $\mathbb{R}^2$ the TPS provides the interpolant to scattered function values that minimises a 2-norm of second derivatives.
- $\phi(r) = (r^2 + c^2)^{1/2}$, called multiquadric (MQ), first introduced by Hardy [4] to interpolate geophysical data. Franke [5] compared over 30 two-dimensional interpolation methods and found that MQ performed best, followed by TPS.

4. Numerical experiments

Example 1. Let

$$\Omega = \{(x, y) \mid x^2 + 4y^2 \leq 4\},$$
$$u = 0 \quad \text{on } \partial\Omega.$$

The domain Ω is discretised using 16 linear boundary elements evenly placed on the ellipse, and 17 internal nodes, as in [6].

Radial basis function	$1+r$	TPS	MQ (c=0.0008)
Percentage error	0.95 %	0.95 %	0.94 %

Table 1. $u(0,0) = 0.8$ with $f(x,y) = -2$.

Radial basis function	$1+r$	TPS	MQ (c=1.0556)
Percentage error	2.79 %	2.67 %	2.49 %

Table 2. $u(0.3,0) = 0.083$ with $f(x,y) = -x$.

Radial basis function	$1+r$	TPS	MQ (c=0.0056)
Percentage error	6.38 %	5.35 %	5.04 %

Table 3. $u(0,0) = 0.136$ with $f(x,y) = -x^2$.

Radial basis function	$1+r$	TPS	MQ (c=0.0556)
Percentage error	0.55 %	0.49 %	0.001%

Table 4. The temperature at the centre point.

Free parameter c	0.5	0.05	0.0556
Percentage error	1.48	0.11 %	0.001%

Table 5. The solution with MQ for different values of c.

Example 2. We consider the NAFEMS thermal analysis benchmark in [6]. This problem is governed by the Poisson equation

$$\Delta u(x,y) = -\frac{10^6}{52} \quad \text{in } \Omega, \tag{8}$$

where $\Omega = \{(x,y) \,|\, |x| \leq 0.3, |y| \leq 0.2\}$. The temperature u on the boundary is maintained at 0°C. The temperature at the centre point should be 310.1°C for $K_x = K_y = 52$. The domain Ω is discretised using 20 linear boundary elements evenly placed on the rectangle and 1 internal node at the centre.

The solution obtained with MQ is highly dependent on the value of the free parameter c, as shown in Table 5.

5. Concluding remarks

Numerical results indicate that the MQ performs better than both $1 + r$ and the TPS on the examples chosen. But results with the MQ are highly dependent on the value of the free parameter c. However, the mechanism to find out the optimal choice of the parameter c for a given problem remains an open question.

As to the relationship between the error in f and the error in u, we notice that the inhomogeneous term may be better approximated by the TPS, but the MQ gives better BIEM results. For the same number of internal nodes, smaller errors are obtained with the MQ. But results appear to have converged for the MQ, although the maximum error in f is still large, while the other RBF give poorer results for a much smaller error in the inhomogeneous term f.

References

1. M.J.D. Powell, *The theory of radial basis function approximation in 1990*, Adv. Numer. Anal. **2**, Oxford Science Publications, Oxford, 1992.
2. D. Nardini and C.A. Brebbia, A new approach to free vibration analysis using boundary elements, *Appl. Math. Modelling* **1** (1983), 157–162.
3. J. Duchon, Spline minimizing rotation-invariant seminorms in Sobolev spaces, in *Constructive theory of functions of several variables*, Lecture Notes in Math. **571**, Springer-Verlag, Berlin, 1977.
4. R.L. Hardy, Multiquadratic equations of topography and other irregular surfaces, *J. Geophys. Res.* **26** (1971), 1905–1915.
5. R. Franke, Scattered data interpolation: tests of some methods, *Math. Comp.* **38** (1982), 181–199.
6. P.W. Partridge, C.A. Brebbia and L.C. Wrobel, *The dual reciprocity boundary element method*, Comput. Mech. Publ., Elsevier, London, 1992.

School of Computer Science and Mathematics, University of Portsmouth, Mercantile House, Hampshire Terrace, Portsmouth PO1 2EG, UK

E-mail: charafi@sms.port.ac.uk

I. CHUDINOVICH and A. LYTOVA

Boundary equations in mixed problems for the Maxwell system

1. Introduction

It is very difficult to overestimate the significance of potential theory methods both in the theoretical study of static and quasi-static problems of diffraction of electromagnetic waves and in solving them numerically. There are numerous papers and monographs on these questions. The interest in analogous methods for the approximate solving of essentially non-stationary problems is now on the increase. But in order to promote the further development of corresponding numerical methods, it is necessary to have a sound base that is a mathematically rigorous and sufficiently complete theory of non-stationary boundary equations in electromagnetic problems. The fact is that the boundary equations in the non-stationary case differ essentially from those in the static and quasi-static cases with some very important properties. And it is these properties that influence the convergence and stability of their numerical solution algorithms.

In [1]–[3] the mixed problem for the Maxwell system was reduced to the mixed problem for the vector wave equation. Surface retarded potentials were introduced and the solvability of the corresponding systems of boundary equations was proved in functional spaces of Sobolev type. In this paper we construct the boundary equation theory directly for the Maxwell system, without any transition to the vector wave equation.

2. Formulation of the problem

Let $\Omega \subset \mathbb{R}^3$ be an exterior domain bounded by a smooth closed surface Γ. The standard diffraction problem may be reduced to finding the electromagnetic field $\{\mathbf{E}(x,t), \mathbf{H}(x,t)\}$ satisfying in $G = \Omega \times \mathbb{R}_+$, $\mathbb{R}_+ = (0,\infty)$, the Maxwell system

$$\begin{aligned} \varepsilon\partial_t \mathbf{E}(x,t) - \operatorname{curl} \mathbf{H}(x,t) &= 0, \\ \mu\partial_t \mathbf{H}(x,t) + \operatorname{curl} \mathbf{E}(x,t) &= 0, \end{aligned} \qquad (x,t) \subset G,$$

where ε and μ are the dielectric constant and magnetic permeability of the medium, respectively. We assume that the initial conditions are homogeneous:

$$\mathbf{E}(x,0+) = \mathbf{H}(x,0+) = 0, \quad x \in \Omega.$$

In the case when the boundary surface Γ is an ideal conductor, the boundary condition has the form

$$\mathbf{E}_\tau(x,t) = \mathbf{f}_\tau(x,t), \quad (x,t) \in \Sigma = \Gamma \times \mathbb{R}_+,$$

where $\mathbf{E}_\tau$ is the component of the electric field $\mathbf{E}$ tangent to Γ and $\mathbf{f}_\tau$ is a vector field given on Σ and tangent to Γ. It should be pointed out that the homogeneity of the initial conditions does not lead to a loss of generality because existing nonhomogeneities may easily be transferred to the boundary condition.

3. Electromagnetic surface potentials

The first step in constructing the electromagnetic surface potentials is to write down the fundamental solution for the Maxwell system. This is a 6×6-matrix which we denote by

$$\begin{pmatrix} \Phi(x,t) \\ \Psi(x,t) \end{pmatrix}.$$

Here Φ and Ψ are the 3×6-matrices satisfying the equations

$$\begin{pmatrix} \varepsilon\partial_t & -\mathrm{curl} \\ \mathrm{curl} & \mu\partial_t \end{pmatrix} \begin{pmatrix} \Phi(x,t) \\ \Psi(x,t) \end{pmatrix} = \delta(x,t)\mathrm{I}, \quad (x,t) \in \mathbb{R}^4, \quad \Phi(x,t) = \Psi(x,t) = 0, \quad t < 0,$$

where δ is the Dirac distribution and I is the unit 6×6-matrix. We do not give here the explicit form of the fundamental solution. Instead, we give the explicit form of the retarded potentials.

Let $\alpha_\tau(x,t)$ be a vector field defined on Σ and tangent to Γ, and let $\Phi_j, j = 1, \ldots, 6$, be the columns of the matrix Φ. The surface electromagnetic potential of the first kind

$$(V\alpha_\tau)(x,t) = \{(V\alpha_\tau)_{\mathbf{E}}(x,t), (V\alpha_\tau)_{\mathbf{H}}(x,t)\}$$

is introduced by the formula

$$(V\alpha_\tau)_j(x,t) = \int\limits_\Sigma (\Phi_j(x-y, t-\theta), \alpha_\tau(y,\theta))_{\mathbb{R}^3}\, ds_y d\theta, \quad j = 1, \ldots, 6.$$

The first three of its components represent the electric field and the last three are the components of the magnetic field. The explicit form of this potential is

$$\begin{pmatrix} (V\alpha_\tau)_{\mathbf{E}}(x,t) \\ (V\alpha_\tau)_{\mathbf{H}}(x,t) \end{pmatrix} = \begin{pmatrix} \dfrac{\mu}{4\pi} \displaystyle\int\limits_\Gamma \frac{(\partial_t \alpha_\tau)\left(y, t - \dfrac{|x-y|}{a}\right)}{|x-y|}\, ds_y - \frac{1}{4\pi\varepsilon} \nabla_x \int\limits_0^t d\theta \int\limits_\Gamma \frac{(\mathrm{div}\,\alpha_\tau)\left(y, \theta - \dfrac{|x-y|}{a}\right)}{|x-y|}\, ds_y \\ -\dfrac{1}{4\pi} \mathrm{curl}_x \displaystyle\int\limits_\Gamma \frac{\alpha_\tau\left(y, t - \dfrac{|x-y|}{a}\right)}{|x-y|}\, ds_y \end{pmatrix},$$

where $a^2 = (\varepsilon\mu)^{-1}$. It is easy to check that the limiting value $(V\alpha_\tau)^-(x,t)$ of this potential as $(x,t) \to \Sigma$ from G is connected with its direct value $(V\alpha_\tau)^0(x,t)$, $(x,t) \in \Sigma$, by the jump formula

$$(V\alpha_\tau)^-(x,t) = \frac{1}{2} \begin{pmatrix} \varepsilon^{-1}\nu(x) \displaystyle\int\limits_0^t (\mathrm{div}\,\alpha_\tau)(x,\theta)\, d\theta \\ \nu(x) \wedge \alpha_\tau(x,t) \end{pmatrix} + (V\alpha_\tau)^0(x,t), \quad (x,t) \in \Sigma,$$

where $\wedge$ denotes the vector product in $\mathbb{R}^3$ and $\nu(x)$ is the inward unit normal to Γ. We see that the tangent component of the electric field and the normal component of the magnetic field are continuous. On the contrary, the tangent component of the magnetic field and the normal component of the electric field have jumps.

The electromagnetic potential of the second kind

$$(W\beta_\tau)(x,t) = \{(W\beta_\tau)_{\mathbf{E}}(x,t), (W\beta_\tau)_{\mathbf{H}}(x,t)\}$$

of density β_τ tangent to Γ is defined by

$$\begin{pmatrix} (W\beta_\tau)_{\mathbf{E}}(x,t) \\ (W\beta_\tau)_{\mathbf{H}}(x,t) \end{pmatrix}$$

$$= \begin{pmatrix} \dfrac{1}{4\pi}\operatorname{curl}_x \displaystyle\int_\Gamma \dfrac{\beta_\tau\left(y, t - \dfrac{|x-y|}{a}\right)}{|x-y|}\, ds_y \\ \dfrac{\varepsilon}{4\pi}\displaystyle\int_\Gamma \dfrac{(\partial_t\beta_\tau)\left(y, t-\dfrac{|x-y|}{a}\right)}{|x-y|}\, ds_y - \dfrac{1}{4\pi\mu}\nabla_x \displaystyle\int_0^t d\theta \int_\Gamma \dfrac{(\operatorname{div}\beta_\tau)\left(y, \theta-\dfrac{|x-y|}{a}\right)}{|x-y|}\, ds_y \end{pmatrix}.$$

The jump formula for this potential has the form

$$(W\beta_\tau)^-(x,t) = \frac{1}{2}\begin{pmatrix} \beta_\tau(x,t)\wedge\nu(x) \\ \mu^{-1}\nu(x)\displaystyle\int_0^t (\operatorname{div}\beta_\tau)(x,\theta)\, d\theta \end{pmatrix} + (W\beta_\tau)^0(x,t), \quad (x,t)\in\Sigma.$$

Here again $(W\beta_\tau)^-(x,t)$ and $(W\beta_\tau)^0(x,t)$ are the limiting value and the direct value of this potential, respectively. We see that the tangent component of the electric part and the normal component of the magnetic part have jumps, while the tangent component of the magnetic field and the normal component of the electric field are continuous. Representing the solution of the diffraction problem as an electromagnetic potential of the first kind, we obtain the system of boundary equations

$$\nu(x)\wedge\left[\int_\Gamma \frac{(\partial_t\alpha_\tau)\left(y, t-\dfrac{|x-y|}{a}\right)}{|x-y|}\, ds_y - a^2\nabla_x\int_0^t d\theta\int_\Gamma \frac{(\operatorname{div}\alpha_\tau)\left(y,\theta-\dfrac{|x-y|}{a}\right)}{|x-y|}\, ds_y\right]$$

$$= \frac{4\pi}{\mu}\nu(x)\wedge\mathbf{f}_\tau(x,t), \quad (x,t)\in\Sigma,$$

or, for short,

$$R_1\alpha_\tau = \mathbf{f}_\tau. \tag{1}$$

The representation of the solution as an electromagnetic potential of the second kind leads to the system

$$\frac{1}{2}\beta_\tau(x,t) + \frac{1}{4\pi}\nu(x) \wedge \operatorname{curl}_x \int_\Gamma \frac{\beta_\tau\left(y, t - \frac{|x-y|}{a}\right)}{|x-y|} ds_y$$

$$= \nu(x) \wedge \mathbf{f}_\tau(x,t), \quad (x,t) \in \Sigma,$$

or, for short,

$$R_2\beta_\tau = \mathbf{f}_\tau. \tag{2}$$

4. Existence theorems

Let $\mathbf{H}_{m,\kappa}(\Gamma \times \mathbb{R})$ and $H_{m,\kappa}(\Gamma \times \mathbb{R})$ be the standard Sobolev spaces of order $m \in \mathbb{R}$ with weight $\exp(-2\kappa t)$, $\kappa > 0$, which consist of three-component vector fields and scalar functions, respectively, defined on $\Gamma \times \mathbb{R}$. $\mathbf{H}_{r;m,\kappa}(\Sigma)$ and $H_{r;m,\kappa}(\Sigma)$ are their subspaces consisting of elements vanishing for $t < 0$. Denoting by ∂_t^k the fractional derivative of order $k \in \mathbb{R}$ with respect to time, we introduce the spaces $\mathbf{H}_{r;m,k,\kappa}(\Sigma)$ and $H_{r;m,k,\kappa}(\Sigma)$ of vector and scalar functions u, respectively, vanishing for $t < 0$ and such that $\partial_t^k u \in \mathbf{H}_{r;m,\kappa}(\Sigma)$ or $H_{r;m,\kappa}(\Sigma)$, respectively. $\mathbf{H}^\tau_{r;m,k,\kappa}(\Sigma)$ is the subspace of $\mathbf{H}_{r;m,k,\kappa}(\Sigma)$ that consists of the vector fields tangent to Γ. Finally, we denote by $\mathbf{H}^\tau_{r;m,k,\kappa}(\operatorname{div}, \Sigma)$ the space of vector functions $\mathbf{u} \in \mathbf{H}^\tau_{r;m,k,\kappa}(\Sigma)$ such that $\operatorname{div}\mathbf{u} \in H_{r;m+1,k-2,\kappa}(\Sigma)$. The norm of an element $\mathbf{u}$ in this space is defined as the sum of the norms of $\mathbf{u}$ in $\mathbf{H}^\tau_{r;m,k,\kappa}(\Sigma)$ and $\operatorname{div}\mathbf{u}$ in $H_{r;m+1,k-2,\kappa}(\Sigma)$. In fact, the indices m and k show that the corresponding vector or scalar function has generalised derivatives of order m with respect to the space variables changing along the surface Γ, and of order $m + k$ with respect to time, which are square integrable on Σ with weight $\exp(-2\kappa t)$.

Theorem 1. *There exists $\kappa_0 > 0$ such that the systems of boundary equations* (1) *and* (2) *are uniquely solvable for every $\mathbf{f}_\tau \in \mathbf{H}^\tau_{r;1/2,k,\kappa}(\Sigma)$, $k \in \mathbb{R}$, $\kappa \geq \kappa_0$. The resolvent operators*

$$R_1^{-1} : \mathbf{H}^\tau_{r;1/2,k,\kappa}(\Sigma) \mapsto \mathbf{H}^\tau_{r;-1/2,k,\kappa}(\operatorname{div}, \Sigma),$$

$$R_2^{-1} : \mathbf{H}^\tau_{r;1/2,k,\kappa}(\Sigma) \mapsto \mathbf{H}^\tau_{r;1/2,k-2,\kappa}(\Sigma)$$

are continuous for all $k \in \mathbf{R}$, $\kappa \geq \kappa_0$.

5. The case of the impedance boundary condition

We consider the mixed problem that consists in finding a solution

$$\{\mathbf{E}(x,t), \mathbf{H}(x,t)\}, \quad (x,t) \in G,$$

to the Maxwell system which satisfies the homogeneous initial conditions and the impedance boundary condition

$$\mathbf{E}_\tau(x,t) - \chi\nu(x) \wedge \mathbf{H}(x,t) = \mathbf{g}_\tau(x,t), \quad (x,t) \in \Sigma,$$

where $\chi > 0$. This type of boundary condition occurs when the boundary surface Γ is not an ideal conductor. Representing the solution of this problem as an electromagnetic potential of the first kind, we obtain a system of boundary equations which we write symbolically as $J_1\alpha_\tau = \mathbf{g}_\tau$. The representation of the solution as an electromagnetic potential of the second kind leads to a system of the form $J_2\beta_\tau = \mathbf{g}_\tau$. To formulate the result of unique solvability of these systems, we need one more functional space. Let $\mathcal{H}^\tau_{r;m,k,\kappa}(\mathrm{div}, \Sigma)$ be the space of vector functions $\mathbf{u} \in \mathbf{H}^\tau_{r;m,k,\kappa}(\Sigma)$ such that $\mathrm{div}\,\mathbf{u} \in H_{r;m,k-1,\kappa}(\Sigma)$.

Theorem 2. *There exists κ_0 such that the systems of boundary equations $J_1\alpha_\tau = \mathbf{g}_\tau$ and $J_2\beta_\tau = \mathbf{g}_\tau$ are uniquely solvable for every $\mathbf{g}_\tau \in \mathcal{H}^\tau_{r;-1/2,k,\kappa}(\mathrm{div}, \Sigma)$, $k \in \mathbb{R}$. The resolvent operators*

$$J_1^{-1} : \mathcal{H}^\tau_{r;-1/2,k,\kappa}(\mathrm{div}, \Sigma) \mapsto \mathbf{H}^\tau_{r;-1/2,k-2,\kappa}(\mathrm{div}, \Sigma),$$

$$J_2^{-1} : \mathcal{H}^\tau_{r;-1/2,k,\kappa}(\mathrm{div}, \Sigma) \mapsto \mathbf{H}^\tau_{r;1/2,k-4,\kappa}(\Sigma)$$

are continuous for all $\kappa \geq \kappa_0$, $k \in \mathbb{R}$.

References

1. I.Yu. Chudinovich, Mathematical questions of the boundary equation method in electrodynamics, *Proc. Conf. MMET94*, 67–70.
2. I.Yu. Chudinovich, The solvability of boundary equations in mixed problems for the nonstationary Maxwell system, *Math. Methods Appl. Sci.* (to appear).
3. I.Yu. Chudinovich, The solvability of boundary equations in mixed problems for the Maxwell system, *Dokl. Russian Acad. Sci.* (to appear).

Department of Mathematics and Mechanics, Kharkov State University, 4 Svobody Sq., 310077 Kharkov, Ukraine

C. CONSTANDA

Robin-type conditions in plane strain

1. Preliminaries

Let $S \subset \mathbb{R}^2$ be a domain bounded by a simple closed C^2-contour ∂S of unit outward normal vector ν, and suppose that S is occupied by a homogeneous and isotropic material with Lamé constants λ and μ. If $x = (x_1, x_2)^{\mathrm{T}}$ is a generic point in $\mathbb{R}^2$, then the equilibrium equations of plane strain in the absence of body forces can be written in the form

$$A(\partial_x)u(x) = A(\partial/\partial x_1, \partial/\partial x_2)u(x) = 0, \quad x \in S, \tag{1}$$

where

$$A(\xi_1, \xi_2) = \begin{pmatrix} \mu\Delta + (\lambda+\mu)\xi_1^2 & (\lambda+\mu)\xi_1\xi_2 \\ (\lambda+\mu)\xi_1\xi_2 & \mu\Delta + (\lambda+\mu)\xi_2^2 \end{pmatrix}.$$

We also consider the stress boundary operator $T(\partial_x) = T(\partial/\partial x_1, \partial/\partial x_2)$ defined by

$$T(\xi_1, \xi_2) = \begin{pmatrix} (\lambda+2\mu)\nu_1\xi_1 + \mu\nu_2\xi_2 & \mu\nu_2\xi_1 + \lambda\nu_1\xi_2 \\ \lambda\nu_2\xi_1 + \mu\nu_1\xi_2 & \mu\nu_1\xi_1 + (\lambda+2\mu)\nu_2\xi_2 \end{pmatrix}.$$

We assume that $\lambda + \mu > 0$ and $\mu > 0$, which ensures that the system (1) is elliptic [1].

The space of rigid displacements $\mathcal{F}$ is spanned by the columns of the matrix

$$F = \begin{pmatrix} 1 & 0 & x_2 \\ 0 & 1 & -x_1 \end{pmatrix}.$$

We denote by S^+ the finite domain enclosed by ∂S and set $S^- = \mathbb{R}^2 \setminus (S^+ \cup \partial S)$. Let $\mathcal{A}$ be the space of functions in S^- that, as $r = |x| \to \infty$, admit an asymptotic expansion (in terms of polar coordinates) of the form

$$\begin{aligned} u_1(r,\theta) &= r^{-1}(\alpha m_0 \sin\theta + m_1 \cos\theta + m_0 \sin 3\theta + m_2 \cos 3\theta) + O(r^{-2}), \\ u_2(r,\theta) &= r^{-1}(m_3 \sin\theta + \alpha m_0 \cos\theta + m_4 \sin 3\theta - m_0 \cos 3\theta) + O(r^{-2}), \end{aligned}$$

and $\mathcal{A}^* = \mathcal{A} \oplus \mathcal{F}$. Also, let $\sigma(x)$ be a positive definite (2×2)-matrix and f a (2×1)-matrix given on ∂S. The interior and exterior boundary value problems with elastic boundary conditions are formulated as follows:

(R^+) Find $u \in C^2(S^+) \cap C^1(\bar{S}^+)$ such that $Au = 0$ in S^+ and $(Tu + \sigma u)|_{\partial S} = f$.

(R^-) Find $u \in C^2(S^-) \cap C^1(\bar{S}^-) \cap \mathcal{A}^*$ such that $Au = 0$ in S^- and $(Tu - \sigma u)|_{\partial S} = f$.

Using the Betti formula, we can easily show that each of (R^+), (R^-) has at most one solution. In what follows we apply the direct boundary integral equation method to prove that these problems do, in fact, have (unique) solutions.

2. Elastic potentials

We consider the matrix of fundamental solutions

$$D(x,\, y) = (\operatorname{adj} A)(\partial_x)[(8\pi)^{-1}(|x-y|^2 \ln|x-y|)E_2],$$

where E_2 is the identity (2×2)-matrix, and the matrix of singular solutions

$$P(x,\, y) = [T(\partial_y)D(y,\, x)]^{\mathrm{T}}.$$

The elastic single and double layer potentials are defined, respectively, by

$$(V\varphi)(x) = \int\limits_{\partial S} D(x,\, y)\varphi(y)\, ds(y), \quad (W\varphi)(x) = \int\limits_{\partial S} P(x,\, y)\varphi(y)\, ds(y).$$

Theorem 1. [1] (i) *If $\varphi \in C(\partial S)$, then $W\varphi \in \mathcal{A}$ and $V\varphi = M^\infty(p\varphi) + \sigma^{\mathcal{A}}$, where*

$$p\zeta = \int\limits_{\partial S} F^{\mathrm{T}}\zeta\, ds,$$

$\sigma^{\mathcal{A}} \in \mathcal{A}$, and M^∞ is the (2×3)-matrix defined in [1].

(ii) *If $\varphi \in C(\partial S)$, then $V\varphi$ and $W\varphi$ are analytic in $S^+ \cup S^-$ and*

$$A(V\varphi) = A(W\varphi) = 0 \quad in \quad S^+ \cup S^-.$$

(iii) *If $\varphi \in C^{0,\alpha}(\partial S)$, $\alpha \in (0,1)$, then the direct values $V_0\varphi$ and $W_0\varphi$ of $V\varphi$ and $W\varphi$ on ∂S exist (the latter as principal value), the functions*

$$\mathcal{V}^+(\varphi) = (V\varphi)|_{\bar{S}^+}, \quad \mathcal{V}^-(\varphi) = (V\varphi)|_{\bar{S}^-}$$

are of class $C^\infty(S^+)\cap C^{1,\alpha}(\bar{S}^+)$ and $C^\infty(S^-)\cap C^{1,\alpha}(\bar{S}^-)$, respectively, and

$$T\mathcal{V}^+(\varphi) = (W_0^* + \tfrac{1}{2}I)\varphi, \quad T\mathcal{V}^-(\varphi) = (W_0^* - \tfrac{1}{2}I)\varphi,$$

where W_0^* is the adjoint of W_0 and I is the identity operator.

(iv) *If $\varphi \in C^{1,\alpha}(\partial S)$, $\alpha \in (0,1)$, then the functions*

$$\mathcal{W}^+(\varphi) = \begin{cases} (W\varphi)|_{S^+} & in\ S^+, \\ (W_0 - \frac{1}{2}I)\varphi & on\ \partial S, \end{cases} \quad \mathcal{W}^-(\varphi) = \begin{cases} (W\varphi)|_{S^-} & in\ S^-, \\ (W_0 + \frac{1}{2}I)\varphi & on\ \partial S \end{cases}$$

are of class $C^\infty(S^+)\cap C^{1,\alpha}(\bar{S}^+)$ and $C^\infty(S^-)\cap C^{1,\alpha}(\bar{S}^-)$, respectively, and

$$T\mathcal{W}^+(\varphi) = T\mathcal{W}^-(\varphi) \quad on \quad \partial S.$$

We define $N_0 : C^{1,\,\alpha}(\partial S) \to C^{0,\,\alpha}(\partial S)$ by $N_0\zeta = T\mathcal{W}^+(\zeta) = T\mathcal{W}^-(\zeta)$.

Theorem 2. (i) *For every* $\partial S \in C^2$ *and any* $\alpha \in (0,1)$, *there are a unique* (2×3)-*matrix* Φ *of class* $C^{1,\alpha}(\partial S)$ *and a unique constant* (3×3)-*matrix* $\mathcal{C}$ *such that the columns of* Φ *are linearly independent and*

$$V_0\Phi = F\mathcal{C}, \quad p\Phi = E_3,$$

where E_3 *is the identity* (3×3)-*matrix.*

(ii) *For a continuous* (2×1)-*matrix* ζ *we define*

$$q\zeta = \int\limits_{\partial S} \Phi^{\mathrm{T}}\zeta\, ds.$$

Then $q(V_0\zeta) = \mathcal{C}^{\mathrm{T}}p\zeta$ *and* $q[(W_0+\frac{1}{2}I)\zeta] = 0$. *In particular,* $q(Fk) = k$ *for any constant* (3×1)-*vector* k.

3. The interior problem

Theorem 3. *If* $\sigma \in C^{1,\alpha}(\partial S)$, (R^+) *has a unique solution for any* $f \in C^{0,\alpha}(\partial S)$.

Proof. Let $f \in C^{0,\alpha}(\partial S)$, and let u be a solution of (R^+). By the Somigliana representation formula,

$$u = \mathcal{V}^+(Tu|_{\partial S}) - \mathcal{W}^+(u|_{\partial S}) \quad \text{in} \quad \bar{S}^+. \tag{2}$$

The boundary condition yields $Tu|_{\partial S} = -\sigma u|_{\partial S} + f$, so, from (2),

$$u = -\mathcal{V}^+(\sigma u|_{\partial S}) - \mathcal{W}^+(u|_{\partial S}) + \mathcal{V}^+(f). \tag{3}$$

We write $u|_{\partial S} = \varphi$. Then, by Theorem 1(iii), the representation (3) on ∂S yields the singular boundary integral equation of index zero

$$V_0(\sigma\varphi) + (W_0 + \tfrac{1}{2}I)\varphi = V_0 f. \tag{$\mathcal{R}^+$}$$

Let φ_0 be a solution of the homogeneous equation, that is,

$$V_0(\sigma\varphi_0) + (W_0 + \tfrac{1}{2}I)\varphi_0 = 0. \tag{4}$$

By (4) and Theorem 2(i), the function $u_0 = \mathcal{V}^-(\sigma\varphi_0) + \mathcal{W}^-(\varphi_0) - \mathcal{V}^-(\Phi)p(\sigma\varphi_0)$ satisfies

$$Au_0 = 0 \quad \text{in} \quad S^-,$$
$$u_0|_{\partial S} = V_0(\sigma\varphi_0) + (W_0 + \tfrac{1}{2}I)\varphi_0 - F\mathcal{C}p(\sigma\varphi_0) = -F\mathcal{C}p(\sigma\varphi_0),$$
$$u_0 = M^\infty p(\sigma\varphi_0) - M^\infty(p\Phi)p(\sigma\varphi_0) + u_0^{\mathcal{A}} = u_0^{\mathcal{A}}.$$

This is a Dirichlet problem with the unique solution $u_0 = -F\mathcal{C}p(\sigma\varphi_0) = 0$ in S^-.

Suppose that $\det \mathcal{C} \neq 0$. Then $p(\sigma\varphi_0) = 0$, which implies that $\mathcal{V}^-(\sigma\varphi_0) + \mathcal{W}^-(\varphi_0) = 0$ in $\bar{S}^-$. Applying the T operator to this equality, we find that, by Theorem 1(iii),

$$(W_0^* - \tfrac{1}{2}I)(\sigma\varphi_0) + N_0\varphi_0 = 0. \tag{5}$$

Now by Theorem 1(iii), (4) and (5), the function $v_0 = \mathcal{V}^+(\sigma\varphi_0) + \mathcal{W}^+(\varphi_0)$ satisfies

$$Av_0 = 0 \quad \text{in} \quad S^+,$$

$$\begin{aligned}(Tv_0 + \sigma v_0)|_{\partial S} = [(W_0^* + \tfrac{1}{2}I)(\sigma\varphi_0) + N_0\varphi_0] + \sigma[V_0(\sigma\varphi_0) + (W_0 - \tfrac{1}{2}I)\varphi_0] \\ = \sigma\varphi_0 - \sigma\varphi_0 = 0.\end{aligned}$$

Since $(\mathcal{R}^+)$ has at most one solution, it follows that $\mathcal{V}^+(\sigma\varphi_0) + \mathcal{W}^+(\varphi_0) = 0$ in S^+. Applying the T operator here as well, we arrive at

$$(W_0^* + \tfrac{1}{2}I)(\sigma\varphi_0) + N_0\varphi_0 = 0. \tag{6}$$

Finally, subtracting (5) and (6) leads to $\sigma\varphi_0 = 0$, which implies that $\varphi_0 = 0$. Hence, by the Fredholm Alternative, $(\mathcal{R}^+)$ has a unique solution $\varphi \in C^{0,\alpha}(\partial S)$. In fact, arguments similar to those in [2] show that $\varphi \in C^{1,\alpha}(\partial S)$. In view of (3), a direct application of Theorem 1(iii),(iv) now completes the proof.

4. The exterior problem

Theorem 4. *If* $\sigma \in C^{1,\alpha}(\partial S)$, (R^-) *has a unique solution for any* $f \in C^{0,\alpha}(\partial S)$.

Proof. Let $u \in \mathcal{A}^*$ be a solution of (R^-), that is, $u = u^{\mathcal{A}} + Fk$, where k is some constant (3×1)-vector. By the Somigliana representation formula,

$$u - Fk = -\mathcal{V}^-(T(u - Fk)|_{\partial S}) + \mathcal{W}^-((u - Fk)|_{\partial S}) \quad \text{in} \quad \bar{S}^-,$$

so, from the boundary condition,

$$u = -\mathcal{V}^-(\sigma u|_{\partial S} + f) + \mathcal{W}^-(u|_{\partial S}) + Fk \quad \text{in} \quad \bar{S}^-. \tag{7}$$

We write $u|_{\partial S} = \varphi$. Then (7) on ∂S yields the singular integral equation

$$V_0(\sigma\varphi + f) - (W_0 - \tfrac{1}{2}I)\varphi - Fk = 0 \tag{$\mathcal{R}^-$}$$

of index zero. Applying the operator q to this equality, we find that, by Theorem 2(ii),

$$\mathcal{C}^{\mathrm{T}} p(\sigma\varphi + f) + q\varphi = k.$$

Once again, suppose that $\det \mathcal{C} \neq 0$, and let $k = q\varphi$. Then every solution of $(\mathcal{R}^-)$ satisfies $p(\sigma\varphi + f) = 0$, so, by (7), it generates a function $u \in \mathcal{A}^*$.

Let φ_0 be a solution of the homogeneous equation, that is,

$$V_0(\sigma\varphi_0) - (W_0 - \tfrac{1}{2}I)\varphi_0 - Fq\varphi_0 = 0. \tag{8}$$

By (8), the function $v_0 = \mathcal{V}^+(\sigma\varphi_0) - \mathcal{W}^+(\varphi_0) - Fq\varphi_0$ satisfies

$$Av_0 = 0 \quad \text{in} \quad S^+,$$
$$v_0|_{\partial S} = V_0(\sigma\varphi_0) - (W_0 - \tfrac{1}{2}I)\varphi_0 - Fq\varphi_0 = 0.$$

This interior Dirichlet problem has the unique solution

$$v_0 = \mathcal{V}^+(\sigma\varphi_0) - \mathcal{W}^+(\varphi_0) - Fq\varphi_0 = 0 \quad \text{in} \quad \bar{S}^+. \tag{9}$$

Applying the T operator to this equality leads to

$$(W_0^* + \tfrac{1}{2}I)(\sigma\varphi_0) - N_0\varphi_0 = 0. \tag{10}$$

By (8) and (10), the function $u_0 = \mathcal{V}^-(\sigma\varphi_0) - \mathcal{W}^-(\varphi_0) - Fq\varphi_0$ satisfies

$$Au_0 = 0 \quad \text{in} \quad S^-,$$
$$\begin{aligned}(Tu_0 - \sigma u_0)|_{\partial S} &= [(W_0^* - \tfrac{1}{2}I)(\sigma\varphi_0) - N_0\varphi_0] - \sigma[V_0(\sigma\varphi_0) - (W_0 + \tfrac{1}{2}I)\varphi_0 - Fq\varphi_0] \\ &= -\sigma\varphi_0 + \sigma\varphi_0 = 0,\end{aligned}$$
$$u_0 = M^\infty p(\sigma\varphi_0) + u^{\mathcal{A}} - Fq\varphi_0 = u^{\mathcal{A}^*}.$$

Since (R^-) has at most one solution, it follows that $u_0 = 0$ in $\bar{S}^-$, so $Fq\varphi_0 = 0$ and, by (9), $\mathcal{V}^-(\sigma\varphi_0) - \mathcal{W}^-(\varphi_0) = 0$. Applying the T operator, we find that

$$(W_0^* - \tfrac{1}{2}I)(\sigma\varphi_0) - N_0\varphi_0 = 0. \tag{11}$$

As before, from (10) and (11) we obtain $\varphi_0 = 0$, and the argument continues as in the proof of Theorem 3.

Remark 1. A non-homogeneous system (1) can be reduced to the homogeneous case by means of a Newtonian potential.

Remark 2. If $\det \mathcal{C} = 0$, $D(x,y)$ is replaced by $D^H(x,y) = D(x,y) + F(x)HF^{\mathrm{T}}(y)$, where H is a constant (3×3)-matrix such that$H^{\mathrm{T}} = H$ and $\det(H + \mathcal{C}) \neq 0$ [3].

References

1. C. Constanda, The boundary integral equation method in plane elasticity, *Proc. Amer. Math. Soc.* **123** (1995), 3385–3396.
2. C. Constanda, *A mathematical analysis of bending of plates with transverse shear deformation*, Pitman Res. Notes Math. Ser. **215**, Longman, Harlow, 1990.
3. C. Constanda, Integral equations of the first kind in plane elasticity, *Quart. Appl. Math.* **53** (1995), 783–793.

Department of Mathematics, University of Strathclyde, Glasgow, U.K.

E-mail: c.constanda@strath.ac.uk

R.F. DAY

Boundary integral simulation of a finite inviscid jet

1. Introduction

The classical jet solution is Rayleigh's infinite cylindrical jet. Instabilities of wavelength $\lambda \approx 4.5d$ on an infinite cylindrical column of fluid of diameter d dominate others and grow until their amplitude is high enough to cause the jet to break up into drops [1]. An engineering application of this mechanism of drop formation is the operation of a continuous inkjet printer. Briefly, these printers work by squirting a continuous jet from a tank under constant pressure, and occasionally deflecting drops on to a surface to write dot-matrix characters. Most drops are not deflected and get collected and returned to the tank.

These printers print very rapidly, but are however quite complicated to manufacture and operate, so the industry is interested to study how to spit a single drop when required. This is the basic idea of a common office inkjet printer which runs relatively slowly and is known as drop-on-demand. The fluid dynamics of this type of printer is called a finite jet, which is analogous to the transient startup of the Rayleigh jet with varying driving pressure. In particular, this work is a model of a finite jet of an inviscid fluid from an axisymmetric orifice, as shown in Fig. 1.

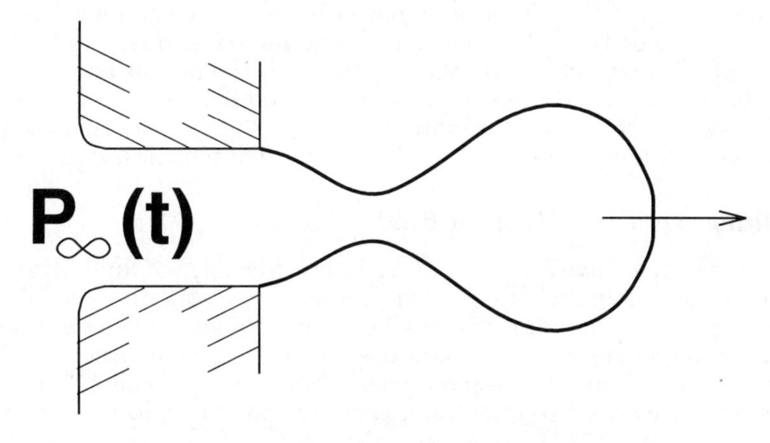

Fig. 1. Finite jet issuing from a nozzle.

The model that has been developed assumes incompressibility and irrotationality, hence potential flow. The following are the non-dimensional equations, using the unsteady Bernoulli relation, and taking surface pressure equal to $\gamma\kappa$, where κ is curvature and γ is surface tension. The equations have been non-dimensionalised

This work is partly supported by Domino UK Ltd., Bar Hill, Cambridge.

using γ, density ρ, and nozzle radius a:

$$\mathbf{u} = \nabla\phi,$$
$$\nabla^2\phi = 0,$$
$$\frac{D\phi}{Dt} - \tfrac{1}{2}|\nabla\phi|^2 + \kappa = P_\infty(t).$$

The difficult equation to solve is Laplace's equation, $\nabla^2\phi = 0$, which is solved using the Boundary Integral Method, a well-known numerical method described in many books such as [2]. The method is also used for studying Hele-Shaw, electrostatics, and elasticity problems. Given the distribution of the potential ϕ, one can solve for the normal derivative of the potential $d\phi/dn$ using the BIM. In this case, $d\phi/dn$ represents the velocity normal to the surface, which has velocity components dz/dt and dR/dt. Other numerical tools used are the common Runge-Kutta timesteppers for integrating the ODEs for the node coordinates z and R, and the potential ϕ. The surface is represented by quartic splines so the curvature is a quadratic approximation.

Various numerical techniques are used to overcome some difficulties with the BIM. The surface integrals of the Green's function $\psi = -4\pi/|\mathbf{x} - \mathbf{x}_0|$ become expressions involving complete elliptic integrals of the first and second kind, $K(m)$ and $E(m)$, because of axisymmetry. These involve $\log(m)$ terms which are singular when m goes to zero. These singularities are removed by subtracting the offending terms and adding them back after integrating analytically. The remaining regular terms are integrated using the usual Gaussian quadrature.

In this manner, the shape and potential on a given boundary is evolved by calculating the velocity of the nodes on the boundary and the material derivative of the potential, $D\phi/Dt$, advancing the position and the potential for a time increment Δt, and then repeating the process. Any shape and potential can be imposed, and given a backpressure $P_\infty(t)$, one can evolve into highly non-linear shapes, which is not possible with long-wavelength and perturbative schemes.

2. Dynamics of a free drop of fluid

The first shapes to evolve are those describing a free drop of fluid, drawn in the upper half of the (z, R)-plane, where the axisymmetry is about the z-axis. Perturbations to a sphere are created using various Legendre polynomials added to a circle, and higher order shapes resembling dog-bones are created using other (z, R)-curves. The oscillation of such shapes is observed over many periods. This is both a check of the numerics, and a method of modelling the behaviour of odd-shaped ink drops after ejection.

Rayleigh [3] found that a linearly perturbed sphere oscillates at a frequency

$$\omega_0 = [N(N-1)(N+2)]^{1/2},$$

where N is the index of the perturbation, the Legendre polynomial $P_N\cos(\theta)$. A perturbation analysis by Tsamopoulos and Brown in 1983 [4] calculated the correction term for the frequency for oscillations of moderate amplitude. The given initial distortion required to get a perfectly periodic oscillation is not trivial, as one would see when observing the "out of phase" wobble of an ellipsoid with $R = 1 + \epsilon P_2(\cos\theta)$, where $\epsilon \approx O(1)$. They found the amount of P_2 and P_4 required to give a periodic frequency to second order. The given initial shape found for $N = 2$ was

$$R(\theta) = 1 + \epsilon P_2(\cos\theta) + \epsilon^2 \left(-\tfrac{1}{5} + \tfrac{1}{21} P_2(\cos\theta) + \tfrac{72}{178} P_4(\cos\theta)\right).$$

The resulting clean periodic oscillation has a frequency $\omega(\epsilon) = \omega_0 - 0.5854\omega_0\epsilon^2$, where ω_0 is defined above. The coefficient of the ϵ^2 term is a corrected value taken from their paper from 1984 [5]. Initial shapes defined as above were given to our numerical model, and the resulting frequencies for various ϵ are plotted in Fig. 2. Agreement is good for ϵ up to about 0.6.

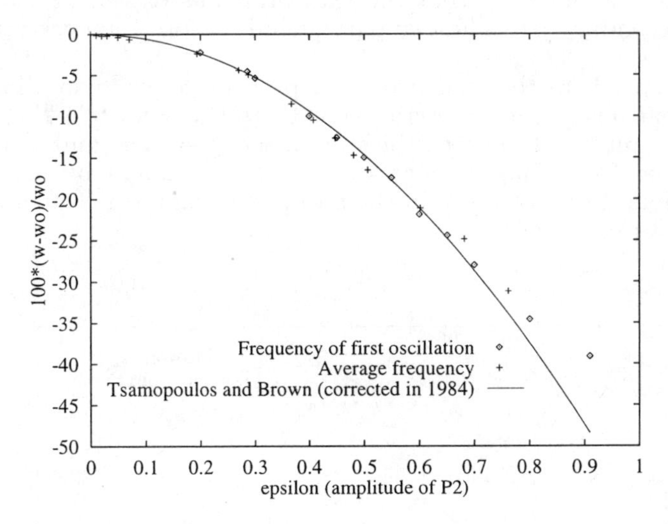

Fig. 2. The change in frequency versus aspect ratio—initial shapes as in [5].

Naturally, the usefulness of this numerical scheme is that arbitrary shapes can be used as initial conditions. This technique would be useful in determining whether an elongated drop with a neck would in fact remain in one piece after being ejected from the ink-jet printer.

3. Dynamics of a finite jet

The next application of this model is to evolve a surface which is emerging from an axisymmetric orifice. This has applications in ink-jet printing, fuel injection, manufacturing of printed circuit boards and printing of biological material to make chemical sensors.

The boundary of the control volume now has three segments: the free surface emerging from the nozzle (as in the free drop case), the rigid wall where $d\phi/dn = 0$, shown as the cross-hatched region in Fig. 1, and an arc of constant radius to close off the region back down to the axis of symmetry. Along the third segment, the flow is assumed to be sink flow, and the sole unknown is the flowrate Q.

Apart from the description of the boundary, the numerics are essentially the same. The input backpressure $P_\infty(t)$ is now significant because it is the driving force for the evolution of the jet. Given an initial shape of a spherical cap sitting on the edge of the nozzle, a number of $P_\infty(t)$ functions were imposed. The resulting jet evolution is calculated and can be visualised in 3D using a program called Geomview, obtained from http://www.geom.umn.edu for free.

Three simple $P_\infty(t)$ functions were imposed. A non-dimensional pressure $P = 4$ was tried, equal to twice that required to balance a spherical cap of radius 1. The resulting jet, which would eventually be a straight circular cylinder, was found to begin as a growth of a sphere of increasing radius, and the flowrate Q was found to be almost exactly linear until the sphere pulled away from the nozzle. The jet appears to consist of two parts: a lead drop of roughly the volume of the sphere when it detached from the nozzle, and a secondary jet like a straight cylinder of radius equal to the nozzle radius. A neck rapidly forms in the region joining these two parts, and it was observed that the lead drop was travelling too slowly to get away from the secondary jet. The two parts collide before the neck reaches a radius near zero.

In order to speed up the lead drop and prevent collision to allow the neck to reach a radius near zero, another pressure tried was $P = 8$ for $0 < T < 1$, then $P = 4$ for $T \geq 1$. The result was to detach the spherical cap earlier (and thus with smaller volume) and to send it off fast enough to avoid being caught by the secondary jet. The radius was found to go to zero in the neck, allowing for a cleaner pinch-off than with the constant pressure case.

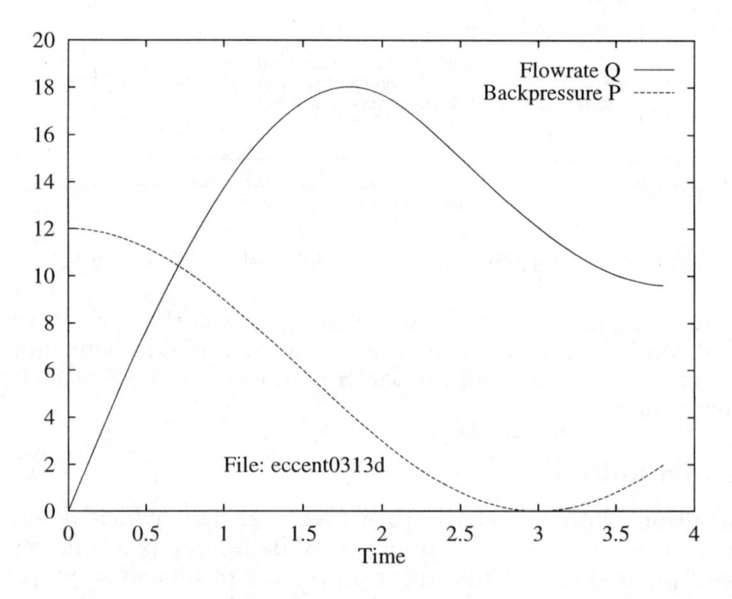

Fig. 3. Flowrate and pressure versus time.

The third regime of pressure functions explored was to use smooth pressure changes, such as $A\sin(\omega t)$ and $A\cos(\omega t)$. With this form for the pressure, there are two dimensionless parameters defined as Aa/γ and $\sqrt{\rho a^3\omega^2/\gamma}$, which are dimensionless amplitude and frequency, respectively. Two plots to describe the evolution of a jet under an $A\cos(\omega t)$ pressure are shown in Figs. 3 and 4: the applied pressure and the resulting flowrate Q are graphed in Fig. 3, and Fig. 4 is a multi-exposure picture to illustrate the evolution of the shape at equal time intervals. The reduction of the neck radius to zero is visible in the last frame, despite the coarse grid there. The lead drop is almost twice the diameter of the nozzle. We notice that the maximum flowrate occurs roughly where the sphere detaches from the nozzle. The long secondary jet is not a desirable feature as it is not likely to get retracted back into the nozzle.

4. Conclusions

A model for the evolution of a boundary for inviscid fluid has been developed using the numerical scheme known as the Boundary Integral Method.

The code was verified by finding agreement between the frequency of oscillation of linearly and moderate amplitude perturbed spheres and the frequency of oscillation from analytic theories in the literature. Moderate amplitude oscillations were found to agree up to $\epsilon \approx 0.6$.

The evolution of a finite jet from an orifice was modelled by certain changes to the numerics, and trials were performed using pressure determined by two dimensionless parameters, Aa/γ and $\sqrt{\rho a^3 \omega^2/\gamma}$.

The numerical scheme to model the motion of an elongated free drop can be used to determine the fate of an odd-shaped lead drop after being ejected from an ink-jet printer.

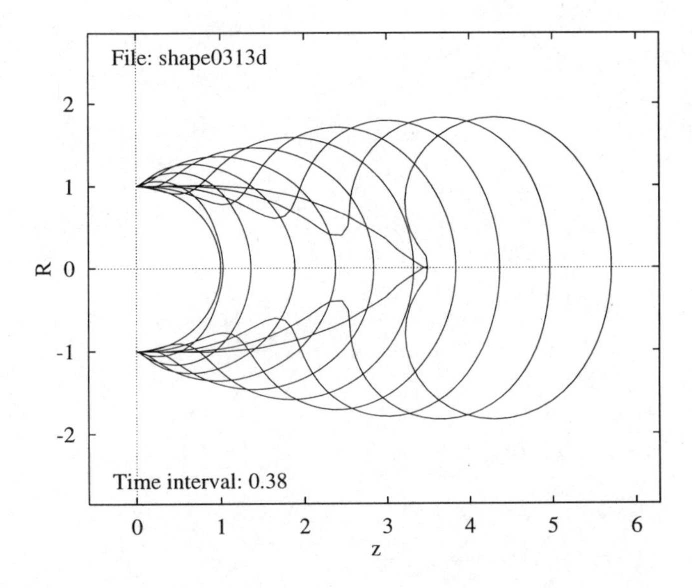

Fig. 4. Multi-exposure plot of jet evolution.

References

1. Lord Rayleigh, On the capillary phenomena of jets, *Proc. R. Soc. Lond. A* **29** (1879), 71–97.

2. C. Pozrikidis, *Boundary integral and singularity methods for linearized viscous flow*, Cambridge University Press, Cambridge, 1992.

3. S.H. Lamb, *Hydrodynamics*, 6th ed., Cambridge Math. Lib., Cambridge University Press, Cambridge, 1993, 471–475.
4. J.A. Tsamopoulos and R.A. Brown, Nonlinear oscillations of inviscid drops and bubbles, *J. Fluid Mech.* **127** (1983), 519–537.
5. J.A. Tsamopoulos and R.A. Brown, Resonant oscillations of inviscid charged drops, *J. Fluid Mech.* **147** (1984), 373–395.

Department of Applied Mathematics and Theoretical Physics, University of Cambridge, Silver Street, Cambridge, U.K.

P.B. DUBOVSKI

A uniqueness theorem for a Boltzmann equation arising in coagulation–fragmentation dynamics

1. Preliminaries

We examine the general coagulation-fragmentation equation which can be written as

$$\frac{\partial c(x,t)}{\partial t} = \tfrac{1}{2}\int_0^x K(x-y,y)c(x-y,t)c(y,t)dy + \int_0^\infty F(x,y)c(x+y,t)dy$$
$$- c(x,t)\int_0^\infty K(x,y)c(y,t)dy - \tfrac{1}{2}c(x,t)\int_0^x F(x-y,y)dy, \quad (1)$$
$$c(x,0) = c_0(x) \geq 0. \quad (2)$$

Equation (1) describes the distribution $c(x,t) \geq 0$ of particles of mass $x \geq 0$ at time $t \geq 0$ whose change in mass is governed by the non-negative reaction rates K and F called, respectively, the coagulation and fragmentation kernels. The coagulation kernel K models the rate at which particles of size x coalesce with those of size y, while F expresses the rate at which particles of size $(x+y)$ fragment into others of sizes x and y. From a physical point of view it is clear that K and F must be non-negative and symmetric, that is, $K(x,y) = K(y,x) \geq 0$ and $F(x,y) = F(y,x) \geq 0$ for all $0 \leq x,\ y < \infty$; all functions in (1) and (2) must be non-negative, and a solution $c(x,t)$ must have a bounded first moment

$$\int_0^\infty xc(x,t)dx < \infty,$$

which is equal to the total mass of the particles. The first two integrals in (1) describe the growth of the number of particles of size x due to coagulation and fragmentation, respectively, while the other integrals describe the reverse of these processes. A brief physical interpretation of the integrals on the right-hand side of (1) can be found, for example, in [1]. Applications of (1) can be found in many problems in chemistry (reacting polymers), physics (aggregation of colloidal particles, growth of gas bubbles in solids), astrophysics (formation of stars and planets) and meteorology (merging of water drops in atmospheric clouds). Many papers have analysed the existence of solutions for the above problem (see the references in [1]). However, uniqueness results are fewer.

This work was partially supported by the Russian Foundation for Basic Research, grants 97-01-00275, 97-01-00333 and 97-01-00389.

A uniqueness theorem for the above problem was proved by Melzak [2] for bounded coagulation and fragmentation kernels with the additional condition that

$$\int_0^x F(x-y,y)dy \leq \text{const.}$$

In [3] we established uniqueness for coagulation kernels with linear growth at infinity, that is,

$$K(x,y) \leq k(1+x+y), \quad F \leq \text{const} \tag{3}$$

in a class of functions with exponential decay.

In [4] it was shown that for kernels satisfying $K(x,y) \leq k(1+x^\alpha y^\alpha)$, $\alpha < 1$, and $F = 0$ our problem can have at most one solution in thc class of functions integrable with the weight $\exp(\lambda x^\alpha)$, $\lambda > 0$. In [5] uniqueness is proved for kernels such that

$$K(x,y) \leq k_1\sqrt{1+x}\sqrt{1+y}, \quad \int_0^x \sqrt{1+y}F(x-y,y)dy \leq k_2\sqrt{1+x} \tag{4}$$

in the natural class of functions with bounded first moment.

2. Uniqueness theorem

Suppose that the coagulation kernels are symmetric, that for all $x \geq 0$ there exists $X(x) \geq 1$ such that

$$K(x,y) = a(x)y + b(x,y) \quad \text{if} \quad y \geq X(x), \tag{5}$$

and that there exist positive constants λ and G such that

$$\sup_{0\leq y\leq X(x)} K(x,y) + \sup_{y\geq X(x)} b(x,y) + a(x)X(x) \leq G\exp(\lambda x), \quad x \geq 0. \tag{6}$$

The functions a and b must be non-negative. The class (5), (6) includes many physically reasonable coagulation kernels. In particular, this class contains a large subset of coagulation kernels satisfying (3) or (4). Also, the class (5) includes bounded coagulation kernels considered in [2], linear kernels of type (3) and multiplicative kernels $K = (Ax+B)(Ay+B)$, which are discussed, for example, in [6]–[8]. In addition, this class includes kernels of the form

$$K(x,y) = \alpha(x,y) + \beta(y)x + \beta(x)y + \gamma(x,y),$$

where

$$\gamma(x,y) = \begin{cases} g_1(x)x + g_2(x)y + g_3(x)xy & \text{for } y \geq x, \\ g_1(y)y + g_2(y)x + g_3(y)xy & \text{for } y \leq x. \end{cases}$$

The functions α, β and g_i, $i = 1,2,3$, are non-negative and bounded. Hence, the functions (5) constitute a large class that includes most of the important coagulation kernels.

We consider fragmentation kernels that are non-negative and symmetric and satisfy condition

$$\int_0^x F(x-y,y)\exp(-\mu y)dy \le A, \quad x \ge 0, \tag{7}$$

with some positive constants μ and A. This class includes bounded and many other fragmentation kernels (for example, $F(x,y) = (x+y)^{-1}$).

We introduce the class Y of non-negative continuous functions of $(x,t) \in [0,\infty) \times [0,\infty)$ with the same first moment, that is, which satisfy

$$\int_0^\infty x c_1(x,t)dx = \int_0^\infty x c_2(x,t)dx < \infty, \quad t \ge 0, \tag{8}$$

for any c_1, $c_2 \in Y$. Consequently, the space $C(\mathbb{R}_+^2)$ of continuous functions is partitioned into many classes Y, depending on the behaviour of their first moment.

Theorem. *The initial value problem* (1), (2) *with a coagulation kernel of class* (5) *and a fragmentation kernel* (7) *has at most one non-negative continuous solution in any class* Y.

The proof is based on the decomposition

$$\int_X^\infty a(x)y(c_1(y,t) - c_2(y,t))dy = -\int_0^X a(x)y(c_1 - c_2)dy,$$

where c_1 and c_2 are two solutions of (1), (2) in the same class Y, and the following statement.

Lemma. *Let* $v(q,t)$ *be a real continuous function with continuous partial derivatives* v_q *and* v_{qq} *on* $D = \{0 < q_0 \le q \le q_1,\ 0 \le t \le T\}$, *and suppose that* $\alpha(q), \beta(q,t), \gamma(q,t)$ *and* $\theta(q,t)$ *are real continuous functions on* D *with continuous first partial derivatives with respect to* q. *Also, let* v, v_{qq}, β *and* γ *be non-negative and* v_q, α_q, β_q, γ_q *and* θ_q *non-positive functions on* D, *and suppose that in* D

$$v(q,t) \le \alpha(q) + \int_0^t (-\beta(q,s)v_q(q,s) + \gamma(q,s)v(q,s) + \theta(q,s))\,ds, \tag{9}$$

$$v_q(q,t) \ge \alpha_q(q) + \int_0^t \frac{\partial}{\partial q}(-\beta(q,s)v_q(q,s) + \gamma(q,s)v(q,s) + \theta(q,s))\,ds. \tag{10}$$

If $c_0 = \sup_{q_0 \le q \le q_1} \alpha$, $c_1 = \sup_D \beta$, $c_2 = \sup_D \gamma$ *and* $c_3 = \sup_D \theta$, *then*

$$v(q,t) \le c_0 \exp(c_2 t) + (c_3/c_2)(\exp(c_2 t) - 1)$$

in any region $R \subset D$ *defined by*

$$R = \{(q,t) : 0 \le t \le T';\ q_0 + c_1 t \le q \le q_1 - \varepsilon + c_1 t,\ 0 < \varepsilon < q_1 - q_0,\},$$

where $T' = \min\{T, \varepsilon/c_1\}$.

3. Conclusions

(i) The condition (8) is very natural. In fact, if we multiply (1) by x and integrate it over $x \in [0, \infty)$, then we get the mass conservation law

$$\frac{d}{dt} \int_0^\infty xc(x,t)dx = 0.$$

In deriving this important equality we have assumed that all the integrals are bounded. Mass conservation takes place, for example, for coagulation kernels with linear growth at infinity and bounded fragmentation kernels [3]. In this case the equality (8) holds [3].

(ii) If K does not satisfy (3), then the mass conservation law may be violated. This phenomenon is discussed, for example, in [6] and [7] for the important multiplicative case $K = xy$, $F = 0$. In this case the behaviour of the total mass (expressed by the first moment of the solution) is uniquely defined. The condition (8) is also valid, and we obtain global uniqueness. For the Flory-Stockmayer discrete model of polymerization with $K_{i,j} = (Ai + 2)(Aj + 2)$, Ziff and Stell [8] found the value of the first moment of the solution for all $t \geq 0$, which also leads to a global uniqueness result.

(iii) In the well-known example of non-uniqueness with $K \equiv 0$, $F \equiv 2$, $c_0(x) = (\lambda + x)^{-3}$ and $\lambda > 0$ (see [5] and [9]), there are two solutions

$$c_1(x,t) = \frac{\exp(\lambda t)}{(\lambda + x)^3}, \quad c_2(x,t) = \exp(-tx)\left(c_0(x) + \int_x^\infty c_0(y)[2t + t^2(y - x)]dy\right).$$

The first one does not satisfy the mass conservation law, while the second one does. Hence, the condition (8) does not hold. Therefore, this non-uniqueness does not contradict the theorem.

(iv) Our theorem remains valid for the discrete form of the problem (1), (2) and for the case that includes sources and efflux terms, which, mathematically, means adding $h(x,t) - f(x,t)c(x,t)$, with $f(x,t) \leq \mathrm{const}(1 + x)$, to the right-hand side in (1). The theorem also includes the conditions for existence mentioned in [10], thus offering a solid basis for the analysis of well-posedness of many problems.

In addition, we can extend the uniqueness theorem in [5] to obtain uniqueness for cases to which the corresponding existence theorem in [11] applies, but which are not covered by the results in [5].

(v) There is a large intersection class of kernels satisfying both our assumptions and those that guarantee existence [4]. This class includes many unbounded kernels modelling the fast interaction of particles with approximately equal masses ($x \approx y$). As an example of a function $K(x,y)$ satisfying both (5) and the existence conditions in [4] we mention

$$K(x,y) = \alpha(x,y) + \begin{cases} \exp(\nu(2y - x)) & \text{for } y \leq x, \\ \exp(\nu(2x - y)) & \text{for } y \geq x, \end{cases} \qquad 0 \leq \nu \leq \lambda,$$

where λ is defined in (6) and the function α is bounded. Therefore, our theorem yields new uniqueness results for these kernels.

(vi) Uniqueness has also been proved for bounded coagulation kernels in a special coagulation model [12], although the boundedness is not necessary in the proof of existence. Our approach extends such results to kernels of type (5).

References

1. P.B. Dubovski, *Mathematical theory of coagulation*, GARC-KOSEF, Seoul, 1994.
2. Z.A. Melzak, A scalar transport equation, *Trans. Amer. Math. Soc.* **85** (1957), 547–560.
3. P.B. Dubovski and I.W. Stewart, Existence, uniqueness and mass conservation for the coagulation-fragmentation equation, *Math. Methods Appl. Sci.* **19** (1996), 571–591.
4. V.A. Galkin and P.B. Dubovski, Solution of the coagulation equation with unbounded kernels, *Differential Equations* **22** (1986), 504–509 .
5. I.W. Stewart, A uniqueness theorem for the coagulation-fragmentation equation, *Math. Proc. Cambridge Philos Soc.* **107** (1990), 573–578.
6. M.H. Ernst, R.M. Ziff and E.M. Hendriks, Coagulation processes with a phase transition, *J. Colloid and Interface Sci.* **97** (1984), 266–277.
7. V.A. Galkin, On the solution of the kinetic coagulation equation with the kernel $\Phi = xy$, *Meteorology and Hydrology* (1984), no. 5, 33–39 (Russian).
8. R.M. Ziff and G. Stell, Kinetics of polymer gelation, *J. Chem. Phys.* **73** (1980), 3492–3499.
9. R.M. Ziff and E.D. McGrady, The kinetics of cluster fragmentation and depolymerisation, *J. Phys. A* **18** (1985), 3027–3037.
10. J.L. Spouge, An existence theorem for the discrete coagulation-fragmentation equations, *Math. Proc. Cambridge Philos. Soc.* **96** (1984), 351–357; **98** (1985), 183–185.
11. I.W. Stewart, A global existence theorem for the general coagulation-fragmentation equation with unbounded kernels, *Math. Methods Appl. Sci.* **11** (1989), 627–648.
12. O. Bruno, A. Friedman and F. Reitich, Asymptotic behavior for a coalescence problem, *Trans. Amer. Math. Soc.* **338** (1993), 133–158.

Institute of Numerical Mathematics, Russian Academy of Sciences, 8 Gubkin St., 117333 Moscow, Russia

E-mail: dubp@inm.ras.ru

V.A. GEYLER and A.V. POPOV

Hofstadter's butterfly for a periodic array of quantum dots

1. Introduction

Beginning with the classical works of Azbel, Hofstadter, and Wannier (see [1]–3]), unusual spectral properties of two-dimensional periodic systems in a uniform magnetic field attract ever increasing attention. The main results here are concerned with theoretical explanations of the quantum Hall effect discovered by K. von Klitzing [4] (Nobel Prize, 1985). From the mathematical point of view this subject is of considerable interest because of its relations to a number of modern areas of mathematics: theory of characteristic classes, non-commutative geometry, operator extension theory, fractal geometry, etc. (see, for example, the review articles [5]–[7]). The most interesting properties of periodic systems with a magnetic field are conditioned by the presence of two natural geometric scales, namely, the magnetic length and the size of an elementary cell of the period lattice. The commensurability (or incommensurability) of the scales leads to such a pecularity of the systems as the transition from a band structure of the spectrum to a fractal one. This transition is described by the flux–energy diagram known as the "Hofstadter butterfly" (see [2] and [3]). Nevertheless, because the number of flux quanta through a unit cell of a crystal lattice is very small for experimentally accessible values of the magnetic field strength, no energy spectrum of the Hofstadter type is observable in usual Hall systems. However, artificial two-dimensional periodic-modulated systems (so-called periodic arrays of quantum dots) have recently been produced in which the above geometric scales are comparable and, consequently, an experimental observation of the Hofstadter butterfly has become possible [8].

It should be borne in mind that the Hofstadter-type spectrum is obtained either in the framework of tight-binding approximation or in that of nearly-free-electron approximation. If we denote by η the number of magnetic flux quanta through an elementary cell of the period lattice of the system, then the Hofstadter butterfly is a diagram describing the η-dependence of the spectrum of the Harper operator, that is, of the difference operator $h(\alpha)$ which is defined in the space $l^2(\mathbb{Z})$ by the formula

$$(h(\alpha)\phi)_m = \phi_{m+1} + \phi_{m-1} + 2\cos(2\pi m\alpha + \nu)\phi_m. \tag{1}$$

Here $\phi = (\phi_m)_{m\in\mathbb{Z}} \in \lambda^2(\mathbb{Z})$, ν is the Bloch parameter, $\alpha = \eta$ in the case of tight-binding approximation, and $\alpha = \eta^{-1}$ in the case of nearly-free-electron approximation. The flux–energy diagram for the initial Schrödinger operator with a magnetic field is more complicated and less beautiful than the Hofstadter butterfly. For example, the numerical results for sine-shaped potentials presented in [9] and [10] show a very irregular structure of the flux–energy diagram. It is pertinent to note that this irregularity can be considered as a manifestation of chaos in quantum billiard systems (see [10] and [11]). In connection with the foregoing discussion we consider an explicitly solvable model of a magneto-Bloch electron in the periodic array of

This work was partly supported by a grant from the Russian Foundation for Basic Research.

quantum dots proposed in [12] and [13]. This model admits a strict analytic investigation, which allows us to perform a numerical determination of the flux–energy diagram for magneto-Bloch bands.

2. Description of the model

Here we describe the considered model of quantum dot array in a uniform magnetic field (on the physical level of rigour it was studied in [12] and [13]). For simplicity, we restrict ourselves to a square lattice of dots and use a system of units in which the physical constants $\hbar$, c, and e (the particle charge) are equal to 1 and $m = 1/2$ (m is the mass of the particle). We choose the single-particle Hamiltonian for a two-dimensional quantum dot in the form $H_d = H_0 + V(r)$, where $V(r) = \omega_0^2 r^2$ is the confining potential of the dot and

$$H_0 = (i\partial/\partial x - \pi\xi y)^2 + (i\partial/\partial y + \pi\xi x)^2$$

is the Hamiltonian of the particle in the (x, y)-plane subjected to a uniform magnetic field with flux density ξ. Suppose that the dots of the array are placed at the nodes of a square lattice Λ, and let $\mathbf{a}_1$, $\mathbf{a}_2$ be some fixed primitive linearly independent vectors of Λ. The state space of our model is the direct sum $\mathcal{H} = \sum_{\lambda\in\Lambda}^{\oplus} \mathcal{H}_\lambda$, where each space $\mathcal{H}_\lambda$ coincides with $L^2(\mathbb{R}^2)$. Let $H_\lambda = H_d$ for every $\lambda \in \Lambda$; the direct sum $H^0 = \sum_{\lambda\in\Lambda}^{\oplus} H_\lambda$ is the unperturbed Hamiltonian of the model. This operator is the Hamiltonian of a set of isolated quantum dots. To take into account the charge carrier tunneling between dots we use the "restriction–extension procedure" of operator extension theory [14]. For every $\lambda \in \Lambda$ we denote by $\mathcal{D}_\lambda$ the set of functions f from the domain $D(H_d)$ which vanish in a neighbourhood of zero. Let $\mathcal{D} = \sum_{\lambda\in\Lambda}^{\oplus} \mathcal{D}_\lambda$. We consider the symmetric operator S that is the restriction of H^0 to the domain $\mathcal{D}$. We seek the "true" Hamiltonian H of the quantum dot array among the self-adjoint extensions of S. Let $R(\zeta)$ be the resolvent of the desired operator: $R(\zeta) = (H - \zeta)^{-1}$, and $R^0(\zeta) = (H^0 - \zeta)^{-1}$. Then R and R^0 are connected by the Krein resolvent formula [14]

$$R(\zeta) = R^0(\zeta) - \gamma(\zeta)[Q(\zeta) + A]^{-1}\gamma^*(\bar{\zeta}), \tag{2}$$

where the operator-valued holomorphic functions $\gamma(\zeta)$ and $Q(\zeta)$ are the so-called Krein Γ- and $\mathcal{Q}$-functions, respectively, and A is a self-adjoint operator in the deficiency space of S which parametrizes the self-adjoint extensions of S. In our case we can rewrite the formula (2) in a more detailed form. First we represent the state space $\mathcal{H}$ in the form of a tensor product $\mathcal{H} = L^2(\mathbb{R}^2) \otimes l^2(\Lambda)$. Then $R^0(\zeta) = R_d(\zeta) \otimes I$, where $R_d(\zeta)$ is the resolvent of H_d. Let $G_d(x, y; \zeta)$ be the Green's function of H_d, that is, the integral kernel of $R_d(\zeta)$. It is easy to prove that the term $\gamma(\zeta)[Q(\zeta) + A]^{-1}\gamma^*(\bar{\zeta})$ in (2) has the form $K(\zeta) \otimes T(\zeta)$, where $K(\zeta)$ is the integral operator in $L^2(\mathbb{R}^2)$ with kernel $K(x, y; \zeta) = G_d(x, 0; \zeta)G_d(0, y; \zeta)$ and $T(\zeta)$ is an operator in $l^2(\Lambda)$ having the diagonal matrix $T_{\lambda\mu}(\zeta) = q(\zeta)\delta_{\lambda\mu}$. To describe the function $q(\zeta)$, we write $\omega_c = 4\pi|\xi|$ (ω_c is known as the cyclotron frequency), $\Omega = \sqrt{\omega_c^2 + \omega_0^2}$, and $\omega_{1,2} = (\Omega \pm \omega_c)/2$ (the hybrid frequencies). Then $q(\zeta) = -(1/4\pi)[\psi(1/2 - \zeta/\Omega] + \log(\Omega/4\pi) + 2C_E]$. Here ψ is the logarithmic derivative of the Euler Γ-function and C_E is the Euler constant. The operator A must be invariant with respect to a unitary representation of the magnetic translation group in the space $l^2(\Lambda)$ [7], hence, the matrix $A_{\lambda\mu}$ of the operator A is fully determined only by the elements $A_{\lambda 0}$ according to the formula

$$A_{\lambda-\gamma,\mu-\gamma} = \exp[\pi i \xi(\gamma \wedge (\lambda - \mu))] A_{\lambda\mu}. \tag{3}$$

3. The main results

Next, we consider the most interesting case of the nearest-neighbour hopping operator A. More precisely, we choose

$$A_{\lambda 0} = g[\delta(\lambda_1, 0)(\delta(\lambda_2, 1) + \delta(\lambda_2, -1)) + \delta(\lambda_2, 0)(\delta(\lambda_1, 1) + \delta(\lambda_1, -1))],$$

where g is a real non-zero constant (the so-called "coupling constant"); here we write $\delta(k,l)$ instead of δ_{kl}, for convenience. In this case we are able to describe the spectrum of H completely. For this purpose we denote by $\chi(x)$ the inverse function to $\psi(x)$, $x \in \mathbb{R}$; χ is a multivalued real-analytic function defined for all x in $\mathbb{R}$ and having continuous single-valued branches $\chi_n(x)$ $(n = 0, 1, \ldots)$ with values in the intervals $(0, +\infty)$, $(-1, 0)$, $(-2, -1), \ldots$. We denote by τ_n the functions

$$\tau_n(x) = \Omega[1/2 - \chi_n(4\pi g x - \log(\Omega/4\pi) - 2C_E)].$$

We recall that the spectrum of H_d (hence, of H^0) is a pure point spectrum and consists of the eigenvalues

$$E_{mn} = \omega_1(m + 1/2) + \omega_2(n + 1/2) \qquad (m, n = 0, 1, 2, \ldots),$$

known as the "Fock-Darwin" levels.

The following assertion holds.

Theorem. *Suppose that the matrix of the operator A has the form* (3). *Then the spectrum of H consists of three parts:*

(I) *the Fock–Darwin levels E_{mn} for which $m \neq n$;*

(II) *the Fock–Darwin levels of the form E_{nn}, provided that E_{nn} is a degenerate point of the spectrum of H_d (this part of the spectrum may be empty);*

(III) *the magneto-Bloch bands B_n. The band B_n is the set of all values of the function $E = \tau_n(x)$ on the spectrum of the Harper operator $h(\eta)$, where $\eta = \xi \mathbf{a}_1 \wedge \mathbf{a}_2$; B_n lies below the level E_{nn} and (if $n \geq 1$) above the level $E_{n-1,n-1}$.*

Needless to say, the band spectrum of H (that is, (III)) is the most interesting part of the spectrum. Fig. 1 shows the flux-energy diagram for the four lowest bands of this part obtained by means of numerical calculations. In our case the parameter g has the value $1/10$. The bands B_1, B_2, B_3 look like a slightly deformed Hofstadter butterfly, while B_0 is highly dissimilar to one. These pecularities of the band shapes are caused by the difference in the behaviour of the functions τ_n for $n = 0$ and $n > 0$, respectively. Roughly speaking, the behaviour of $\tau_n(x)$ for $n > 0$ is similar to that of the function $\tan^{-1}(x)$, while the behaviour of τ_0 is similar to $\exp(x)$. A comparison of our diagrams to those in [9] and [10] shows that they are similar to the diagrams in [10] rather than those in [9]. In connection with this we mention that the

Hamiltonians in [9] and [10] have the form $H_0 + U$, where $U(x,y) = a(\cos bx \cos by)^k$ in [9] and $U(x,y) = a(\cos bx + \cos by)$ in [10].

The above theorem means that the "Fock–Darwin" levels of E_{nn}-type are broadened in a magneto-Bloch band. This effect is discussed in [12].

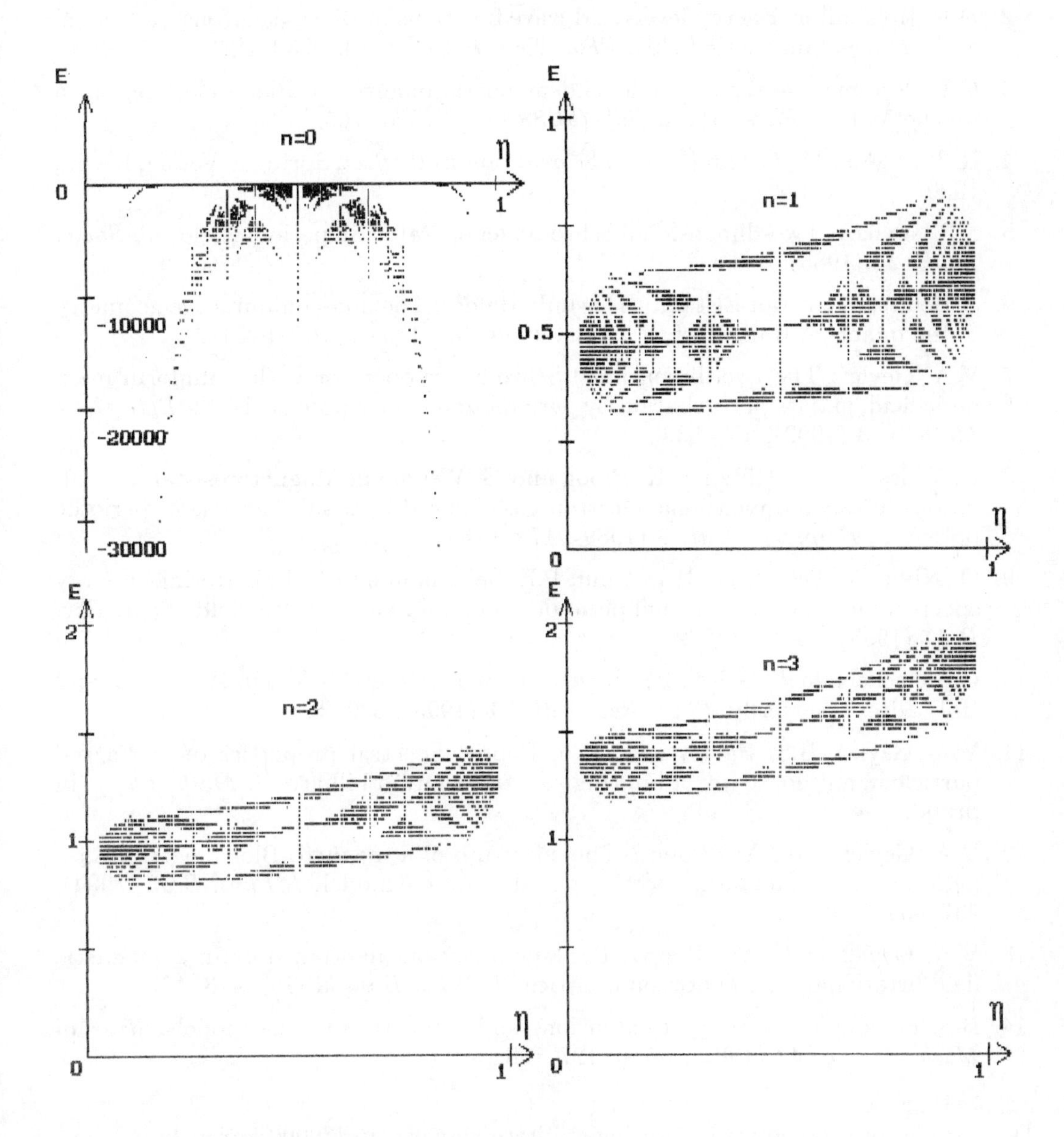

Fig. 1. View of the four lowest magneto-Bloch bands.

References

1. M.Ya. Azbel, Energy spectrum of a conductivity electron in a magnetic field, *Soviet Phys. ZETP* **19** (1964), 634–657 .
2. D.R. Hofstadter, Energy levels and wave functions of Bloch electrons in rational and irrational magnetic fields, *Phys. Rev. B* **14** (1976), 2239–2249.
3. G.H. Wannier, A remark independent on rationality for Bloch electrons in a magnetic field, *Phys. Status Sol. (b)* **88** (1978), 757–765.
4. R. Prange and S. Girvin (eds.), *The quantum Hall effect*, Springer-Verlag, Berlin, 1990.
5. S.P. Novikov, Two-dimensional Schrödinger operators in periodic fields, *J. Soviet Math.* **28** (1985), 3–32.
6. J. Bellissard, A. van Elst and H. Schulz-Baldes, The non-commutative geometry of the quantum Hall effect, *J. Math. Phys.* **35** (1994), 5373–5451.
7. V.A. Geyler, The two-dimensional Schrödinger operator with a uniform magnetic field, and its perturbation by periodic zero-range potentials, *St. Petersburg Math. J.* **3** (1992), 489–532.
8. D. Weiss, K. von Klitzing, K. Ploog and G. Weimann, Magnetoresistance oscillations in a two-dimensional electron gas induced by a submicrometer periodic potential, *Europhys. Lett.* **8** (1989), 179–184.
9. O. Kühn, V. Fessatidis, H.L. Chui, P.E. Selbmann and N.J.M. Horing, Energy spectrum for two-dimensional periodic potential in a magnetic field, *Phys. Rev. B* **47** (1993), 13019–13022.
10. G. Petschel and T. Geisel, Bloch electron in a magnetic field: classical chaos and Hofstadter's butterfly, *Phys. Rev. Lett.* **71** (1993), 239–242.
11. V.A. Geyler, B.S. Pavlov and I.Yu. Popov, Spectral properties of a charged particle in antidot array: a limiting case of quantum billiard, *J. Math. Phys.* (in press).
12. V.A. Geyler and I.Yu. Popov, The spectrum of a magneto-Bloch electron in a periodic array of quantum dots: explicitly solvable model, *J. Phys. B* **93** (1994), 437–439.
13. V.A. Geyler and I.Yu. Popov, Periodic array of quantum dots in a magnetic field: irrational flux; honeycomb lattice, *J. Phys. B* **98** (1995), 473–477.
14. B.S. Pavlov, The theory of extensions and explicitly-solvable models, *Russian Math. Surveys* **42** (1987), no. 6, 127–168.

Department of Mathematics, Mordovian State University, 430000 Saransk, Russia

S. HEIKKILÄ

Existence results for first order discontinuous differential equations of general form

1. Hypotheses and main results

Given a real interval $J = [t_0, t_1]$, consider the implicit initial value problem (IVP)

$$f(t, x(t), x'(t)) = 0 \text{ a.e. on } J, \quad x(t_0) = x_0, \tag{1.1}$$

where the function $f : J \times \mathbb{R}^2 \to \mathbb{R}$ is allowed to be discontinuous in all its arguments. We say that a function $x : J \to \mathbb{R}$ which belongs to the set

$$AC(J) = \{x : J \to \mathbb{R} \mid x \text{ is absolutely continuous on } J\}$$

is a *lower solution* of the IVP (1.1) if

$$f(t, x(t), x'(t)) \leq 0 \text{ a.e. on } J \text{ and } x(t_0) \leq x_0,$$

and an *upper solution* of (1.1) if the opposite inequalities hold. If equalities hold, we say that x is a *solution* of (1.1). For $x \in AC(J)$, we define $x'(t) = 0$ if x is not differentiable at $t \in J$.

We define a partial ordering on $AC(J)$ by

$$y \leq z \text{ iff } y(t) \leq z(t) \text{ for all } t \in J \text{ and } y'(t) \leq z'(t) \text{ for a.a. } t \in J.$$

Using integral methods, we construct a generalised iteration method and the method of upper and lower solutions, in which the conditions

(f0) (1.1) has a lower solution y, an upper solution z, and $y \leq z$,

(f1) there is $\mu : J \times \mathbb{R}^2 \to (0, \infty)$ such that $\mu \cdot f$ is sup-measurable, and the function $(u, v) \mapsto v - (\mu \cdot f)(t, u, v)$ is increasing in u and in v for a.a. $t \in J$

ensure that the IVP (1.1) has extremal solutions in the order interval $[y, z]$ of $AC(J)$, decreasing with respect to f and increasing with respect to x_0.

We also study the existence of upper and lower solutions and the convergence of successive approximations in some special cases, and give examples to illustrate the obtained results.

2. Preliminaries

Let $L^1(J)$ denote the space of Lebesgue integrable functions $v : J \to \mathbb{R}$ with a.e. pointwise partial ordering $\preceq$. We first convert problem (1.1) to an operator equation in $L^1(J)$.

Lemma 2.1. *Suppose that the conditions* (f0) *and* (f1) *hold, and let* $y, z \in AC(J)$ *be the lower and upper solutions of* (1.1), *assumed in* (f0). *Then the equation*

$$Gv(t) = v(t) - (\mu \cdot f)(t, x_0 + \int_{t_0}^{t} v(s)ds, v(t)), \quad t \in J, \tag{2.1}$$

defines an increasing mapping G *from the order interval* $[y', z']$ *of* $L^1(J)$ *to* $[y', z']$. *Moreover,* x *is a solution of the IVP* (1.1) *in* $[y, z]$ *if and only if* $x(t) = x_0 + \int_{t_0}^{t} v(s)ds$, $t \in J$, *where* v *is a fixed point of* G.

Proof. From condition (f1) it follows that equation (2.1) defines a measurable mapping Gv for each $v \in L^1(J)$. Condition (f1) also implies that

$$Gv(t) \leq G\hat{v}(t) \text{ a.e. on } J \text{ whenever } v \preceq \hat{v} \text{ in } L^1(J). \tag{a}$$

If y, $z \in AC(J)$, $y \leq z$, are the lower and upper solutions of (1.1), assumed in (f0), it is easy to see that $y' \preceq Gy'$ and $Gz' \preceq z'$. Let $v \in [y', z']$ be given. This means that $y'(t) \leq v(t) \leq z'(t)$ a.e. on J. In view of the above results, we have $y'(t) \leq Gv(t) \leq z'(t)$ for a.a. $t \in J$. This implies that $|Gv(t)| \leq |y'(t)| + |z'(t)|$ for a.a. $t \in J$. Because Gv is measurable, we then have $Gv \in L^1(J)$. Thus, Gv belongs to $[y', z']$, whence $G[y', z'] \subset [y', z']$. From (a) it follows that G is increasing. The last conclusion is an easy consequence of the given notation and definitions.

As a special case of Theorem 1.2.2 of [2] we obtain the following result.

Lemma 2.2. *Let* $[\alpha, \beta]$ *be a non-empty order interval in* $L^1(J)$. *If a mapping* $G : [\alpha, \beta] \to [\alpha, \beta]$ *is increasing, then* G *has a least fixed point* v_* *and a greatest fixed point* v^*, *and*

$$\begin{cases} v_* = & \max C = \min\{w \in [\alpha, \beta] \mid Gw \preceq w\}, \\ v^* = & \min C' = \max\{w \in [\alpha, \beta] \mid w \preceq Gw\}, \end{cases} \tag{2.2}$$

where C (C') *is a countable chain of* G*-iterations of* α (β).

Proof. Let $(v_n)_{n=o}^{\infty}$ be a monotone sequence in $[\alpha, \beta]$. Because G is increasing, the sequence $(Gv_n)_{n=o}^{\infty}$ is monotone in $[\alpha, \beta]$. Applying the dominated convergence theorem, we see that $(Gv_n)_{n=o}^{\infty}$ converges in $(L^1(J), \| \cdot \|_1)$. Thus, the conditions in Theorem 1.2.2 of [2] hold, whence the least and the greatest fixed points v_*, v^* of G exist and (2.2) holds.

3. Existence of extremal solutions and dependence on data

We are now ready to prove our main existence result.

Theorem 3.1. *Suppose that* $f : J \times \mathbb{R}^2 \to \mathbb{R}$ *satisfies conditions* (f0) *and* (f1). *Then the IVP* (1.1) *has extremal solutions* x_* *and* x^* *in* $[y, z]$ *in the sense that if* $x \in [y, z]$ *is a solution of* (1.1), *then* $x \in [x_*, x^*]$.

Proof. The results of Lemma 2.1 and Lemma 2.2 imply that the operator G, defined by (2.1), has a least fixed point v_* and a greatest fixed point v^*. By the last conclusion in Lemma 2.1, the functions $x_*, x^* \in AC(J)$ defined by

$$x_*(t) = x_0 + \int_{t_0}^{t} v_*(s)ds, \quad x^*(t) = x_0 + \int_{t_0}^{t} v^*(s)ds, \quad t \in J, \tag{3.1}$$

are the least and the greatest solutions of (1.1) in $[y, z]$.

Applying the formulae (2.2), we easily establish the following result for the dependence of the extremal solutions of (1.1) on the function f and on the initial value x_0.

Proposition 3.1. *If conditions* (f0) *and* (f1) *hold, then the extremal solutions of the IVP* (1.1) *in* $[y, z]$ *are decreasing with respect to* f *and increasing with respect to* x_0.

Remark 3.1. It can be shown that (1.1) has at most one solution on J if the function $f : J \times \mathbb{R}^2 \to \mathbb{R}$ satisfies the following conditions:

(fa) $f(t, u, v) \leq g(t, u)\varphi(v)$ for a.a. $t \in J$ and for all $u, v \in \mathbb{R}$, where $g : J \times \mathbb{R} \to (0, \infty)$, $\varphi : \mathbb{R} \to \mathbb{R}$ and $\varphi^{-1}[\mathbb{R}_+] \subseteq \mathbb{R}_+$;

(fb) for each $z_0 \in \mathbb{R}$ there are p, λ with $p/(1-\lambda) \in L^1_+(J)$, ν $J \times \mathbb{R}^2 \to (0, \infty)$ and $r > 0$ so that $(\nu \cdot f)(t, u+\delta u, v+\delta v) - (\nu \cdot f)(t, u, v) \geq (1-\lambda(t))\delta v - p(t)\delta u$ for all $u, v \in \mathbb{R}$ and $\delta u, \delta v \in \mathbb{R}_+$ satisfying $z_0 \leq u < u + \delta u \leq z_0 + r$, and for a.a. $t \in J$.

4. Special cases

Next, consider the IVP

$$x'(t) = g(t, x(t), x'(t)) \quad \text{a.e. on} \quad J = [t_0, t_1], \quad x(t_0) = x_0, \tag{4.1}$$

where $g : J \times \mathbb{R}^2 \to \mathbb{R}$ is assumed to satisfy the following conditions:

(g0) g is sup-measurable, and there is $c \in L^1_+(J)$ such that $g(t, u, v) + c(t)v$ is increasing in u and in v for a.a. $t \in J$;

(g1) $|g(t, u, v)| \leq p(t)\psi(|u|) + \lambda(t)\,|v|$ for a.a. $t \in J$ and for all $u, v \in \mathbb{R}$, where $p/(1-\lambda) \in L^1_+(J)$, $\psi : \mathbb{R}_+ \to (0, \infty)$, ψ is increasing and $\int_0^\infty dv/\psi(v) = \infty$.

Proposition 4.1. *If conditions* (g0) *and* (g1) *hold, then for each* $x_0 \in \mathbb{R}$ *the IVP* (4.1) *has extremal solutions* x_* *and* x^*, *and all the solutions of* (4.1) *belong to the order interval* $[-z, z]$, *where* z *is the solution of the IVP*

$$z'(t) = \frac{p(t)\psi(z(t))}{1 - \lambda(t)} \quad \text{a.e. on } J, \quad z(t_0) = |x_0|. \tag{4.2}$$

Moreover, the extremal solutions x_* *and* x^* *are increasing with respect to* g *and* x_0.

Proof. The conditions (g0) and (g1) imply that (f0) and (f1) hold when

$$f(t, u, v) = v - g(t, u, v), \quad \mu(t, u, v) = \frac{1}{1 + c(t)}, \quad t \in J, \; u, v \in \mathbb{R}, \tag{4.3}$$

and when $y = -z$. Thus, the IVP (1.1), or, equivalently, the IVP (4.1), has by Theorem 3.1 extremal solutions x_* and x^* in $[-z, z]$. It is also easy to show that if $x \in AC(J)$ is a solution of (4.1), then it belongs to $[-z, z]$, so that x_* and x^* are the least and the greatest of all the solutions of (4.1). The last assertion is a consequence of Proposition 3.1.

5. Convergence of successive approximations and examples

Lemma 2.2 implies that the extremal solutions of (4.1) ((1.1)) can be obtained as the last elements of countable chains of functions in $AC(J)$ (cf. [2]). When one-sided continuity hypotheses hold for the function g in its last two arguments, then one of these chains is reduced to a sequence of successive approximations. This is the content of the next result.

Proposition 5.1. *Suppose that $g : J \times \mathbb{R}^2 \to \mathbb{R}$ satisfies conditions* (g0) *and* (g1). *Then for each $x_0 \in \mathbb{R}$ the sequence of the successive approximations y_n, $n \in \mathbb{N}$, defined by the recurrent relations*

$$\begin{aligned} v_{n+1}(t) &= \frac{1}{1+c(t)}(g(t, y_n(t), v_n(t)) + c(t)v_n(t)), \quad t \in J, \\ y_{n+1}(t) &= x_0 + \int_{t_0}^{t} v_{n+1}(s)ds, \quad t \in J, \end{aligned} \tag{5.1}$$

converges in $AC(J)$ with respect to the norm $\|x\|_{01} = \max_{t \in J}|x(t)| + \int_{t_0}^{t_1} |x'(s)|ds$ to

a) the maximal solution of the IVP (4.1) *if $g(t, u_n, v_n) \to g(t, u, v)$ for a.a. $t \in J$, whenever $u_n \searrow u$ and $v_n \searrow v$, and if $y_0 = z$ and $v_0 = z'$, where z is the solution of* (4.2);

b) the minimal solution of the IVP (4.1) *if $g(t, u_n, v_n) \to g(t, u, v)$ for a.a. $t \in J$, whenever $u_n \nearrow u$ and $v_n \nearrow v$, and if $y_0 = -z$ and $v_0 = -z'$.*

Example 5.1. We define the functions $h : J \to \mathbb{R}$ and $q : \mathbb{R} \to \mathbb{R}$ by

$$h(t) = \sum_{m=1}^{\infty}\sum_{k=1}^{\infty} \frac{2 + [k^{1/m}t] - k^{1/m}t}{(km)^2}\left(2 + \sin\left(\frac{1}{1 + [k^{1/m}t] - k^{1/m}t}\right)\right), \quad t \in J,$$

and

$$q(s) = \sum_{m=1}^{\infty}\sum_{k=1}^{\infty} \frac{\tan^{-1}([k^{1/m}s])}{(km)^2}, \quad s \in \mathbb{R},$$

where $[v]$ denotes the greatest integer not exceeding v. Consider the IVP

$$x'(t) = h(t) + q(t + x(t)) + \tan^{-1}([x'(t)]) \text{ a.e. on } J, \quad x(t_0) = x_0. \tag{5.2}$$

It is easy to see that the conditions (g0) and (g1) hold. Thus, by Proposition 4.1, the IVP (5.2) has extremal solutions x_* and x^* for each $x_0 \in \mathbb{R}$. Since the function $v \mapsto [v]$ is right-continuous, it follows from Proposition 5.1 that x^* is obtained as the limit of successive approximations.

Example 5.2. Consider the IVP

$$x'(t) = \frac{[3 - 2t + x(t)]}{2} + \frac{[x'(t)]}{2 + 2|[x'(t)]|} \text{ a.e. on } J, \quad x(t_0) = x_0. \tag{5.3}$$

From Proposition 4.1 it follows that the IVP (5.3) has extremal solutions, and that the maximal solution is obtained by a method of successive approximations. These approximations can be calculated by computers, and in some special cases they even help to find the exact maximal solutions. For instance, the maximal solution of (5.3) on $J = [0, 5]$ with initial condition $x(0) = 1$ is

$$x(t) = \begin{cases} 1 + \frac{7}{3}t, & 0 \leq t < 3, \\ 8 + \frac{17}{6}(t-3)t, & 3 \leq t < \frac{21}{5}, \\ \frac{57}{5} + \frac{27}{8}(t - \frac{21}{5}), & \frac{21}{5} \leq t < \frac{271}{55}, \\ \frac{762}{55} + \frac{31}{8}(t - \frac{271}{55}), & \frac{271}{55} \leq t \leq 5. \end{cases}$$

Another solution of this problem is $x(t) = \begin{cases} \frac{7}{4}t + 1, & 0 \leq t < 4, \\ \frac{5}{4}(t-4) + 8, & 4 \leq t \leq 5. \end{cases}$

Remark 5.1. The functions $\mu \cdot f$ and g satisfy the sup-measurability condition assumed in (f1) and (g0) if they are Shragin functions, in particular, if they are Carathéodory functions or equivalent to Borel functions (cf. [1] and [3]).

References

1. J. Appel and P.P. Zabreiko, *Non-linear superposition operators*, Cambridge University Press, Cambridge, 1990.
2. S. Heikkilä and V. Lakshmikantham, *Monotone iterative techniques for discontinuous nonlinear differential equations*, Marcel Dekker, New York, 1994.
3. I.V. Shragin, On the Carathéodory conditions, *Russian Math. Surveys* **34** (1979), no. 3, 183–189 .

Department of Mathematical Sciences, University of Oulu, 90570 Oulu, Finland

J.V. HORÁK

On solvability and approximation of 1D models of coupled thermoelasticity

1. Introduction

In this paper we briefly present the origin and some mathematical formulations of the "extended" model problem (for beam or plate strip) for quadruples of unknown functions $\mathbf{U} = \{\{u_1, u_2\}, \{\vartheta_1, \vartheta_2\}\}$ within the framework of the *linearised theory of coupled thermoelasticity*. Thus, we introduce a *weak* formulation of the model problem as well as semidiscrete (in time) and complete discrete (in time and space variables) formulations. According to a special, but still real, physical situation, we may split the extended model into two independent and simplified models representing stretching and bending of the beam. For the sake of brevity, we restrict ourselves here only to the bending effect, which is a more interesting case from a mathematical point of view.

The Rothe method of discretisation in time (see [1] for more details) is used for a priori estimations of a semidiscrete solution set and their time derivatives. Approximate properties of the Rothe functions and their convergence to the weak solution of the problem, as well as the continuous dependence of this solution on the given data, are shown. Finally, approximation of Rothe's vector functions in space variable by FEM is introduced, and the convergence of the complete discretised solution is proved.

1.1. Assumption and limitation

Here we discuss only the simplest situation: an elastic and thermally homogeneous isotropic material, first order theory and linearisation of the temperature, and heat sources change along the height of the cross section. We use the "elementary theory" of beam (plate strip) bending, that is, we adopt the Bernoulli-Euler hypothesis; the Timoshenko (Mindlin-Reissner) model (including the effect of transverse shear deformation) is not discussed. The model problem is formulated for a prismatic cross-section of the beam with constant height H and unit width, steady sources and load and classical boundary conditions, and under the assumption of positive material coefficients: $\alpha > 0, k > 0, c > 0$ (see, for example, [2]).

1.2. Origin of the problem

General governing equations of the linearised theory of coupled thermoelasticity can be derived from thermodynamic laws, assumptions on behaviour of the material (*constitutive relations, properties of materials*) and course of evolution of the thermodynamic process (all details for the general situation can be found in [3], or, for our problem, in [2], for example). The basic equations of the linearised thermoelastic theory consist of the Duhamel-Neumann law, the Cauchy equation of motion and the energy equation. In the quasi-static case we assume zero inertia forces: $\ddot{\mathbf{U}} = \mathbf{0}$.

This work was partly supported by grant No 3110 3002 from Palacký University Olomouc, Czech Republic.

Then the complete system of field equations for the *linearised quasi-static theory of coupled thermoelasticity* in 3D (our starting point) has the form

$$\operatorname{div}\mathbf{C}[\nabla\mathbf{U}] + \operatorname{div}(\vartheta\mathbf{M}) + \mathbf{b} = \mathbf{0},$$
$$\operatorname{div}(\mathbf{K}\nabla\vartheta) + \theta_0\mathbf{M}\cdot\nabla\dot{\mathbf{U}} + r = c\dot{\vartheta},$$

where the coupling term $\theta_0\mathbf{M}\cdot\nabla\dot{\mathbf{U}}$ comes from the first thermodynamic law, while the standard term $\operatorname{div}(\vartheta\mathbf{M})$ has its origin in the Duhamel-Neumann law.

2. Model problem formulation

The basic assumptions on the course of the temperature ϑ and source r along the height of the cross-section are that

$$\vartheta(x,y,z,t) \cong \vartheta_1(x,t) + y\vartheta_2(x,t), \quad r(x,y,z,t) \cong r_1(x,t) + yr_2(x,t).$$

As a consequence of the *"technical"* theory of bending ($u_1 = u_1(x,t)$ is axial and $u_2 = u_2(x,t)$ is a vertical displacement), we have $\mathbf{u} = \{u_1 - yDu_2, u_2, 0\}$, $\operatorname{div}\dot{\mathbf{u}} = D_t(Du_1) - yD_t(D^2u_2)$, where $\mathbf{U} = \{\mathbf{u}, \boldsymbol{\vartheta}\}$, $\mathbf{u} = \{u_1, u_2\}$, $\boldsymbol{\vartheta} = \{\vartheta_1, \vartheta_2\}$, $D \equiv \partial/\partial x$, $D_t \equiv \partial/\partial t$. Since the material is homogeneous and isotropic, (in 3D) $\mathbf{M} = m\mathbf{1}$, $\mathbf{K} = k\mathbf{1}$ and $m = -3(\lambda+2\mu)\alpha$, where $m = -E\alpha$ (for a beam) and $m = -E\alpha/(1-2\nu)$ (for an infinite plate strip).

2.1. Resulting set of equations for the model problem

The complete set of four governing equations of the linearised theory of coupled thermoelasticity for our model problem has the form

$$D^2\vartheta_1 - (\alpha_h + \alpha_d)(kH)^{-1}\vartheta_1 + (\alpha_h - \alpha_d)(2k)^{-1}\vartheta_2 - \theta_0 E\alpha k^{-1}D_t(D^2u_1) + \tilde{r}_1 = aD_t\vartheta_1, \tag{1}$$
$$D^2\vartheta_2 - (12H^{-2} + 3(\alpha_h + \alpha_d)(kH)^{-1})\vartheta_2 - 6(\alpha_h - \alpha_d)k^{-1}H^{-2}\vartheta_1 + \theta_0 E\alpha k^{-1}D_t(D^2u_2) + \tilde{r}_2 = aD_t\vartheta_2, \tag{2}$$
$$D(EHDu_1) - D(\alpha EH\vartheta_1) = \tilde{q}_1, \tag{3}$$
$$D^2(EJD^2u_2) + D^2(\alpha EJ\vartheta_2) = \tilde{q}_2, \tag{4}$$

where $\tilde{r}_1 = r_1 + (\alpha_h\vartheta_h + \alpha_d\vartheta_d)/(kH)$ and $\tilde{r}_2 = r_2 + 6(\alpha_h\vartheta_h - \alpha_d\vartheta_d)/(kH^2)$.

In the case $\alpha_h = \alpha_d = \bar{\alpha}$, the model problem can be essentially simplified. Equations (1) and (2) are then uncoupled and problem (1)–(4) can be decomposed into two mutually independent problems:

the stretching effect (in the direction of the axis of the beam)

$$D^2\vartheta_1 - 2\bar{\alpha}(kH)^{-1}\vartheta_1 - \theta_0 E\alpha k^{-1}D_t(Du_1) + \tilde{\tilde{r}}_1 = aD_t\vartheta_1, \tag{5}$$
$$D(EHDu_1) - D(\alpha EH\vartheta_1) = \tilde{q}_1; \tag{6}$$

the bending effect (our model problem equations after simplification)

$$a_1 D_t\vartheta = D^2\vartheta - a_2\vartheta + a_3D^2(D_tu) + r, \tag{7}$$
$$D^4u + \alpha D^2\vartheta = q, \tag{8}$$

where $\tilde{\tilde{r}}_1 = r_1 + (\bar{\alpha}/(kH))(\vartheta_h + \vartheta_d)$, $\vartheta \equiv \vartheta_2$ (temperature gradient), $u \equiv u_2$ (vertical displacement), $r \equiv r_2 + (6\overline{\alpha})/(kH^2)(\vartheta_h - \vartheta_d)$, $q = \tilde{q}_2/(EJ)$, $a_1 = c/k$, $a_2 = 12/H^2 + (6\overline{\alpha})/(kH)$, $a_3 = \theta_0 E\alpha/k$. Next we set $I = (0,T)$, $T \in \mathbb{R}^+$, $\Omega = (0,L)$, $L \in \mathbb{R}^+$, $\partial\Omega = \{0, L\}$, $\Gamma = \partial\Omega \times I$, $\Omega_0 = \Omega \times \{0\}$ and $Q = \Omega \times I$.

In what follows we discuss only the last case: equations (7) and (8) in Q with the boundary conditions $u = 0$ and $\vartheta = 0$ on Γ, $Du = 0$ on $\{0\} \times I$, $D^2u = 0$ on $L \times I$, and $\vartheta = 0$ in Ω_0. This simplified model problem for two of the unknown functions $\{u, \vartheta\}$ is called (P_B).

2.2. Method of factorisation and special cases

The problem (P_B) can be even further simplified in view of the type of prescribed boundary conditions. Thus, for some boundary conditions, the fourth order problem (P_B) can be decomposed into two uncoupled second order problems (see [2], for example). But this is not the case in general, and we will present the complete problem (P_B).

2.3. Weak formulation of the model problem: the continuous case

First, as usual, we define the linear space $\mathcal{V} = \{v \in H_0^1(\Omega) \cap H^2(\Omega) \mid Dv(0) = 0\}$ of kinematically admissible functions, where $H^k(\Omega)$ is the Sobolev space. The Cartesian product of test function spaces is $\mathcal{H} = \mathcal{V} \times H_0^1(\Omega)$,

$$\mathbf{a}(u,v) = \int_\Omega D^2u(x)D^2v(x)dx, \quad u, v \in H^2(\Omega),$$

$$\mathbf{b}(u,v) = \int_\Omega Du(x)Dv(x)dx, \quad u, v \in H^1(\Omega),$$

are bilinear forms defined on $H^k(\Omega) \times H^k(\Omega)$, $k = 1, 2$,

$$\mathcal{A}(\mathbf{U},\mathbf{V}) = \mathbf{a}(u,v) + \mathbf{b}(\vartheta,\eta) + a_2(\vartheta,\eta)_{L_2(\Omega)}, \quad \mathbf{U},\mathbf{V} \in \mathcal{H},$$
$$\mathcal{B}(\mathbf{U},\mathbf{V}) = a_1(\vartheta,\eta)_{L_2(\Omega)} + a_3\mathbf{b}(u,\eta), \quad \mathbf{U},\mathbf{V} \in \mathcal{H},$$
$$\mathcal{C}(\mathbf{U},\mathbf{V}) = \alpha\mathbf{b}(\vartheta,v), \quad \mathbf{U},\mathbf{V} \in \mathcal{H},$$

are bilinear forms defined on $\mathcal{H} \times \mathcal{H}$, where $\mathbf{U} = \{u, \vartheta\}$ and $\mathbf{V} = \{v, \eta\}$, $u, v \in \mathcal{V}$, $\vartheta, \eta \in H_0^1(\Omega)$, and $\mathcal{F}(\mathbf{V}) = (q, v) + \langle r, \vartheta \rangle$, $\mathbf{V} \in \mathcal{H}$, is a linear form defined on $\mathcal{H}$.

Then we define the weak solution of (P_B) as follows. Suppose that a couple of functions $\{q, r\} \in L_2(I; \mathcal{V}^*) \times L_2(I; H^{-1})$ are given. An abstract function $\mathbf{U} = \mathbf{U}(t) : I \to \mathcal{H}$, $\mathbf{U} \in L_2(I; \mathcal{H}) \cap AC(I; \mathcal{V} \times L_2(\Omega))$, $D_t\mathbf{U} \in L_2(I; \mathcal{V} \times L_2(\Omega))$, $(P_2\mathbf{U})(0) = 0$ in $C(I; L_2(\Omega))$ such that

$$\int_I \mathcal{A}(\mathbf{U}(t), \mathbf{V}(t))dt - \int_I \mathcal{C}(\mathbf{U}(t), \mathbf{V}(t))dt$$
$$+ \int_I \mathcal{B}(D_t\mathbf{U}(t), \mathbf{V}(t))dt = \int_I \mathcal{F}(\mathbf{V}(t))dt \quad \text{for any } \mathbf{V} \in L_2(I; \mathcal{H}) \qquad (\mathcal{P})$$

is said to be a *weak solution* of the problem (P_B).

3. Approximation

3.1. Discretisation in time

For a semidiscrete formulation of the problem $(\mathcal{P})$ we define

$$p \in \mathbb{N}, \quad p^{(n)} = 2^{(n-1)}p, \quad l^{(n)} = T/(p^{(n)}),$$
$$\mathcal{D}^{(n)} = \{t_j^{(n)}\}_{j=1}^{p^{(n)}}, \quad t_j^{(n)} = jl^{(n)}, \quad j = 0, 1, \ldots, p^{(n)}.$$

Then a function $\mathbf{Z}_j^{(n)} = \{z_j^{(n)}, \xi_j^{(n)}\} \in \mathcal{V} \times H_0^1(\Omega)$ such that

$$\mathbf{a}(z_j^{(n)}, v) - \alpha\mathbf{b}(\xi_j^{(n)}, v) = (q_j^{(n)}, v) \quad \text{for any } v \in \mathcal{V},$$
$$a_1\left(\frac{\xi_j^{(n)} - \xi_{j-1}^{(n)}}{l^{(n)}}, \eta\right)_{L_2(\Omega)} + \mathbf{b}(\xi_j^{(n)}, \eta) + a_2(\xi_j^{(n)}, \eta)_{L_2(\Omega)} \qquad (\mathcal{P}_j^{(n)})$$
$$+ a_3\mathbf{b}\left(\frac{z_j^{(n)} - z_{j-1}^{(n)}}{l^{(n)}}, \eta\right) = \langle r_j^{(n)}, \eta\rangle \quad \text{for any } \eta \in H_0^1(\Omega),$$

where $\{q_j^{(n)}, r_j^{(n)}\} \in \mathcal{V}^* \times L_2(\Omega)$, is said to be an "instant" weak *semidiscrete* solution in time $t_j^{(n)} \in \bar{I}$. For $j = 0$ we put $\{z_0, 0\}$, where $a(z_0, v) = (q, v)$, $v \in \mathcal{V}$.

3.2. Discretisation in the space variable

Using a standard definition of the finite dimensional (FEM) spaces $\mathcal{H}_h = \mathcal{V}_h \times S_h$ and kinematically admissible spaces $\mathcal{H}_h^o = \mathcal{V}_h^o \times S_h^o$, we can define the space approximation of the solution as follows: by formal substitution of $\beta Z_{j,h}^{(n)}$ in $(\mathcal{P}_j^{(n)})$ instead of $\{\mathbf{Z}_j^{(n)}\}$ and $\mathcal{H}_h^o$ instead of $\mathcal{H}_h$, we obtain the definition of a weak *discrete* solution of the problem $(\mathcal{P}_{j,h}^{(n)})$ in time $t_j^{(n)} \in \bar{I}$.

4. Construction of Rothe's functions

Solving the problem $(\mathcal{P}_j^{(n)})$ $((\mathcal{P}_{j,h}^{(n)}))$ gives the semidiscrete solution $\mathbf{Z}_j^{(n)} \in \mathcal{H}$ (discrete solution $\mathbf{Z}_{j,h}^{(n)} \in \mathcal{H}_h$). Then we can construct Rothe's function $\mathbf{U}^{(n)}$ (its space approximation $\mathbf{U}_h^{(n)}$) as a piecewise linear interpolation (in time) of the semidiscrete solutions $\mathbf{Z}_j^{(n)}$ $(\mathbf{Z}_{j,h}^{(n)})$. We have

$$\mathbf{U}^{(n)}(t)\,\big|_{I_j^{(n)}} = \mathbf{Z}_{j-1}^{(n)} + \frac{\mathbf{Z}_j^{(n)} - \mathbf{Z}_{j-1}^{(n)}}{l^{(n)}}(t - t_{j-1}^{(n)}), \quad t \in I_j^{(n)},$$
$$\mathbf{U}_h^{(n)}(t)\,\big|_{I_j^{(n)}} = \mathbf{Z}_{j-1,h}^{(n)} + \frac{\mathbf{Z}_{j,h}^{(n)} - \mathbf{Z}_{j-1,h}^{(n)}}{l^{(n)}}(t - t_{j-1}^{(n)}), \quad t \in I_j^{(n)}.$$

5. Results

Theorem 1. (Existence and uniqueness of $\mathbf{Z}_j^{(n)}$.) *Suppose that the functions* $\{q, r\} \in \mathcal{V}^* \times L_2(\Omega)$ *are given. Then for any given partition* $\mathcal{D}^{(n)}$ *of the interval* $\bar{I}$ *there exists a unique finite set* $\{\mathbf{Z}_j^{(n)}\}_{j=1}^{p^{(n)}} \in [\mathcal{V} \times H_0^1(\Omega)]^{p^{(n)}}$ *of weak solutions of the problems* $(\mathcal{P}_j^{(n)})$, $j = 1, \dots, p^{(n)}$.

Theorem 2. (Existence of $\mathbf{Z}_{j,h}^{(n)}$.) *Suppose that the functions* $\{q, r\} \in \mathcal{V}^* \times L_2(\Omega)$ *are given. Then for any* $h \in (0, h_o\rangle$ *and any* $n \in \mathbb{N}$ *there exists a unique weak solution* $\mathbf{Z}_{j,h}^{(n)} \in \mathcal{H}_h^o$, *of the problem* $(\mathcal{P}_{j,h}^{(n)})$, $j = 1, \dots, p^{(n)}$.

Thus, Theorem 1 justifies the definition of the Rothe functions and Theorem 2 justifies the definition of the approximation of the Rothe functions.

Theorem 3. (Convergence of $\mathbf{Z}_{j,h}^{(n)}$ as $h \to 0+$.) *The sequence of discrete solutions* $\mathbf{Z}_{j,h}^{(n)}$ *converges strongly in* $\mathcal{H}_h^0$, *as* $h \to 0+$, *to the weak semidiscrete solution* $\mathbf{Z}_j^{(n)}$ *of the problem* $(\mathcal{P}_j^{(n)})$, $j = 1, \dots, p^{(n)}$.

Corollary 1. (Convergence of $\mathbf{U}_h^{(n)}$ as $h \to 0+$.) *The sequence of approximations of the Rothe functions* $\{\mathbf{U}_h^{(n)}\}$ *converges strongly to the Rothe function* $\mathbf{U}^{(n)}$ *as* $h \to 0+$, *for any* $n \in \mathbb{N}$.

Theorem 4. (Existence and uniqueness of $\mathbf{U}$: main result.) *Suppose that the functions* $\{q, r\} \in \mathcal{V}^* \times L_2(\Omega)$ *are given. Then there exists a unique solution of the problem* $(\mathcal{P})$.

Proof. We use a constructive method which consists of (i) the replacement of the continuous formulation of the problem by a finite set of time-independent problems $((\mathcal{P}) \sim \{(\mathcal{P}_j^{(n)})\}_{j=1}^{p^{(n)}})$, and (ii) the construction of a weak solution as a limit element of Rothe functions. This is possible because of the a priori estimates (details can be found, for example, in [2])

$$\|\mathbf{Z}_j^{(n)}\|_{[L_2(\Omega)]^2} \le \|\mathbf{Z}_j^{(n)}\|_{\mathcal{H}} \le \max\left((\bar{c})^2, \frac{2(1+\alpha\bar{c})}{\min(1, a_2)}\right) \|\mathcal{F}\|_{H^*} = C,$$

$$\left\| \frac{\mathbf{Z}_j^{(n)} - \mathbf{Z}_{j-1}^{(n)}}{l^{(n)}} \right\|_{H^2 \times L_2} \le \left(\frac{\bar{c}}{a_1} \frac{\sqrt{(a_1 + \alpha a_3)\alpha}}{\sqrt{a_3}} + \frac{\sqrt{a_1 + \alpha a_3}}{\sqrt{a_1^3}} \right) \|r\|,$$

and because the Rothe functions satisfy the estimates

$$\mathbf{U}^{(n)} \in L_\infty(I; H^2(\Omega) \times H^1(\Omega)) \qquad \text{for} \quad \mathcal{D}^{(n)},\ n \in \mathbb{N},$$
$$\mathbf{U}^{(n)} \in H^1(I; H^2(\Omega) \times L_2(\Omega)) \qquad \text{for} \quad \mathcal{D}^{(n)}\ ,\ n \in \mathbb{N}.$$

Theorem 5. (Convergence of $\mathbf{U}^{(n)}$ as $n \to \infty$.) *The sequence of the Rothe functions* $\{\mathbf{U}^{(n)}\}$ *converges strongly in* $C(\bar{I}; H^1(\Omega) \times L_2(\Omega))$, *as* $n \to \infty$, *to the weak solution* $\mathbf{U}$ *of the model problem* $(\mathcal{P})$.

Corollary 2. (Convergence of $\mathbf{U}_h^{(n)}$ as $n \to \infty$ and $h \to 0+$.) *The sequence of approximations of the Rothe functions* $\{\mathbf{U}_h^{(n)}\}$ *converges strongly to the weak solution* $\mathbf{U}$ *of the model problem* $(\mathcal{P})$ *as* $n \to \infty$ *and* $h \to 0+$.

References

1. K. Rektorys, *Method of dicretisation in time and partial differential equations*, TKI, SNTL, Praha, 1985 (Czech).
2. J.V. Horák, On solvability of one special problem of coupled thermoelasticity. I, *Acta Univ. Palack. Olomuc. Fac. Rerum Natur. Math. XXXV* **113**, 1996.
3. D.E. Carlson, Linear thermoelasticity, in *Encyclopedia of Physics, VIa/2, Mechanics of Solids II,* Springer-Verlag, Berlin, 1972.

Department of Mathematical Analysis and Applied Mathematics, Faculty of Natural Science, Palacký University Olomouc, Tomkova 40, 779 00 Olomouc, Czech Republic

A. KIRJANEN

Stabilisation of a linear control system via input independent of the present time

1. Introduction

Consider a linear control system

$$\dot{x} = Ax + Bu, \tag{1}$$

where $x \in \mathbb{R}^n$, $u \in \mathbb{R}^r$ and A, B are matrices with constant coefficients.

Suppose that A is not a Hurwitz matrix, so that the differential equation $\dot{x} = Ax$ is not asymptotically stable.

Let the input $u(t)$ be of the form

$$u(t) = \sum_{k=1}^{m} C_k x(t - kh), \quad h > 0, \quad C_k = \text{const}. \tag{2}$$

Here $u(t)$ does not depend on $x(t)$, that is, on the present time t. In reality we often have such a problem when we do not know the phase variables at the present time or they are known with errors.

If we substitute (2) into (1), we obtain the feedback system

$$\dot{x}(t) = Ax(t) + B \sum_{k=1}^{m} C_k x(t - kh), \tag{3}$$

and we are looking for conditions for the asymptotical stability of this system.

2. Method of solving the problem

We assume that the pair (A, B) is controllable, that is, $\operatorname{rank}(B, AB, ..., A^{n-1}B) = n$. In this case system (1) may be reduced to the form

$$u = y^{(n)} + P_1 y^{(n-1)} + ... + P_n y, \tag{4}$$

where $y \in \mathbb{R}^r$ and P_i, $i = 1, ..., n$, are $r \times n$ matrices [1]. Then we can seek the input u of the form

$$u = Q_{(n-1)}\left(\frac{d}{dt}\right) y(t - h) + \ldots + Q_0 y(t - nh), \tag{5}$$

where $Q_i\left(\frac{d}{dt}\right)$, $i = \overline{0, n-1}$, are polynomial matrices with constant coefficients and the differential operator d/dt as argument. We will obtain conditions for the polynomial matrices such that the trivial solution of equation (4) is asymptotically stable under the input (5). Since all the above transformations are nonsingular, the feedback system (3) will also be asymptotically stable. The coefficients of the matrices C_k can be chosen via the coefficients of matrices Q_i.

The characteristic equation of the feedback system (3) and the system (4) with input (5) can be written in the form

$$z^n + \sum_{i=1}^{n} P_i z^{n-1} = \sum_{j=1}^{n} Q_{n-j}(z) e^{-jhz}, \tag{6}$$

where the coefficients P_i depend only on the matrix A.

Let $a_1, \ldots, a_n$ be the roots of the equation

$$z^n + \sum_{i=1}^{n} P_i z^{n-1} = 0.$$

They depend only on the eigenvalues of A. We construct the polynomial matrices Q_{n-j} so that equation (6) may be expressed as

$$(z - a_1 + \tilde{q}_1 e^{-h_1 z}) \cdot \ldots \cdot (z - a_n + \tilde{q}_n e^{-h_n z}) = 0$$

with unknown coefficients $\tilde{q}_1, \ldots, \tilde{q}_n$.

Hence, the problem of stabilisation of the feedback system reduces to finding necessary and sufficient conditions for the roots of the quasipolynomials

$$H_k(z) = -ze^z + (p_k + i\gamma_k) + q_k + ir_k,$$
$$p_k + i\gamma_k = a_k h_k, \; q_k + ir_k = -\tilde{q}_k h_k, \; k = \overline{1,n},$$

to lie in the open left half-plane. For the case $\gamma_k = r_k = 0$ necessary and sufficient conditions were obtained by [2], and for $\gamma_k = 0$ in [3].

3. The main result

Theorem 1. [4] *All the roots of the quasipolynomial*

$$H(z) = -ze^z + (p + i\gamma)e^z + q + ir \tag{7}$$

have negative real parts if and only if the following conditions are satisfied:

1) $p < \cos^2 \omega_0$ *or* $\cos^2 \omega_l^{(1)} < p < \cos^2 \omega_l^{(2)}$, *where* ω_0, $\omega_l^{(1)}$, $\omega_l^{(2)}$ *are the roots of the equation*

$$\frac{\omega}{\cos^2 \omega} = \tan \omega + \frac{\gamma}{p}. \tag{8}$$

If $|\gamma/p| < \pi$, *then* (8) *has only one root* $\omega_0 \in \left(-\frac{\pi}{2}, \frac{\pi}{2}\right)$; *if* $|\gamma/p| \in (\pi k, \pi(k+1))$, *then* (8) *has the root* ω_0 *as above and* $2k$ *roots* $\omega_l^{(1)} \in \left(\pi l - \frac{\pi}{2}, \pi l\right)$, $\omega_l^{(2)} \in \left(\pi l, \pi l + \frac{\pi}{2}\right)$, $l = \overline{1,k}$.

2) *Let* $\gamma \in \left(\pi m - \frac{\pi}{2}, \pi m + \frac{\pi}{2}\right)$. *Then the parameter* r *must satisfy the inequality*

$$\min\,(\gamma-\omega)\cos\omega + p\sin\omega < r < \max\,(\gamma-\omega)\cos\omega + p\sin\omega$$

for $\omega \in (\pi(m-1), \pi(m+1))$.

3) *The parameter* q *must satisfy the inequality*

$$\sqrt{(\omega_2-\gamma)^2+p^2-r^2}\,\operatorname{sgn} p < -q < \sqrt{(\omega_{1,3}-\gamma)^2+p^2-r^2},$$

where $\omega_1 < \omega_2 < \omega_3$ *are the roots of the equation* $r + (\gamma-\omega)\cos\omega + p\sin\omega = 0$ *in the interval* $(\pi(m-1), \pi(m+1))$. *On the right-hand side of the inequality for* q, $\omega_{1,3}$ *stands for that of* ω_1 *or* ω_3 *for which the expression* $(\omega-\gamma)^2$ *is smaller.*

Proof. Let $z = i\omega$ and $H(i\omega) = F(\omega) + iG(\omega)$. The quasipolynomial $H(z)$ has all the roots in the open left half-plane if and only if all the roots of the trigonometric polynomials $F(\omega)$ and $G(\omega)$ are real, simple, alternating, and for all ω [5]

$$G'(\omega)F(\omega) - F'(\omega)G(\omega) > 0. \tag{9}$$

$F(\omega)$ and $G(\omega)$ are first degree polynomials in ω and $\cos\omega, \sin\omega$. Hence, all their roots are real, simple and alternating only if there are exactly $4k+1$ roots of these functions on the segment $[-2\pi k, 2\pi k]$. First we consider the equation $G(\omega) = r$, or $\tan\omega = \frac{\omega}{p} - \frac{\gamma}{p}$. If $p \le 0$, then there are $4k+1$ roots on $[-2\pi k, 2\pi k]$. For $p > 0$, in order to obtain $4k+1$ roots, the straight line $y = \frac{\omega}{p} - \frac{\gamma}{p}$ must cross one of the branches of the tangent curve $y = \tan\omega$ three times. From this condition we derive the restriction on p (condition 1).

Condition 2) of the theorem means that if $\gamma \in \left(\pi m - \frac{\pi}{2}, \pi m + \frac{\pi}{2}\right)$, then in this interval there is a root of $G'(\omega)$ with the smallest amplitude of $G(\omega)$. So in condition 2) we take such an r that on the interval $(\pi m - \pi, \pi m + \pi)$ there is a root of $G(\omega)$.

Condition 3) corresponds to inequality (9).

This is a short scheme of the proof.

Corollary. *If all the eigenvalues* a_k, $k = \overline{1,n}$, *of the matrix* A *satisfy*

$$h\,\mathrm{Re}\,a_k < \cos^2\omega_0$$

or

$$\cos^2\omega_l^{(1)} < h\,\mathrm{Re}\,a_k < \cos^2\omega_l^{(2)},$$

then system (1) *can be stabilised by the input* (2) *and the coefficients* q_k, r_k *may be chosen from the conditions* 2) *and* 3) *of Theorem* 1.

Remark. If all $a_k, k = 1, ..., n$, are real, then system (1) can be stabilised via the input (2) if $a_k < 1/h$, $k = \overline{1,n}$ [6].

References

1. V. Zubov, *Mathematical methods of investigation of control systems*, St. Petersburg, 1974.
2. N.D. Hayes, Roots of the transcendental equation associated with a certain differential-difference equation, *J. Lond. Math. Soc.* **25** (1950), 221–246.
3. M. Barszcz and A.W. Olbrot, Stability criterion of a linear differential-difference system, *IEEE Trans. Automatic Control* AC–**24** (1979), 368–369.
4. A. Kirjanen, *Stability of systems with aftereffect and their applications*, St. Petersburg, 1994.
5. R. Bellman and K. Cooke, *Differential difference equations*, Academic Press, New York–London, 1963.
6. A. Prasolov, Sufficient conditions of controllability under a special form input with delay, *Differential Equat.* **4** (1982), 713–716.

St. Petersburg State University, Department of Applied Mathematics and Control Processes, Bibliotechnaya Sq. 2, Peterhof, 198904 St. Petersburg, Russia

E-mail: gurjanov@robot.apmath.spb.su or nick@apmath.spb.su

V.P. KOROBEINIKOV and D.V. VOROBIEV

Analysis of the critical states of physical systems by means of integral relations and catastrophe theory

1. Introduction

This paper is devoted to developing the application of mathematical catastrophe theory to the understanding of the behaviour of different physical systems near their critical states. By critical states we mean fast alterations of the system state under the continuous changing of the main parameters. Here we consider the case of ignition and extinction of flame, electrical breakdown and similar phenomena.

In what follows we discuss these problems using catastrophe theory methods. Autonomous dynamical systems with potential vector field on the right-hand sides of the equations are used.

2. N.N. Semenov's problem for a gaseous combustible mixture

We consider the ignition of a premixed gaseous medium in a vessel. This is a well-known problem described in many books (see, for example, [1]). The equation of the thermal theory of ignition in a vessel has the energy conservation form

$$\frac{dT}{dt} = Ae^{-E/(RT)} - B(T - T_v), \tag{2.1}$$

where $A = \text{const}$, T_v is the temperature of the vessel walls and E is the energy of activation.

We introduce a potential function $V(T, E, T_0) = V_{ET_V}(T) : \mathbb{R}^1 \times \mathbb{R}^2 \to \mathbb{R}$ by means of the integral

$$V(T, E, T_0) = \int_{T_0}^{T} \left[Ae^{-E/(R\theta)} - B(\theta - T_v)\right] d\theta. \tag{2.2}$$

It is easy to study equation (2.1) by means of catastrophe theory [3]. The critical points T_{cr} of the potential V correspond to the condition

$$\frac{dV}{dT} = Ae^{-E/(RT)} - B(T - T_v) = 0. \tag{2.3}$$

The degenerate critical points are found from (2.4) and

$$\frac{d^2V}{dT^2} = \frac{E}{RT^2}Ae^{-E/(RT)} - B = 0. \tag{2.4}$$

To find the bifurcation set, we have to eliminate the variable T from (2.3) and (2.4).

In the vicinity of T_{cr}, the potential $V(E, T_v, T)$ has the asymptotic form $V_0 + y^3$ (if $T < E/(2R)$), and we have a catastrophe of fold type from Thom's classification [3]. Near a critical point the temperature will grow fast.

In the case when $T_v = E/(4R)$, $T_{\text{cr}} = E/(2R)$, we have $V'''(T_{\text{cr}}) = 0$, and the potential has the expression

$$V(E, T_v, T) = V_0 + C(T - T_{\text{cr}})^4 + \mathcal{O}\left[(T - T_{\text{cr}})^5\right].$$

Therefore, this case seems to have a cusp catastrophe locally.

3. Two-phase mixture of a gas with solid particles

In this section we discuss the situation of a combustible mixture of air, gaseous (volatile) fuel, and combustible and inert particles placed in a vessel of volume Ω. We make assumptions analogous to those in [2]. They allow us to write the energy balance equations for each of the three components:

$$\begin{aligned}
M_g C_g \frac{dT_g}{dt} &= Q_g A_g e^{-E_g/(RT_g)} + N_s \alpha_1 S_s (T_s - T_g) - \alpha_2 S_v (T_g - T_v) \\
&\qquad + N_s^{\text{in}} \alpha_1 S_s^{\text{in}} (T_s^{\text{in}} - T_g), \\
N_s M_s C_s \frac{dT_s}{dt} &= N_s Q_s A_s e^{-E/(RT_s)} - \alpha_1 S_s N_s (T_s - T_g) - \epsilon_1 \sigma S_s N_s T_s^4, \\
N_s^{\text{in}} M_s^{\text{in}} C_s^{\text{in}} \frac{dT_s^{\text{in}}}{dt} &= -\alpha_2 S_s^{\text{in}} N_s^{\text{in}} (T_s^{\text{in}} - T_g) - \epsilon_2 \sigma S_s^{\text{in}} N_s^{\text{in}} (T_s^{\text{in}})^4.
\end{aligned} \tag{3.1}$$

The meaning of these equations is clear.

We now consider the right-hand sides of equations (3.1) as the components of a certain vector field $\mathbf{f}$.

The chemical reaction here is supposed to be heterogeneous ($A_g = 0$). Now it is very easy to check that $\mathbf{f} = (f_1, f_2, f_3)$ has a potential of the form

$$\begin{aligned}
\Phi = \alpha_1 N_s S_s T_s T_g + \alpha_2 S_v T_v T_g + \alpha_1 N_s^{\text{in}} S_s^{\text{in}} T_s^{\text{in}} T_g - \tfrac{1}{2} T_g^2 \\
- (\alpha_1 N_s S_s + \alpha_2 S_v + \alpha_1 S_s^{\text{in}} N_s^{\text{in}}) - \tfrac{1}{2} \alpha_1 N_s^{\text{in}} S_s^{\text{in}} (T_s^{\text{in}})^2 - \tfrac{1}{2} \alpha_1 N_s S_s T_s^2 \\
- \tfrac{1}{5} \epsilon_1 \sigma S_s N_s T_s^5 - \tfrac{1}{5} \alpha_1 \sigma S_s^{\text{in}} N_s^{\text{in}} (T_p^{\text{in}})^5 - N_s A \int_{T_0}^{T_s} e^{-E/(R\xi)}\, d\xi.
\end{aligned}$$

As a first step in our analysis, we consider only two-phase media: air and combustible dust. In this case, in the system (3.1) we set $A_g = 0$ and $N_s^{\text{in}} = 0$. We have to construct the catastrophe manifold M and the singular manifold Σ (see [3]):

$$M = \{(T_g, T_s, E, T_v) \in \mathbb{R}^n \times \mathbb{R}^r,\ n = 2,\ r = 2 : \nabla\Phi(T_g, T_s, E, T_v) = 0\}, \tag{3.2}$$

$$\Sigma = \{(T_s, E, T_v) \in M : \det(H\Phi) = 0\}. \tag{3.3}$$

As a simple consequence of the equations describing M and Σ, we can write out the fifth-order polynomial equation

$$4\gamma\epsilon_1\sigma T_s^5 - \frac{E}{R}\gamma\epsilon_1\sigma T_s^4 + \alpha_1\alpha_2 S_v T_s^2 - \frac{E}{R}\alpha_1\alpha_2 S_v T_s + \frac{E}{R}\alpha_1\alpha_2 S_v T_v = 0, \qquad (3.4)$$

where $\gamma = (\alpha_1 N_s S_s + \alpha_2 S_v)$.

Using Newton's method, we have found the roots of this equation for several pairs (E, T_v) of the values of the activation energy and vessel temperature, respectively. Typical results are as follows: $T_v = 500K$, $E = 30\,000 J/\text{mole}$, $T_{\text{cr}}^{\text{up}} = 496K$, $T_{\text{cr}}^{\text{down}} = 982K$; $T_v = 2\,400K$, $E = 80\,000 J/\text{mole}$, $T_{\text{cr}}^{\text{up}} = 993K$, $T_{\text{cr}}^{\text{down}} = 2320K$. The upper positive root may correspond to burning, and the lower one to a slow oxidation process. For the behaviour of the integral curves to be shown, several numerical solutions of the Cauchy problem for equations (3.1) were carried out with prescribed parameters T_v and E and initial values of T_s, T_g and T^{in}. A similar numerical result was produced and compared with experimental values in [4].

4. Mathematical study of breakdown in gases

Suppose now that the domain **D** in Fig. 1 is filled with a gaseous mixture. We assume that the electric field intensity $\mathcal{E}$ is sufficiently strong. The discharge corresponds to an abrupt increase in the electron concentration and electrical current due to an avalanche ionisation process. We consider the spatially averaged diffusion-drift equation in simplified form, namely

$$\frac{dn_e}{dt} = \left(\nu_i - \nu_a - \frac{\mu_e \mathcal{E}}{\Lambda}\right) n_e - \frac{D n_e}{\Lambda^2}. \qquad (4.1)$$

Here n_e is the electron number density averaged over the volume $V \sim \Lambda^3$, Λ is the diffusion length, μ_e the electron mobility, $\nu_i(\mathcal{E}/n_n)$ the ionisation frequency of the neutral molecules of gas with number density n_n due to electron impacts in the electrical field $\mathcal{E}$, and ν_a is the electron attachment frequency. We also assume that

$$p = p_n + p_e + p_i = k(n_n T_n + n_e T_e + n_i T_i), \quad \nu_i = \mu_e \mathcal{E} A p_n e^{-Bp_n/\mathcal{E}}, \qquad (4.2)$$

where k is the Boltzmann constant. Considering n_e as the state variable and $\mathcal{E}$, D, A, B, Λ as the monitoring parameters for the system governed by equation (4.1), we obtain the set M of the catastrophe, determined by the equation

$$\left(\nu_i(\mathcal{E}, n_e) - \nu_a(\mathcal{E}, n_e) - \frac{\mu_e \mathcal{E}}{\Lambda} - \frac{D(T_e)}{\Lambda^2}\right) n_e = 0. \qquad (4.3)$$

The second derivative of the potential function for (4.1) leads to the equation for the case of an electron avalanche development under the assumption that $\nu_a = D = 0$: $\mathcal{E} = Bp_n$. The singular set is defined by these equations and corresponds to the critical values of the parameters, that is, it describes the formation of discharge (breakdown) during the electron avalanche stage. We can also find the bifurcation set. For example, in air $B = 51{,}2V/\text{cm} \cdot \text{torr}$, as reported by Skanavi [5], and the critical value of the breakdown field for the pressure $p_n = 760$ torr is equal to

$\mathcal{E}_c = 38,9kV/\text{cm}$. However, the results ought to be related only to the electron avalanche (streamer) stage. Due to the ion motion in the streamer channel, the electric current increases with its density vector $\vec{j} = \sigma\vec{\mathcal{E}}$ and Joule dissipation $\sigma\mathcal{E}^2$.

The study of the energy equation is carried out in the same manner as for equation (2.1).

5. Integral relations for spatial distribution and integral equations

The previous method was applied to the spatially averaged equations of diffusion or heat transfer processes. We now discuss briefly the spatial distribution of the discharge formation. We consider the same problem as in the previous section. For the sake of simplicity, we examine only the one-dimensional (plane) case for the diffusion equation, that is,

$$\frac{\partial n_e}{\partial t} + \frac{\partial}{\partial x}(n_e v_e + I_e) = \dot{n}_e, \tag{5.1}$$

where v_e is the velocity, $I_e = -n_e \mu_e E - D\,\partial n_e/\partial x$, and μ_e is the mobility of the electrons. Suppose that the distribution of the electric fields is uniform along the electrodes, and that the initial gas state has constant parameters. We divide the space between electrodes into N strips. If we integrate (5.1) from x_i to x_{i+1}, we get a system of ordinary differential equations for $n_{ei} = n_e(x_i)$. We take $N = 2$, $i = 0, 1, 2$, and assume that $(v_e)_i = 0$. Then

$$\begin{aligned}
\frac{dn_{e0}}{dt} &= -F_{e1} + \frac{4}{d}(I_{e0} - I_{e1}) + \dot{n}_{e1} + \dot{n}_{e0},\\
\frac{dn_{e1}}{dt} &= F_{e1} = \frac{4}{d}(2I_{e1} + n_{e1}\mu_{e1}\mathcal{E} - \frac{4}{d}Dn_{e1}) + \dot{n}_{e1},\\
& n_{e0}|_{t=0} = n_e^*, \quad n_{e1}|_{t=0} = n_{e1}^*,\\
& I_{e1} = m_1 n_{e1} + m_0 n_{e0} + M.
\end{aligned}$$

The boundary condition on the walls is $I_{ew} = h(T_e)(n_e - n_{ew})$, where $h(T_e)$ is the coefficient of electron exchange and m_0, m_1 and M are constants. Thus, we have a system of two ordinary differential equations that can be completely analysed by means of well-known methods, for example, they can be studied as in the above method of catastrophe theory.

When we consider the problems of body heating by radiation, we have a case of an integro-differential equation. Before studying this complicated case, it is useful to consider a simpler one. We consider an integral equation of the form

$$f(x) = \int_a^b K(x, y) f(y)\, dy + g(x), \tag{5.2}$$

where

$$K(x, y) = \sum_{i=1}^{r} \phi_i(x)\overline{\psi_i(y)}, \quad f(0) = g(0) = 0.$$

The functions $\phi_1, \ldots, \phi_r$ and $g(x)$ are assumed analytic at the origin.

Proposition 1. *The solution $f(x)$ of equation* (5.2) *has a critical point of order N at zero if and only if $WG_N(0)\vec{\xi} = \vec{h}(0)$.*

Proposition 2. *If the function $g(x)$ has a critical point of the order $r-1$ at zero, then the solution $f(x)$ of equation* (5.2) *cannot typically have a critical point of the same order at zero.*

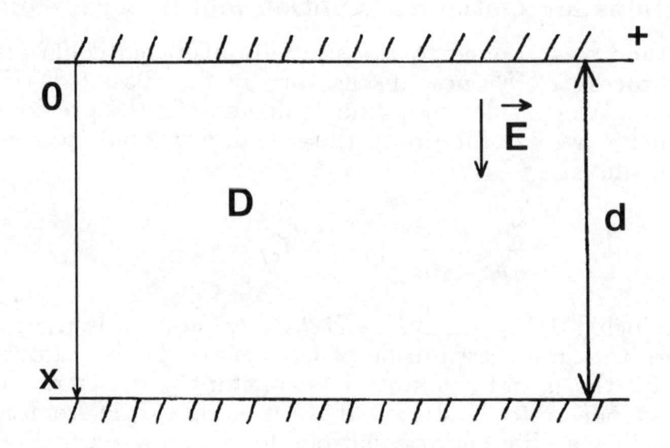

Fig. 1. Domain of discharge.

6. Conclusion and acknowledgments

This paper presents preliminary results in the investigation of the theory of catastrophe methods in various cases of dynamical systems. This work is now in progress and further results will be published later. The authors wish to thank P. Wolanski, R. Klements, V. Markov and their colleague Yu. K. Bobrov for useful discussions and practical assistance. D.V. Vorobiev is indebted to the Russian Foundation for Basic Research (travel grant no. 96-01-10601) and to the Moscow Institute of Physics and Technology for the financial support that made his visit to Oulu possible.

References

1. R. Strehlov, *Combustion fundamentals*, McGraw-Hill, New York, 1984.
2. W. Kordylewski, *Thermal explosion theory and its applications*, Wroclaw, 1985 (Polish).
3. T. Poston and I. Stewart, *Catastrophe theory and its applications*, Pitman, London, 1978.
4. S.W. Baek, M Sichel and C.W. Kaufmann, Asymptotic analysis of the shock wave ignition of dust particles, *Combustion and Flame* **81** (1990), 219–228.
5. G.I. Skanavi, *Physics of dielectrics*, Fizmatlit, Moscow, 1958 (Russian).

Institute for Computer Aided Design, 2nd Brestskaya 19/18, 123056 Moscow, Russia

A. KOSHCHII

Some non-classical dual variational principles and a posteriori error estimates for elliptic boundary value problems

1. Introduction

The importance of conservation laws is well known. Recently, practical requirements have stimulated the construction of modified variational principles and the development of dual variational methods for the solution of elliptic boundary value problems. In the numerical realisation of dual methods, both the direct (original) extremal problem and its dual are solved. After obtaining approximate solutions by means of the corresponding direct and dual variational formulations, we can determine both upper and lower estimates for the exact value of the extremal of the original problem, and a posteriori energy estimate for the accuracy of the approximate solution.

It should be pointed out that a similar procedure may be used to analyse the results independently of how they have been obtained.

As a rule, dual variational formulations are defined on classes of vector functions satisfying certain restrictions of a differential nature in the domain of the problem (for example, by the equilibrium equations in Castigliano's principle). At the same time, for some special narrow classes of problems there are dual variational principles that are free of this restriction (for example, in problems of stationary heat conduction in the presence of heat sources depending on temperature, or in bending of thin plates lying on an elastic base). From an abstract point of view, the existence of such variational principles is connected with the presence of easily invertible operators in the additive structure of the operators of these problems.

The investigations carried out in recent years show that dual variational principles with similar characteristics exist for much wider classes of problems. In this paper we prove such principles for a boundary value problem for a fourth order scalar differential equation. Due to space limitations, results are not presented in their most general form.

2. Formulation of the problem

We consider the boundary value problem

$$\Delta^2 u - \operatorname{div} T \operatorname{grad} u + cu = f, \quad x \in \Omega,$$

$$u = \frac{\partial u}{\partial n} = 0, \quad x \in \Gamma.$$

Here Ω is a bounded domain in $\mathbb{R}^m$, $m \geq 2$, with a sufficiently smooth (for example, Lipschitz continuous) boundary Γ, n is the outward unit normal to Γ, and T is an $m \times m$ matrix with elements t_{ij}, $i, j = 1, \ldots, m$.

We assume that $f \in L_2(\Omega)$, $c, t_{i,j} \in L_\infty(\Omega)$, $i, j = 1, \ldots, m$, T is a symmetric and non-negative matrix ($T \geq 0$) and $c \geq 0$ almost everywhere in Ω. Then the boundary value problem has a unique generalised solution u_e, which at the same time is the solution of the direct extremal problem

$$J^0(u) \to \inf,$$
$$u \in U = \left(u \in W_2^2(\Omega) : u = \frac{\partial u}{\partial n} = 0 \text{ on } \Gamma \right),$$

where

$$J^0(u) = \frac{1}{2} \int_\Omega \left((\Delta u)^2 + (T \text{ grad } u, \text{ grad } u) + cu^2 - 2fu \right) d\Omega$$

is the so-called energy functional and $(\cdot, \cdot)$ is the standard inner product on $\mathbb{R}^m$.

In the case $m = 2$, our boundary value problem describes the bending of a thin isotropic plate in the presence or absence of an elastic base ($c \geq \text{const} > 0$ or $c = 0$ almost everywhere in Ω, respectively) and of forces in the middle plane ($T > 0$ or $T = 0$, respectively). $T > 0$ means that the matrix T is uniformly positive definite almost everywhere in Ω, and $T = 0$ means that the matrix T is equal to zero.

Suppose that the functional J_0 and the set Λ are known and such that

$$\inf_{u \in U} J^0(u) = \sup_{\lambda \in \Lambda} J_0(\lambda).$$

Then we can estimate the difference between the approximate solution u and the unknown exact solution u_e of the direct problem in the energy norm by means of the approximate solution λ of the dual extremal problem

$$J_0(\lambda) \to \sup,$$
$$\lambda \in \Lambda.$$

This estimate is given by the classical inequality

$$[u - u_e]^2 \leq 2 \left(J^0(u) - J_0(\lambda) \right),$$

where

$$[v]^2 = \int_\Omega \left((\Delta v)^2 + (T \text{ grad } v, \text{ grad } v) + cv^2 \right) d\Omega$$

is the energy norm.

Remark. If the dual problem has been constructed, then special error estimates can be obtained usually, in particular, exact hypercircle estimates.

It is well known that the form of classical dual problems depends essentially on the parameters of the original problem. In the case $T = 0$, the dual problems are defined on classes of scalar functions satisfying restrictions of a differential nature in the domain of the given problem $c \geq 0$. If $T \geq 0$, these restrictions depend on T and c.

In the next section we present a new non-classical dual problem under general assumptions on the parameters $T \geq 0$, and $c \geq 0$.

3. A non-classical dual problem

Theorem 1. *Suppose that* $\alpha \in [C^1(\overline{\Omega})]^m$ *and* $a = \text{const}$ *are such that* $S = T + aE$, *where* E *is the unit* $m \times m$ *matrix, is uniformly positive definite, almost everywhere in* Ω *and that*

$$\delta = \mathrm{c} + \operatorname{div} \alpha - \tfrac{1}{2} a^2 - (S^{-1}\alpha, \alpha) \geq \text{const} > 0.$$

Then the dual problem is the extremal problem for the functional

$$J_0(\lambda, \mu) = -\frac{1}{2} \int_\Omega \left[\lambda^2 + \frac{1}{\delta} g^2(\lambda, \mu) + \left(S^{-1}(\mu + \gamma \text{ grad } \lambda),\ \mu + \gamma \text{ grad } \lambda\right) \right] d\Omega$$

with $\lambda \in W_2^{k(\gamma)}(\Omega)$ *and* $\mu \in \left[W_2^1(\Omega)\right]^2$. *Here*

$$g(\lambda, \mu) = f + \operatorname{div} \mu - (1 - \gamma)\Delta\lambda - \tfrac{1}{2} a\lambda - \left(S^{-1}\alpha, \mu + \gamma \text{ grad } \lambda\right),$$

γ *is an arbitrary constant and* $k(\gamma) = \begin{cases} 1, & \gamma = 1, \\ 2, & \gamma \neq 1. \end{cases}$

The equality $J^0(u) = J_0(\lambda, \mu)$ *is attained when*

$$u = u_e,\ \lambda = \lambda_e = \Delta u_e + \tfrac{1}{2} a u_e,$$
$$\mu = \mu_e = S \text{ grad } u_e - \gamma \text{ grad } \lambda_e + u_e \alpha,$$
$$u_e = \frac{1}{\delta} g(\lambda_e, \mu_e).$$

These relations connect the solution u_e of the direct problem with the solution (λ_e, μ_e) of the dual one. After computing approximate solutions for the direct and dual extremal problems, we can use these formulae to obtain additional information about the exact solutions of these problems, namely

$$\lambda_p = \Delta u + \tfrac{1}{2}\, a u,\ \mu_p = S \text{ grad } u - \gamma \text{ grad } \lambda_p + u\alpha,$$
$$u_d = \frac{1}{\delta} g(\lambda, \mu).$$

Here (λ_p, μ_p) is the approximate solution of the dual problem obtained by means of the approximate solution of the direct problem, and u_d is the approximate solution of the direct problem obtained from the approximate solution of the dual problem.

It should also be noted that after finding the approximate solutions, we can obtain the so-called hypercircle approximate solutions

$$u_h = \tfrac{1}{2}(u + u_d), \quad \lambda_h = \tfrac{1}{2}\,(\lambda + \lambda_p), \quad \mu_h = \tfrac{1}{2}\,(\mu + \mu_p).$$

The next assertion indicates effective formulae for choosing the functional parameters a and α.

Theorem 2. *If the closure of Ω lies inside the circle of radius R with the centre at the origin, then the constant a may be chosen anywhere in the interval*

$$0 < a < \frac{m^2}{R^2}.$$

In this case, the vector function α may be taken of the form

$$\alpha = kx = k(x_1, x_2, \dots x_m),$$

where the constant k is such that

$$\frac{ma}{2R^2}\left[1 - \left(1 - \frac{R^2}{m^2}a\right)^{1/2}\right] < k < \frac{ma}{2R^2}\left[1 + \left(1 - \frac{R^2}{m^2}a\right)^{1/2}\right].$$

Theorem 3. *The hypercircle solutions of the given problem satisfy the estimate*

$$\tfrac{1}{2}\left[J^0(u) - J_0(\lambda, \mu)\right] = \int_\Omega \left[(\lambda_h - \lambda_e)^2 + \delta(u_h - u_e)^2 + \left(S^{-1}(\mu_h - \mu_e + \gamma \operatorname{grad}(\lambda_h - \lambda_e)), \mu_h - \mu_e + \gamma \operatorname{grad}(\lambda_h - \lambda_e)\right)\right] d\Omega.$$

Remark. The construction of non-classical dual problems with auxiliary functional parameters was suggested by V.I. Kalinichenko (1980) for two-dimensional stationary heat problems with homogeneous boundary conditions. This idea had also been mentioned earlier by V.L. Rvachev in a different context.

4. Conclusion

We see that the nature of the above non-classical dual problem remains the same for all permissible versions of the original problem (that is, of the matrix T and function c). Its solution is realised in classes of vector functions that do not have to satisfy any differential restriction in Ω. This distinguishes the non-classical dual problem from classical ones.

Corresponding results have already been established for boundary value problems for the stationary heat equation, for the Lamé system in elasticity theory, in problems of bending of thin plates and in some other problems (see [1]–[5]). The nature of the results does not depend on the presence, nor on the absence, of easily invertible operators in the additive structure of the operators of the original problems. In the boundary value problems for bending of plates in Kirchhoff's model are also independent of the presence or absence of forces in the middle plane.

References

1. A.F. Koshchii, V.I. Kalinichenko and V.A. Shcheglov, A dual method for solving fourth-order equations of elliptic type (Russian), *Dokl. Akad. Nauk. Ukrain. SSR Ser. A* **1987**, no. 6, 26–29.

2. A.F. Koshchii, V.I. Kalinichenko and A.I. Ropavka, *Numerical solutions of heat-conduction problems*, Vishcha Shkola, Kharkiv, 1987 (Russian).
3. A.F. Koshchii, P.G. Bova and V.A. Shcheglov, A posteriori estimates for the solution in problems of the theory of thin plates, in *Numerical methods for solving problems of elasticity and plasticity theory*, Novosibirsk, 1988, 30–36 (Russian).
4. A.F. Koshchii and A.I. Ropavka, A posteriori error estimates for the solutions of some boundary value problems in heat conduction, in *Numerical experiment in problems of heat-mass transfer and heat transmission*, vol 9., part 2, ANB, Minsk, 1992, 92–95 (Russian).
5. A.F. Koshchii and B.S. Elkin, On some non-classical dual variational methods and a posteriori error estimates, in *Non-linear boundary value problems of mathematical physics and their applications*, Kiev, 1995, 125–127 (Russian).

Department of Mathematics and Mechanics, Kharkov State University,
4 Svobody Sq., 310077 Kharkov, Ukraine

A. KRASNIKOVS, J. VARNA and L.A. BERGLUND

Transverse crack opening displacement calculations in cross-ply and angle-ply laminated specimens

1. Introduction

Laminated fibre composites are widely used in different branches of industry because of their high strength and stiffness-to-weight ratios. During the fabrication and use of such laminates, the earliest damaging phenomenon that occurs is the formation of transverse cracks parallel to the fibre direction. At later stages, transverse cracks lead to delamination, fibre fractures and whole laminate bearing facilities degradation. Experimental studies of the transverse cracking effect on thermo-elastic properties of cross-ply laminates show (see [1] and [2]) that in laminates with very similar elastic properties of the constituents (unidirectional plies) the reduction of elastic properties (Poisson's ratio) during cracking may be very different. In the present study different analytic and numerical micromechanical method predictions are compared with experimental data for $[0_m,\ 90_n]_S$ cross-ply and angle-ply laminates.

The commonly used theoretical methods that describe this phenomenon are based on the shear lag assumption (see [3]–[5]), the principle of minimum complementary energy (see [6]–[9]) and the finite element method (FEM) calculations [10]. In our paper only Std GF/EP experimental test data are discussed. Such material is elastic and brittle. This is the reason why only linear elastic behaviour is taken into account in theoretical predictions.

2. Theory

In what follows we consider the representative volume of the laminated plate, called laminate, with the geometry shown in Fig. 1. The elastic properties of the laminate are reduced during transverse cracking. The laminate is made of orthotropic layers and subjected to mechanical loading in the x-direction. The layer's thermo-elastic properties of interest are E_1, E_2, ν_{12}, ν_{23}, α_1 and α_2. Thermal stress occurs during fabrication (cooling from production to room temperature). The geometric parameters are h, l_0, b and d (see Fig. 1). We denote the stress components in the undamaged laminate, obtained from laminate theory, by σ_{x0}^{0}, σ_{y0}^{0}, σ_{x0}^{90} and σ_{y0}^{90}. They contain mechanical and thermal parts. The stress state in the laminate with cracks may be written in the form [10]

$$\sigma_x^{90} = \sigma_{x0}^{90} - \sigma_{x0}^{90} f_1(\bar{x}, \bar{z}), \tag{1}$$

$$\sigma_x^{0} = \sigma_{x0}^{0} + \sigma_{x0}^{0} f_2(\bar{x}, \bar{z}). \tag{2}$$

Here, the second term in each expression represents the stress perturbation caused by the presence of a crack. We introduce the non-dimensional coordinates $\bar{x} = x/d$,

This work was supported by grant N 94 1059 from the Latvian Council of Science.

$\bar{z} = z/d$. The integral equilibrium in the x-direction yields

$$\int_0^1 f_1(\bar{x}, \bar{z}) d\bar{z} = \int_1^{\bar{h}} f_2(\bar{x}, \bar{z}) d\bar{z}, \tag{3}$$

where $\bar{h} = h/d$.

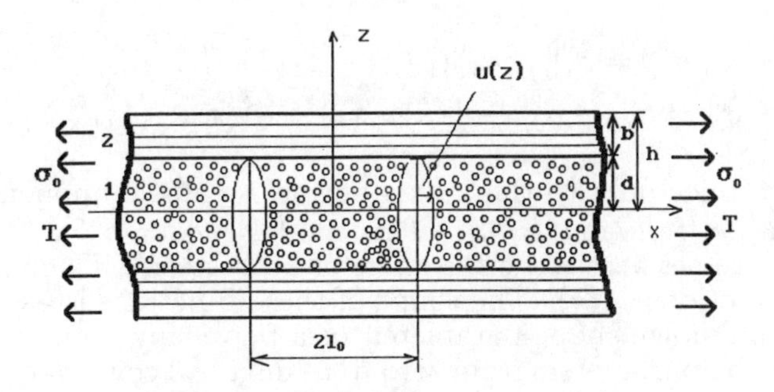

Fig. 1. Lamina geometry.

Since all known analyses neglect the edge effect and are based on the assumption that the strains are not dependent on the y-coordinate, stress and strain averaging is necessary only in the x- and z-directions. Using (3) and (1), (2), we obtain the average stresses in the form

$$\bar{\sigma}_x^{90} = \sigma_{x0}^{90} - \sigma_{x0}^{90} R(\bar{l}_0)(2\bar{l}_0)^{-1}, \tag{4}$$

$$\bar{\sigma}_x^{0} = \sigma_{x0}^{0} + \sigma_{x0}^{90} R(\bar{l}_0)(2\bar{l}_0\bar{b})^{-1}, \tag{5}$$

where $\bar{b} = b/d, \bar{l}_0 = l_0/d$ and

$$R(\bar{l}_0) = \int_{-\bar{l}_0}^{+\bar{l}_0} \int_1^{\bar{h}} f_2(\bar{x}, \bar{z}) d\bar{x} d\bar{z}. \tag{6}$$

For the average crack opening displacement and modulus reduction rate we can obtain the equations [10]

$$\bar{u} = \frac{d}{2}(\varepsilon_{x0}^m + \varepsilon_{x0}^{th}) \left(1 + \frac{E_2 d}{E_1 b}\right) R(\bar{l}_0), \tag{7}$$

$$\frac{E_x}{E_{x0}} = \left(1 + \frac{E_2 d}{2E_1 b} \frac{1}{\bar{l}_0} R(\bar{l}_0)\right)^{-1}, \tag{8}$$

where ε_{x0}^m and ε_{x0}^{th} are the mechanical and thermal parts of strain, respectively. Equations (7) and (8) show that the rate of reduction of the longitudinal modulus is related to the average COD. This relationship is described by the function $R(\bar{l}_0)$. Since $R(\bar{l}_0)$ depends on $\bar{l}_0$, the relationship holds also at high crack density (crack interaction region). The function $R(\bar{l}_0)$ is different for different transverse cracking models. These differences cause different predictions of the crack opening and modulus reduction. In the remainder of this paper we briefly analyse the assumptions of different models and the corresponding expressions for $R(\bar{l}_0)$.

In the model [9] the solution for the stress field is found using the principle of minimum complementary energy. The admissible stress state that satisfies the equilibrium equations is chosen to be

$$\sigma_x^{90} = \sigma_{x0}^{90} - \sigma_{x0}^{90}(\psi(\bar{x}) - \psi_1(\bar{x})\varphi_2''(\bar{z})),$$
$$\sigma_x^{0} = \sigma_{x0}^{0} + \sigma_{x0}^{90}(-\psi(\bar{x})\varphi_1''(\bar{z}) + \psi_1(\bar{x})\varphi_3''(\bar{z})).$$

Here and below we refer only to the x-axis normal stress components that are explicitly present in the function $R(\bar{l}_0)$. This model accounts for the non-uniform distribution of all stresses across the thickness of both layers. The non-uniformity is described by the functions $\varphi_i(\bar{z})$. The shape of these functions is assumed to be exponential with an unknown shape parameter, or a power function with an unknown power. By varying the parameters from zero to infinity, we cover very different stress states, from uniform to very strong boundary effects at the interface. The actual values of these parameters are calculated during the minimisation procedure. The functions that describe the x-distribution are found by solving a system of two fourth order differential equations with constant coefficients. These equations follow from the minimisation routine. For (6) we obtain

$$R(\bar{l}_0) = \int_{-\bar{l}_0}^{+\bar{l}_0} (\psi(\bar{x}) - \psi_1(\bar{x}))d\bar{x}.$$

The calculated analytic expressions of $\psi(\bar{x})$ and $\psi_1(\bar{x})$ are not given here because of their complexity.

The next model [8] is a particular and simplified case of the previous one. The assumptions about the admissible stress state are as follows. 1) In the 90-layer, the x-axis stress distribution in the z-direction is uniform, the shear stress is linear, and the z-axis stress is normal and parabolic. 2) In the 0-layer, the non-uniformity described by an exponent with an unknown shape parameter is included. In this case,

$$R(\bar{l}_0) = \int_{-\bar{l}_0}^{+\bar{l}_0} \psi(\bar{x})d\bar{x}.$$

Hashin's model [6] was historically the first model that used the variational approach to determine the stress state in the cross-ply laminate with cracks. The admissible stress system, which is necessary in order to use the principle of minimum complementary energy, is constructed using the assumption that the x-axis normal stress in the 90- and in the 0-layers is independent of the z-coordinate (uniform stress distribution across the layer thickness). This implies linear distributions for the shear stresses and a parabolic one for the z-axis normal stress.

Models based on the shear lag assumption (see [3]–[5]) are the simplest among those discussed in this paper. There is a large variety of different modifications. However, the basic assumptions are the same. The equation that describes the stress distribution is obtained from integral equilibrium conditions (the equilibrium equations are not satisfied at each point). For this reason the equation is a second order differential equation with constant coefficients. As a result, the zero shear stress condition on the crack surface cannot be satisfied. The solution for the x-axis stress distribution can be expressed as

$$\sigma_x^{90} = \sigma_x^{90} - \sigma_x^{90} \frac{\cosh \alpha \bar{x}}{\cosh \alpha \bar{l}_0}, \quad \sigma_x^0 = \sigma_x^0 + \sigma_x^{90} \frac{\cosh \alpha \bar{x}}{\cosh \alpha \bar{l}_0},$$

where

$$\alpha = \sqrt{k(d/E_2)[1 + (E_2 d)/(E_1 b)]}, \quad R(\bar{l}_0) = (2/\alpha)\tanh(\alpha \bar{l}_0)$$

and k is the fitting parameter.

FEM calculations were realised using the commercial software NISA for 2D and ABAQUS for 3D. During the calculations, the computational accuracy was verified. In contrast to all other approximate methods, FEM also yields the shape of the crack opening.

3. Results and analysis

Predictions of the COD were performed for $[0_2^0, 90_n^0]_S$ standard GF/EP laminates with $E_1 = 45.5\,\text{GPa}$, $E_2 = 12.7\,\text{GPa}$, $G_{12} = 3.45\,\text{GPa}$, $G_{23} = 4.5\,\text{GPa}$, $\nu_{12} = 0.28$, $\nu_{23} = 0.42$, $\alpha_1 = 4 \cdot 10^{-6}(°\text{C}^{-1})$, $\alpha_2 = 17 \cdot 10^{-6}(°\text{C}^{-1}$ and $T = -100°\text{C}$. The actual layer thicknesses were

n	2	4	8
b (mm)	0.25	0.25	0.23
d (mm)	0.25	0.50	0.92

We start the comparison with the low crack density case. This diluted crack distribution corresponds to non-interactive cracks. FEM produces the most accurate stress and displacement distribution. The calculated crack opening displacement distribution is compared with test results. The agreement with Standard GF/EP data is very good. The predictions of COD by different models and test data for laminates with different 90-layer thickness always give the best agreement for FEM. The model [9] slightly overestimates the normalised average COD. The difference between this model and FEM is smaller for thin 90-layers, and increases with increasing 90-layer thickness. This is an indication that the stress state in the thick 90-layer is described sufficiently well by this model. The model [8], as well as Hashin's model, predict rather inaccurate values of COD. In the latter, the deviation from the FEM solution is independent of the 90-layer thickness. The predictions of the model [8] are better for thin 90-layers and worsen for thick 90-layers. For thicker 90-layers, the predictions of this model are comparable to those of Hashin's model. This indicates that the stress non-uniformity in the 0-layer included in the model [8] improves for laminates with thin 90-layers and becomes negligible in the case of thick 90-layers.

The mean values of the normalised average COD measured at 0.65% mechanical strain are shown against the 90-layer thickness in Fig. 2, together with the predictions of all the models discussed. According to the FEM analysis, the calculated normalised average COD is almost independent of the 90-layer thickness. Tests show smaller COD than FEM for thin 90-layers. This may be due to fibres bridging the crack and reducing the opening. For thick 90-layers, experimental COD are larger than in the FEM. The reason may be non-regularities in crack geometry (local delamination, crack branching etc.) that increase the COD. Fiber bridging is less significant in this case because of the much larger size of the crack.

The model [9] describes better the increase of COD as the 90-layer thickness increases. However, the COD values are overestimated by approximately 20%. Certainly, the better slope determination by the model [9] is the only coincidence, caused by the less accurate stress calculation.

The predictions of [8] and Hashin's model yield an approximately 50% higher COD than that observed in the test. The difference in COD caused by the non-linear material behaviour observed in [2] was 20%. For this reason, such models are not appropriate for the analysis of transverse cracking in layers with non-linear material behaviour. The same conclusion also hold for the model [9].

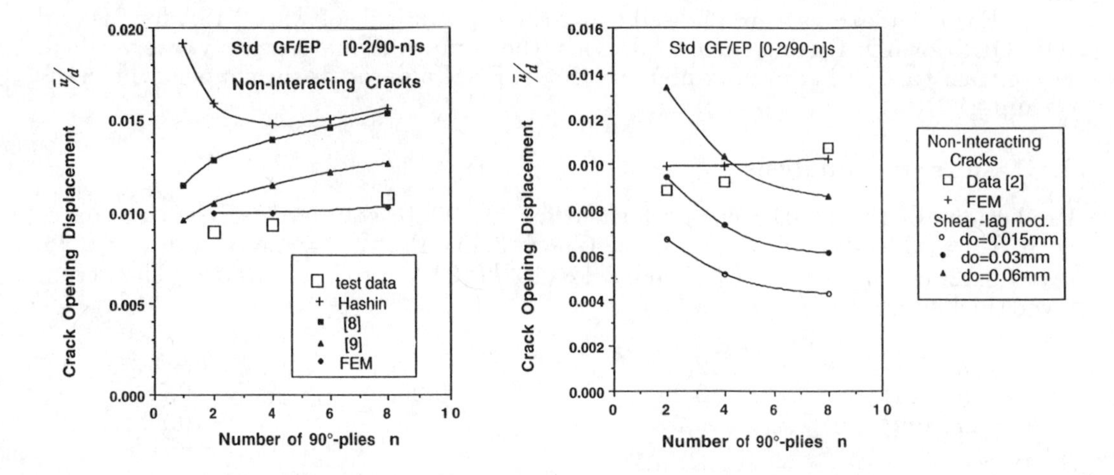

Fig. 2. The average crack opening displacement vs. the number of 90-layers.

Fig. 3. The normalised crack opening displacement vs. the number of 90-layers.

The predictions of the shear lag model are compared with the test data and the FEM results in Fig. 3. The results are given only for the model modification that considers the resin-rich region at the interface. The typical size of the zone is about 0.015 mm. For this value, the predicted COD is very unrealistic. Neither the quantitative values nor the qualitative behaviour are correct.

Finally, we consider the effect on the COD of the crack interaction at high crack densities. Tests show a significant decrease of the COD. The model predictions are rather far from the actual data. Even the FEM analysis fails to predict the crack interaction properly. The actual COD is always lower than the FEM predictions. Since we do not doubt the accuracy of the FEM solution, the reason for the difference may lie in the assumed geometry of the crack, or in the material behaviour. We could not observe any qualitative differences in the geometry between non-interactive and interactive cracks in the inspected crack density region. Similar results have been obtained for angle-ply laminates.

To conclude our analysis, we remark that the predictive ability of every commonly used theoretical model may be restricted, and that its accuracy must be verified before use.

References

1. R. Talreja, S. Yalvac, L.D. Yats and D.G. Wetters, Transverse cracking and stiffness reduction in cross-ply laminates of different matrix toughness, *J. Composite Materials* **26** (1992), 1644–1663.
2. J. Varna, L.A. Berglund, R. Talreja and A. Jakovics, A study of the crack opening displacement of transverse cracks in cross ply laminates, *Int. J. Damage Mech.* **2** (1993), 272–289.
3. Y.M. Han and H.T. Hahn, Ply cracking and property degradation of symmetric balanced laminates under general in-plane loading, *Composite Sci. Technology* **39** (1989), 377–397.
4. P.A. Smith and J.R. Wood, Poisson's ratio as a damage parameter in the static tensile loading of the simple cross-ply laminates, *Composite Sci. Technology* **38** (1990), 85–93.
5. S.G. Lim and C.S. Hong, Prediction of transverse cracking and stiffness reduction in cross-ply laminated composites, *J. Composite Materials* **23** (1989), 695–713.
6. Z. Hashin, Analysis of cracked laminates: a variational approach, in *Mechanics of materials*, vol. 4, North-Holland, 1985, 121–136.
7. J.A. Nairn, The strain energy release rate of composite microcracking: a variational approach, *J. Composite Materials* **23** (1989), 1106–1129.
8. J. Varna and L.A. Berglund, Multiple transverse cracking and stiffness reduction in cross-ply laminates, *J. Composites Technology Res.* **13** (1991), 97–106.
9. J. Varna and L.A. Berglund, Thermo-elastic properties of composite laminates with transverse cracks, *J. Composites Technology Res.* **16** (1994), 77–87.
10. J. Varna, L.A. Berglund, A. Krasnikovs and A. Chikhalenko, Crack opening geometry in cracked composite laminates, *Internat. J. Damage Mech.* **6** (1997), 96–118.

A.K.: Riga Technical University, Kalku St. 1, Riga LV 1058, Latvia

J.V., L.A.B.: Lulea University of Technology, Lulea, Sweden

I.A. KUZIN

A remark on the number of solutions to critical and supercritical semilinear problems with non-zero right-hand sides

1. Results

Consider the "anticoercive" problem

$$\begin{aligned} \Delta u + |u|^{p-2}u &= h(\mathbf{x}) \quad \text{in } \Omega, \\ u|_{\partial\Omega} &= 0, \end{aligned} \tag{1}$$

where Ω is a smooth bounded domain in $\mathbb{R}^N$, $N > 2$, $p \geq 2^*$, $2^* \equiv 2N/(N-2)$ is the critical Sobolev exponent and h is sufficiently regular. This problem has received much attention recently.

It is known that if $h \equiv 0$, then problem (1) has infinitely many distinct solutions $u \in H_1^0(\Omega)$ for $2 < p < 2^*$. However, if $p \geq 2^*$ and the domain Ω is star-shaped (a ball, for instance), then problem (1) with $h \equiv 0$ has no non-zero solution [1].

For a non-zero right-hand side the situation is not that clear, only some partial results being available. It is known that for $2 < p < (2N-2)/(N-2)$ problem (1) has infinitely many distinct solutions $u \in H_1^0(\Omega)$ (see [2]–[4]). The answer can also be obtained for $(2N-2)/(N-2) \leq p < 2^*$ if Ω is a ball and h is radial [4]. Pokhozhaev (see, for example, [5]) has obtained sufficient conditions on h under which the radial problem (1) has no classical sign-defined solutions.

The natural question arises of the multiple solvability of (1) with non-zero right-hand sides for the critical and supercritical cases $p \geq 2^*$. We study this problem when Ω is a ball and h is a radial function.

It has already been shown that, in contrast to the superlinear subcritical case, for all even integers $p \geq 2^*$ and a sufficiently smooth h, with the possible exception of some special cases, the radial problem (1) has at most finitely many distinct classical solutions (exact statements can be found in [5]).

Here we present new and more detailed information on the set of solutions in some typical situations.

In the radial case, problem (1) in the classical setting has the form

$$\begin{aligned} u'' + \frac{N-1}{r}u' + |u|^{p-2}u &= h(r), \quad 0 \leq r \leq R, \\ u'(0) &= 0, \\ u(R) &= 0, \end{aligned} \tag{2}$$

where $r = |\mathbf{x}|$ and $R > 0$ is the radius of the ball Ω.

This work was partly supported by the grants RFBR 96-01-00097 and INTAS 94-2187.

Theorem 1a. *Let $p \geq 2^*$, let $h \in C^1_{\mathrm{loc}} \cap L_\infty([0,+\infty))$, and suppose that there is an r_0 such that $h(r_0) < 0$ and $h'(r) \leq 0$ for $r \geq r_0$. Then there exists an R_* such that for any $R > R_*$ problem (2) has no classical solution.*

Theorem 1b. *Let $p \geq 2^*$, let $h \in C^1_{\mathrm{loc}} \cap L_\infty([0,+\infty))$, and suppose that there is an r_0 such that $h(r_0) > 0$ and $h'(r) \geq 0$ for $r \geq r_0$. Then there exists an R_* such that for any $R > R_*$ problem (2) has no classical solution.*

We now consider the critical case.

Theorem 2a. *Let $p = 2^*$, let $h \in C_{\mathrm{loc}}([0,+\infty))$, and suppose that there is a positive constant ε such that $h(r) < -\varepsilon$ for all $r \geq 0$. Then there exists an R^* such that for any $0 < R < R^*$ problem (2) has at least two positive classical solutions.*

Theorem 2b. *Let $p = 2^*$, let $h \in C_{\mathrm{loc}}([0,+\infty))$, and suppose that there is a positive constant ε such that $h(r) > \varepsilon$ for all $r \geq 0$. Then there exists an R^* such that for any $0 < R < R^*$ problem (2) has at least two negative classical solutions.*

2. Proofs

We study problem (2) by means of the auxiliary Cauchy problem

$$\begin{aligned} u'' + \frac{N-1}{r}u' + |u|^{p-2}u &= h(r), \quad 0 \leq r < +\infty, \\ u'(0) &= 0, \\ u(0) &= a, \end{aligned} \tag{3}$$

with various values of a. A standard argument shows that if $h \in C_{\mathrm{loc}}([0,+\infty))$, $p \geq 2^*$, then problem (3) has a unique solution $u_a \in C^2_{\mathrm{loc}}([0,+\infty))$ for each a.

We prove Theorems 1 and 2 for $p = 2^*$. The case $p > 2^*$ is analogous.

For $p = 2^*$ we need the following assertion.

Lemma 1. *If $p = 2^*$ and $h \in C_{\mathrm{loc}}([0,\infty))$, then for all $0 < r_0 \leq R_0 < \infty$*

$$\lim_{|a|\to\infty} \max_{r_0 \leq r \leq R_0} |u_a(r) - u_0(r)| + |u_a'(r) - u_0'(r)| = 0.$$

The proof of this lemma can be found in [5].

We now prove Theorem 2a. The proof of Theorem 2b is analogous.

Consider the auxiliary Cauchy problem (3) with $a = 0$. It is obvious that $u_0''(0) < 0$ and $u_0'(0) = 0$. Hence, $u_0(r) < 0$ in some neighbourhood of the origin excluding $r = 0$.

Consider problem (3) with $a = 1$. Similarly, $u_1''(0) < 0$ and $u_1'(0) = 0$. Hence, $u_1(r) > 0$, $u_1'(r) < 0$ and $u_1''(r) < 0$ in a neighbourhood of the origin excluding $r = 0$. We claim that $u_1'(r) < 0$ for all $r > 0$ in the neighbourhood of the origin defined by $u_1(r) \geq 0$. Indeed, let r_0 be the least $r > 0$ for which $u_1'(r_0) = 0$, and suppose that $u_1(r) > 0$ for $0 \leq r \leq r_0$. Then $u_1''(r_0) \geq 0$. On the other hand, by (3) and the conditions on h, we arrive at the contradiction

$$u_1''(r_0) = -u^{p-1}(r_0) + h(r_0) < 0.$$

The statement $u_1'(r) < 0$ means that $u_1(R_1) = 0$ at some point R_1 or u_1 has a positive limit as $r \to \infty$. The latter case is easily eliminated by the equation. Hence, $u_1(R_1) = 0$ and $u_1(r) > 0$, $u_1'(r) < 0$ for $0 < r < R_1$.

Since the solution depends continuously on the initial data, for any $0 < r < R_1$ there exists an a, $0 < a < 1$, such that $u_a(r) = 0$.

Now consider problem (3) with $a > 1$. Just as in the case $a = 1$, there exists an R_a such that $u_a(R_a) = 0$, $u_a(r) > 0$ and $u_a'(r) < 0$ for $0 < r < R_a$. On the other hand, by Lemma 1, for any $r > 0$

$$\lim_{a \to +\infty} u_a(r) < 0. \tag{4}$$

This means that

$$\lim_{a \to +\infty} R_a(r) = 0.$$

Hence, the continuous dependence on the initial data implies the existence of a radius R_0 such that for any $0 < R < R_0$ there exists $a > 1$ with $u_a(R) = 0$ and $u_a'(r) < 0$ for $0 < r < R$.

Choosing $R^* = \min\{R_0, R_1\}$, we deduce the desired statement.

We now provere Theorem 1a; the proof of Theorem 1b is analogous.

First we consider the following assertion.

Lemma 2. *Under the conditions in Theorem* 1a, *for each $a_0 > 0$ there is R such that $|a| < a_0$ and $u_a(r) = 0$ (if such an r exists) imply $r < R$.*

Proof. Since the interval $[-a_0, a_0]$ is compact and the solutions depend continuously on the initial data, it suffices to prove the result for a particular a.

It is easy to see that any solution u_a is bounded above. Let A be its maximal value, and let $u_a = v + A$. Then $v \leq 0$. We multiply (3) by v' and integrate from the point R_0 where the maximum is attained, to some $R > R_0$. Taking into account that $v(R_0) = 0$ and $v'(R_0) = 0$, we obtain

$$\frac{1}{2}v'(R)^2 + \frac{1}{p}|v(R) + A|^p - h(R)v(R) - \frac{1}{p}A^P$$
$$= -\int_{R_0}^{R} h'v\,dr - (N-1)\int_{R_0}^{R} \frac{|v'(r)|^2}{r}\,dr. \tag{5}$$

Consider a sequence $\{R_j\}$ consisting of a local maximum of u_a with $v(R_j) > h(r_0)/2$ in increasing order. If this sequence is finite (or empty), then the assertion is true.

Conversely, suppose now that this sequence is infinite. Obviously,

$$\int_{R_0}^{R} \frac{|v'(r)|^2}{r}\,dr = \sum_{j:\, R_j \geq R_0} \int_{R_j}^{R_{j+1}} \frac{|v'(r)|^2}{r}\,dr \geq \sum_{j:\, R_j \geq R_0} \frac{1}{R_{j+1}} \int_{R_j}^{R_{j+1}} |v'(r)|^2\,dr$$
$$\geq \sum_{j:\, R_j \geq R_0} \frac{1}{R_{j+1}(R_{j+1} - R_j)} |\delta v_j|^2,$$

where

$$\delta v_j = \max_{r_1, r_2 \in [R_j, R_{j+1}]} |v(r_1) - v(r_2)|.$$

By the equation, the local minima are bounded above; therefore, $\delta v_j > C$ for some positive constant C. It is easy to see that the sequence $\{R_{j+1} - R_j\}$ is bounded. Thus, the right-hand side in (5) tends to $-\infty$ as $R \to \infty$. Then so does the left-hand side, which implies that $v(R) < h(r_0)/2$ for a sufficiently large R.

The contradiction we have obtained completes the proof of Lemma 2.

We return to the proof of Theorem 1. Since $h' \leq \varepsilon$, we have $h(r) \leq h(r_0) < 0$ for $r > r_0$. Hence, as in the case f (4), there are $r_1 \geq r_0$ and $a_0 > 0$ such that for $a > a_0$ and $r > r_1$ there holds $u_a(r) < 0$. With the appropriate change of notation, the same estimate is analogously true for $a < -a_0$.

Lemma 2 applied to this a_0 now leads to the desired statement.

References

1. S.I. Pokhozhaev, On eigenvalues of $\Delta u + \lambda f(u) = 0$, *Dokl. Acad. Sci. USSR* **165** (1965), 36–39.
2. A. Bahri and P.-L. Lions, Remarks on variational critical point theory and application, *C.R. Acad. Sci. Paris Sér. I* **301** (1985), 145–147.
3. M. Struwe, Infinitely many critical points for functionals which are not even and applications to superlinear boundary value problems, *Manuscr. Math.* **32** (1980), 335–364.
4. A. Bahri and H. Berestycki, A perturbation method in critical point theory and applications, *Trans. Amer. Math. Soc.* **267** (1981), 1–32.
5. S.I. Pohozaev, On quasilinear elliptic problems in $\mathbb{R}^N$ in the supercritical case, *Proc. Steklov Math. Inst.* **201**, 324-341.
6. I.A. Kuzin, On the absence of a countable set of solutions with supercritical exponent of nonlinearity and non-zero right-hand side, *Mat. Zametki* **56** (1996).
7. I.A. Kuzin, On the solvability of the semilinear elliptic equation $\Delta u + |u|^{p-2}u = h(r)$ for $(2N-2)/(N-2) < p < 2N/(N-2)$ (to appear).

Department of Theoretical Problems, Russian Academy of Sciences, Moscow, Russia

M. LAITINEN

Radiation heat transfer in absorbing, emitting and scattering media

1. Introduction

Heat radiation is a significant factor in heat transfer in many cases, typically always when a hot surface is in contact with a transparent or semi-transparent medium with relatively low heat conductivity.

When modelling heat transfer in the presence of radiation we have to examine radiative heat exchange at any point, for each direction and wavelength of the rays. However, to simplify the situation, we restrict ourselves to grey and isotropically scattering materials bounded by grey and diffuse walls. This means that we can disregard the angular and spectral distribution of radiation, and model only the total intensities.

We consider a three-dimensional connected domain Ω occupied by a conductive, semi-transparent and grey material. For definiteness, we assume that Ω is surrounded by an opaque medium and denote the boundary of Ω by Σ. The steady state balance equations for absolute temperature T can be written as

$$-k\Delta T = f + h_{\text{rad}} \quad \text{in } \Omega,$$
$$k\frac{\partial T}{\partial n} = g + q_{\text{rad}} \quad \text{on } \Sigma,$$

where h_{rad} and q_{rad} are the heat source and flux due to radiation, k is the coefficient of heat conductivity and f and g are known data.

We denote by $\bar{R}_i$ the amount of emitted and scattered radiation in the volume, and by $\bar{R}_s$ the corresponding quantity on the surface. We used a superposed bar to distinguish the physical variables from their scaled versions used in the mathematical analysis. The radiative heat source can be written as

$$h_{\text{rad}} = (\kappa + \gamma)(\bar{K}_{ii}\bar{R}_i + \bar{K}_{is}\bar{R}_s) - \bar{R}_i. \tag{1}$$

Here $\bar{K}_{ii}\bar{R}_i + \bar{K}_{is}\bar{R}_s$ is the amount of incoming radiation, $\kappa \geq 0$ is the emission/absorption coefficient, and $\gamma \geq 0$ is the scattering coefficient. $\bar{K}_{ii}$ and $\bar{K}_{is}$ are integral operators with kernels defined on $\Omega \times \Omega$ and $\Omega \times \Sigma$, respectively. On the radiating boundary (interface between the opaque and semi-transparent media) we have the radiative heat flux

$$q_{\text{rad}} = \bar{K}_{si}\bar{R}_i + \bar{K}_{ss}\bar{R}_s - \bar{R}_s, \tag{2}$$

with a similar interpretation.

The radiosities $\bar{R}_i$ and $\bar{R}_s$ depend on the temperature in non-linearly and non-locally. More precisely, the radiosity at a point is the sum of the Stefan-Boltzmann

---

This work was supported by the Academy of Finland.

radiation emitted by the point and the scattered/reflected part of incoming radiation. Thus, we can write

$$\bar{R}_i = 4\kappa\sigma|T|^3T + \gamma(\bar{K}_{ii}\bar{R}_i + \bar{K}_{is}\bar{R}_s)\,, \tag{3}$$

$$\bar{R}_s = \epsilon\sigma|T|^3T + (1-\epsilon)(\bar{K}_{si}\bar{R}_i + \bar{K}_{ss}\bar{R}_s)\,, \tag{4}$$

where $0 \leq \epsilon \leq 1$ is the emissivity coefficient of the surface.

In the following sections we state that, under fairly natural conditions, the above problem is well defined and has a solution. The main difficulties are that $|T|^3T$ is not necessarily integrable on the boundary when T is in $H^1(\Omega)$, and that the problem is not monotone. The proofs can be found in [1].

2. Radiative integral operators

We introduce the radiative operators and recall some of their properties. For the construction of these operators, see [1] and [2].

To abbreviate the notation, we write $\beta = \kappa + \gamma$. Moreover, we denote by n_x the outward unit normal at a point x and by Ξ the visibility factor, that is, $\Xi(x,z) = 1$ if x and z can see each other ($\overline{xz} \cap \Omega = \overline{xz}$) and $\Xi(x,z) = 0$ otherwise. If we set

$$\tau(x,z) = \exp\left[-\int_x^z \beta(s)\,ds\right]\Xi(x,z),$$

then the above the integral formulae can be expressed as

$$(\bar{K}_{ii}\bar{R}_i)(x) = \int_\Omega \frac{\tau(x,z)}{4\pi\|z-x\|^2}\,\bar{R}_i(z)\,dV(z),$$

$$(\bar{K}_{is}\bar{R}_s)(x) = \int_\Sigma \frac{\tau(x,z)\,n_z\cdot(z-x)}{\pi\|z-x\|^3}\,\bar{R}_s(z)\,dS(z),$$

$$(\bar{K}_{si}\bar{R}_i)(x) = \int_\Omega \frac{\tau(x,z)\,n_x\cdot(x-z)}{4\pi\|z-x\|^3}\,\bar{R}_i(z)\,dV(z),$$

$$(\bar{K}_{ss}\bar{R}_s)(x) = \int_\Sigma \frac{\tau(x,z)\,n_x\cdot(z-x)\,n_z\cdot(x-z)}{\pi\|z-x\|^4}\,\bar{R}_s(z)\,dS(z).$$

We now assume that the radiating body is absorbing and emitting at every point of Ω and Σ, so that there exists a constant α such that $\kappa \geq \alpha > 0$ and $\epsilon \geq \alpha > 0$. In addition, we assume that $\kappa, \gamma, \beta \in L^\infty(\Omega)$.

To keep the notation simple, we define the scaled radiosity R by

$$R = \begin{cases} \bar{R}_i(x)/4\beta(x) & \text{if } x \in \Omega, \\ \bar{R}_s(x) & \text{if } x \in \Sigma, \end{cases}$$

and set $\epsilon(x) = \kappa(x)/\beta(x)$ for $x \in \Omega$. Now we rewrite the system (1)–(4) as

$$h_{\text{rad}} = -4\beta(I-K)R, \quad x \in \Omega, \tag{5}$$

$$q_{\text{rad}} = -(I-K)R, \quad x \in \Sigma, \tag{6}$$

$$\epsilon\sigma|T|^3T = (I-(I-E)K)R, \quad x \in \Omega \cup \Sigma, \tag{7}$$

where E is the operator of multiplication by ϵ. The operator K is defined by

$$(KR)(x) = \begin{cases} (K_{ii}R)(x) + (K_{is}R)(x) & \text{if } x \in \Omega, \\ (K_{si}R)(x) + (K_{ss}R)(x) & \text{if } x \in \Sigma, \end{cases}$$

where $K_{ii} = \bar{K}_{ii}\beta$, $K_{is} = (1/4)\bar{K}_{is}$, $K_{si} = 4\bar{K}_{si}\beta$ and $K_{ss} = \bar{K}_{ss}$.

We define a measure μ such that

$$\int R\, d\mu = \int_\Omega 4\beta(x) R\, dx + \int_\Sigma R\, dS,$$

and denote by $L^p_\mu = L^p(\Omega \cup \Sigma; \mu)$ the set of functions $f : \Omega \cup \Sigma \to \mathrm{R}$ whose p-th powers are μ-integrable. In what follows, $\| \cdot \|_p$ is the L^p_μ-norm. The next assertion shows that the measure μ and the formulation (5)–(7) are, in some sense, natural for the problem.

Lemma 1. *Let $1 \leq p \leq \infty$ and suppose that Ω is bounded, connected and Lipschitz. Then the operator K maps L^p_μ into itself compactly and $\|K\|_p \leq 1$, K is positive, that is, it maps positive functions into positive functions, and is self-adjoint in the sense that $\int (Kf)g\, d\mu = \int (Kg)f\, d\mu$.*

This implies that $\|(I - E)K\|_p < 1$; hence, $(I - (I - E)K)$ is invertible. Thus, R can be eliminated from (5) and (6), and the radiative heat source and flux can be expressed in terms of T alone as

$$h_{\text{rad}} = -4\beta G(\sigma |T|^3 T), \quad x \in \Omega,$$
$$q_{\text{rad}} = -G(\sigma |T|^3 T), \quad x \in \Sigma,$$

where the operator G maps L^p_μ into itself and is defined by

$$G = (I - K)(I - (I - E)K)^{-1} E.$$

Lemma 2. *The operator G can be written as $G = I - H$, where H is self-adjoint, positive and $\|H\|_p \leq 1$. G is positive semi-definite in the sense that $\int (Gf) f\, d\mu \geq 0$ for all $f \in L^2_\mu$.*

In some practical cases (an oven with a hatch, for example), a part of the boundary allows radiation to escape the system. In this case we write $\partial\Omega = \Sigma \cup \Sigma_0$, where Σ, as before, is the radiating part of the boundary and Σ_0 is the transparent part (see Fig. 1(b) for a typical example). This does not change the structure of the model derived above. However, additional data terms appear in (1)–(4), namely, the radiation coming from outside of the system. Moreover, the operators K and H become contractive.

Lemma 3. *Let $1 < p < \infty$, and suppose that $\partial\Omega \setminus \Sigma$ has positive surface measure. Then $\|K\|_p < 1$ and $\|H\|_p < 1$.*

3. Conduction-radiation problem

We consider a system that consists of conductive materials occupying $\tilde{\Omega} = \Omega \cup \Omega_0 \cup \Sigma$. In Ω the material is assumed grey and semi-transparent, whereas Ω_0 is opaque. By Σ we denote the common boundary of Ω and Ω_0. The exterior boundary of Ω_0 is denoted by Γ, and the possible exterior boundary of Ω by Σ_0. Two typical cases are shown in Fig. 1. The first one corresponds to radiative heat transfer in an enclosure, while in the second case, a part of the wall Σ_0 is assumed transparent, allowing some radiation to escape.

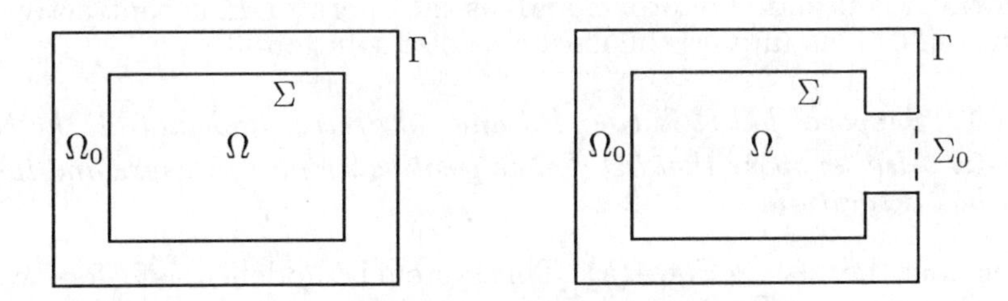

Fig. 1. (a) An enclosure. (b) A system where some radiation can escape.

First we consider the case in Fig. 1(b). The heat balance is governed by the system

$$-k\Delta T = f \quad \text{in } \Omega_0, \tag{8}$$

$$-k\Delta T = f - 4\beta G(\sigma|T|^3 T) + q_i^\infty \quad \text{in } \Omega, \tag{9}$$

$$[k\partial T/\partial n] = -G(\sigma|T|^3 T) + q_s^\infty \quad \text{on } \Sigma, \tag{10}$$

where $[\cdot]$ denotes the jump across Σ. Here q_i^∞ and q_s^∞ denote the additional radiative heat source/flux arriving from outside the system. On Γ we choose the condition

$$k\frac{\partial T}{\partial n} + \alpha(T - T_0) = 0 \quad \text{on } \Gamma. \tag{11}$$

On Σ_0 we have the condition

$$k\frac{\partial T}{\partial n} = 0 \quad \text{on } \Sigma_0, \tag{12}$$

as Σ_0 does not conduct or radiate heat outside the system.

Since $|T|^3 T$ is not necessarily integrable on the boundary when T is in $H^1(\Omega)$, we have to look for a weak solution in the space

$$V = \{v \in H^1(\tilde{\Omega}) : v|_\Sigma \in L^5(\Sigma)\}.$$

Now we can write (8)–(12) in weak form as

$$A(T, v) = a(T, v) + b(T, v) = \langle \tilde{f}, v \rangle \quad \forall v \in V, \tag{13}$$

where

$$a(T,v) = \int_{\Omega\cup\Omega_0} k\nabla T\nabla v + \int_\Gamma \alpha T v,$$
$$b(T,v) = \int_\Omega 4\beta G(\sigma T^4)v + \int_\Sigma G(\sigma T^4)v,$$
$$\langle \tilde{f}, v\rangle = \int_{\Omega\cup\Omega_0} (f + q_i^\infty)v + \int_\Gamma \alpha T_0 v + \int_\Sigma q_s^\infty\, v.$$

The form A is pseudomonotone, and, as the operator H is contractive, it is also coercive in V [1]. This implies the existence of a solution [3].

Theorem 1. *Suppose that $\tilde{\Omega}$ is bounded and connected, and that all the boundaries are Lipschitz. Also, suppose that $\partial\Omega\backslash\Sigma$ has positive surface measure and that $\tilde{f} \in V'$. Then* (13) *has a solution.*

We consider the case in Fig. 1(a). This system is modelled as before, except that now $q_i^\infty = 0$ and $q_s^\infty = 0$. This problem is pseudomonotone, but now coercivness is difficult to prove in V, as H is not a contraction. Hence, we prove existence by assuming the existence of sub- and supersolutions. For this reason, we denote by V^+ the cone of non-negative elements

$$V^+ = \{v \in V \,:\, v \geq 0\}\,.$$

Theorem 2. *Suppose that $\tilde{\Omega}$ is bounded and connected, and that all the boundaries are Lipschitz. Also, let $\tilde{f} \in V'$, and suppose that there exist functions ϕ and ψ in V such that $\phi \leq \psi$ and*

$$A(\phi, w) \leq \langle \tilde{f}, w\rangle \quad \forall w \in V^+,$$
$$A(\psi, w) \geq \langle \tilde{f}, w\rangle \quad \forall w \in V^+.$$

Then our problem has a solution T. Moreover, $\phi \leq T \leq \psi$ in $\tilde{\Omega}$.

Such ϕ and ψ can easily be found, for example, when $f = 0$.

References

1. M. Laitinen and T. Tiihonen, *Heat transfer in conducting, radiating and semi-transparent materials*, report 8, University of Jyväskylä, Department of Mathematics, Laboratory of Scientific Computing, 1996.

2. R. Siegel and J.R. Howell, *Thermal radiation heat transfer*, 3rd ed., Hemisphere, New York, 1992.

3. E. Zeidler, *Nonlinear functional analysis and its applications. II B: Nonlinear monotone operators*, Springer-Verlag, Heidelberg-New York, 1988.

Department of Mathematics, University of Jyväskylä, Box 35, 40351 Jyväskylä, Finland

A.B. MURAVNIK

On weighted norm estimates for the mixed Fourier-Bessel transforms of non-negative functions

1. Introduction

It is proved in [5] that if f is non-negative, then for any $\alpha \in (0,(n-1)/2]$

$$\|r^{\alpha}\sigma(f)\|_{\infty} \leq C\|r^{\alpha-1}\sigma(f)\|_{1}, \tag{1.1}$$

where $\sigma(f)(r)$ is the mean value of $|\hat{f}|^2$ over the sphere of radius r with the centre at the origin, and C depends only on the dimension of the space.

We note that, generally, (1.1) does not hold since we can construct a sequence $\{f_m\}_{m=1}^{\infty}$ such that $\|r^{\alpha-1}\sigma(f_m)\|_1$ does not depend on m but $\sigma(f_m)(1)$ tends to infinity as $m \to \infty$. Thus, the requirement that f be non-negative prohibits the above type of behaviour. Actually, it represents a certain restriction on the shape of the graph of $\hat{f}$.

In this work we investigate the mixed Fourier-Bessel transform, which is applied in the theory of partial differential equations containing singular Bessel operators with respect to a selected variable (called the *special* variable). These equations arise in models of mathematical physics with degenerative space heterogeneities. We prove that estimates of the type (1.1) are valid for weighted spherical means of $|\hat{f}|^2$, but that the weights in both parts of the inequality and the weights of the means themselves are controlled by the relation between the dimension of the space of non-special variables and the index of the Bessel functions from the kernel of the transform (or the parameter at the singularity of the Bessel operator contained in the equation). More exactly, the following statements are valid (for $f \geq 0$): if $n > 1$, then for any $p > -n$ and any $q > -1$ there exists C such that, for any $\alpha \in (0,(n-1)/2)$, any $\beta \in (0,k/2)$ and $(\alpha,\beta) = ((n-1)/2, k/2)$

$$\|r^{\alpha+\beta}\sigma^{p+\alpha,q+\beta}(f)\|_{\infty} \leq C\|r^{\alpha+\beta-1}\sigma^{\alpha-n,\beta-1}(f)\|_{1}. \tag{1.2}$$

If $n = 1$, then for any $p > -1$ and any $q > k/2 - 1$ there exists C such that

$$\|r^{k/2}\sigma^{p,q}(f)\|_{\infty} \leq C\|\eta^{k/2-1}|\hat{f}|_{\xi=0}|^2\|_{1}. \tag{1.3}$$

(Here $\sigma^{p,q}$ denotes the mean value of $|\hat{\cdot}|^2$ with weight $|x|^p y^q$.)

If $n = 0$, then we actually deal with the case of a *pure* Fourier-Bessel transform; this is investigated in [6] and [7].

This paper was written whilst the author worked at the University of Jyväskylä. The author was supported by the Finnish Centre for International Mobility.

2. Preliminaries

In this section we introduce the necessary notations and definitions; we also recall the properties of the Fourier-Bessel transform.

Let $k \stackrel{\text{def}}{=} 2\nu + 1$ be a positive parameter. In what follows, all the absolute constants generally depend on ν and n. We write

$$\mathbb{R}_+^{n+1} \stackrel{\text{def}}{=} \left\{(x,y) \,\middle|\, x \in \mathbb{R}^n,\ y > 0\right\}.$$

$S_+(r)$ denotes the upper hemisphere in $\mathbb{R}_+^{n+1}$ with radius r, centred at the origin; dS_z denotes the surface measure with respect to the (vector) variable z. Also, let

$$L_{p,k}(\mathbb{R}_+^{n+1}) \stackrel{\text{def}}{=} \left\{f \,\middle|\, \|f\| = \left(\int\limits_{\mathbb{R}_+^{n+1}} y^k |f(x,y)|^p \, dx dy\right)^{1/p} < \infty\right\}, \quad p \text{ finite}.$$

The set of infinitely smooth functions with compact support defined on $\mathbb{R}^{n+1}$ is denoted by $C_0^\infty(\mathbb{R}^{n+1})$. We consider the subset of $C_0^\infty(\mathbb{R}^{n+1})$ consisting of even functions with respect to y, and denote by $C_{0,\text{even}}^\infty(\mathbb{R}_+^{n+1})$ the set of restrictions of elements of that subset to $\mathbb{R}_+^{n+1}$. The space $C_{0,\text{even}}^\infty(\mathbb{R}_+^{n+1})$ is known as *the space of test functions.*

Distributions on $C_{0,\text{even}}^\infty(\mathbb{R}_+^{n+1})$ are introduced (following, for example, [1]) with respect to the degenerative measure $y^k dx dy$ by

$$\langle f, \varphi \rangle \stackrel{\text{def}}{=} \int\limits_{\mathbb{R}_+^{n+1}} y^k f(x,y) \varphi(x,y) \, dx dy \quad \text{for any } \varphi \in C_{0,\text{even}}^\infty(\mathbb{R}_+^{n+1}). \tag{2.1}$$

Thus, all linear continuous functionals on $C_{0,\text{even}}^\infty(\mathbb{R}_+^{n+1})$ given by (2.1) (with $f \in L_{1,k,\text{loc}}(\mathbb{R}_+^{n+1})$) are called *regular* (and the corresponding function f is called *ordinary*).

The Fourier-Bessel transform is given by [2]

$$\hat{f}(\xi_1, \ldots, \xi_n, \eta) \stackrel{\text{def}}{=} \mathcal{F}_\text{b}(f)(\xi, \eta) \stackrel{\text{def}}{=} \int\limits_{\mathbb{R}_+^{n+1}} y^k j_\nu(\eta y) e^{-ix \cdot \xi} f(x,y) \, dx dy,$$

where $j_\nu(z) = J_\nu(z)/z^\nu$ is the normalised (in the uniform sense) Bessel function. We note that

$$f(x,y) = C \int\limits_{\mathbb{R}_+^{n+1}} \eta^k j_\nu(\eta y) e^{ix \cdot \xi} \hat{f}(\xi, \eta) \, d\xi d\eta.$$

3. A brief scheme of the proof

The proof is based (as in [5]) on the central idea that the distribution of the weighted spherical averaging (denoted by σ_r in the sequel) is singular, but that its Fourier-Bessel transform is regular and could be computed explicitly.

We have

$$\begin{aligned}\langle\widehat{\sigma_r},\varphi\rangle = C\langle\sigma_r,\check{\varphi}\rangle &= C\int\limits_{S_+(1)}|\xi|^p\eta^q\check{\varphi}(r\xi,r\eta)\,dS_{\xi,\eta}\\ &= \int\limits_{S_+(1)}|\xi|^p\eta^q\int\limits_{\mathbb{R}^{n+1}_+}y^k\varphi(x,y)e^{irx\cdot\xi}j_\nu(ry\eta)\,dxdydS_{\xi,\eta}\\ &= \int\limits_{\mathbb{R}^{n+1}_+}y^k\varphi(x,y)\int\limits_{S_+(1)}|\xi|^p\eta^q e^{irx\cdot\xi}j_\nu(ry\eta)\,dS_{\xi,\eta}dxdy,\end{aligned}$$

where φ is an arbitrary test function. The inner integral in the last expression is equal to

$$Cr^{(2-n)/2-\nu}|x|^{(2-n)/2}\,y^{-\nu}\int\limits_0^1\eta^{q-\nu}(1-\eta^2)^{(n+2p-2)/4}J_{(n-2)/2}(r\sqrt{1-\eta^2}|x|)J_\nu(ry\eta)\,d\eta$$

(see, for example, [8], p. 155). This implies the estimate

$$|\widehat{\sigma_r}(x,y)|\le Cr^{-(n+k-1)/2}|x|^{-(n-1)/2}y^{-k/2} \tag{3.1}$$

(since C now depends on p and q as well).

Next, we observe that $\sigma(f)(r)\overset{\text{def}}{=}\langle\sigma_r,|\hat{f}|^2\rangle=\langle\sigma_r,\hat{f}\,\overline{\hat{f}}\,\rangle=\langle\hat{f}\sigma_r,\hat{f}\rangle$.

On the other hand, $\langle f*\widehat{\sigma_r},f\rangle=C\langle\widehat{f*\widehat{\sigma_r}},\hat{f}\rangle=C\langle\hat{f}\widehat{\widehat{\sigma_r}},\hat{f}\rangle=C\langle\hat{f}\sigma_r(-x,y),\hat{f}\rangle$. But we have just proved that $\widehat{\sigma_r}$ is an ordinary function, even with respect to each non-special variable. Thus, $\sigma(f)(r)=\langle f*\widehat{\sigma_r},f\rangle$.

Using known properties of the generalised shift operator and generalised convolution (see [3] and [4]), we find that in the general case ($n>1$)

$$\begin{aligned}\sigma(f)(r) &= \int\limits_{S_+(1)}|\xi|^p\eta^q|\hat{f}(r\xi,r\eta)|^2\,dS_{\xi,\eta}\\ &\le Cr^{(1-n-k)/2}\langle\mathcal{F}_{\rm b}(|x|^{-(n-1)/2}y^{-k/2}),|\hat{f}|^2\rangle.\end{aligned}$$

Also, $\mathcal{F}_{\mathrm{b}}\big(|x|^{-(n-1)/2}y^{-k/2}\big)$ may be computed explicitly (if $n > 1$ and $k > 0$):

$$\mathcal{F}_{\mathrm{b}}\big(|x|^{(1-n)/2}y^{-k/2}\big) = \int_0^{\infty} y^k j_\nu(y\eta) y^{-k/2}\, dy \int_{\mathbb{R}^n} e^{-ix\cdot\xi}|x|^{(1-n)/2}\, dx = C\eta^{-k/2-1}|\xi|^{-(n+1)/2}$$

(see, for example, [8], p. 155, and [6]).

This leads to (1.2) with $\alpha = \alpha_0 \stackrel{\text{def}}{=} (n-1)/2$ and $\beta = \beta_0 \stackrel{\text{def}}{=} k/2$. To extend (1.2) to the whole claimed intervals, we just apply (as in [5]) the already proved inequality to the new function

$$f_{\gamma,\delta} \stackrel{\text{def}}{=} f * \big(|x|^{\gamma-n}y^{\delta-k-1}\big),$$

where

$$\gamma \stackrel{\text{def}}{=} \frac{\alpha_0 - \alpha}{2} > 0, \quad \delta \stackrel{\text{def}}{=} \frac{\beta_0 - \beta}{2} > 0.$$

In fact, the mixed Fourier-Bessel transform of $|x|^{\gamma-n}y^{\delta-k-1}$ is decomposed (in the same way as the mixed Fourier-Bessel transform of $|x|^{-(n-1)/2}y^{-k/2}$ above) into the product of the Fourier transform of the Riesz kernel $|x|^{\gamma-n}$ and the pure Fourier-Bessel transform of the power function $y^{\delta-k-1}$. From this we conclude that $\widehat{f_{\gamma,\delta}} = C_{\gamma,\delta}\hat{f}|x|^{-\gamma}y^{-\delta}$. Then all that remains is to go back from γ, δ to α, β and obtain the powers in (1.2). We note that $C_{\gamma,\delta}$ appears on both sides of the inequality. Consequently, C in (1.2) does not really depend on α, β.

In the critical case ($n = 1$) x disappears from the right-hand side of (3.1), and the above scheme leads to (1.3).

Remark. Inequalities (1.2) and (1.3) are valid for any non-negative f such that the right-hand side of the inequality converges (in particular, for any non-negative f in $L_{1,k}(\mathbb{R}_+^{n+1}) \cap L_{2,k}(\mathbb{R}_+^{n+1})$). However, even if the right-hand side of the inequality diverges, the inequality still holds formally.

Acknowledgements. The author is indebted to P. Mattila for his guidance and attention, and to P. Sjölin for fruitful discussions. He also wishes to thank the University of Jyväskylä for its hospitality.

References

1. V.V. Katrakhov, On the theory of partial differential equations with singular coefficients, *Sov. Math. Dokl.* **15** (1974), 1230–1234.

2. I.A. Kipriyanov, Fourier-Bessel transforms and imbedding theorems for weight classes, *Proc. Steklov Inst. Math.* **89** (1967), 149–246.

3. I.A. Kipriyanov and A.A. Kulikov, The Paley-Wiener-Schwartz theorem for the Fourier-Bessel transform, *Sov. Math. Dokl.* **37** (1988), 13–17.

4. B.M. Levitan, Expansions in Fourier series and integrals with Bessel functions, *Uspekhi Mat. Nauk* **6** (1951), no. 2, 102–143.

5. P. Mattila, Spherical averages of Fourier transforms of measures with finite energy; dimensions of intersections and distance sets, *Mathematika* **34** (1987), 207–228.

6. A.B. Muravnik, On weighted uniform estimates for pure Fourier-Bessel transform of nonnegative functions, *Reports of Mittag-Leffler Inst.* **33** (1994/1995).

7. A.B. Muravnik, On weighted uniform estimates for the pure Fourier-Bessel transform of nonnegative functions with compact support, preprint **186** (1995), Department of Mathematics, University of Jyväskylä.

8. E.M. Stein and G. Weiss, *Introduction to Fourier analysis on Euclidean spaces*, Princeton Univ. Press, Princeton, 1971.

Department of Applied Mathematics and Mechanics, Voronezh State University, Universitetskaya Sq. 1, 394693 Voronezh, Russia

A. NASTASE

Qualitative analysis of partial differential equations of the three-dimensional compressible boundary layer via spectral solutions

1. Introduction

In our previous papers [1] and [2], some original spectral forms for the velocity components u, v, w inside the three-dimensional, compressible boundary layer (CBL) are proposed, that is,

$$u = u_e \sum_{i=1}^{N} u_i \eta^i, \quad v = v_e \sum_{i=1}^{N} v_i \eta^i, \quad w = w_e \sum_{i=1}^{N} w_i \eta^i. \tag{1a, b, c}$$

Let $\eta = [x_3 - Z(x_1, x_2)]/\delta(x_1, x_2)$ be the coordinate in the boundary layer, u_e, v_e, w_e the edge velocities, $Z(x_1, x_2)$ the equation of the flattened flying configuration, $\delta(x_1, x_2)$ the boundary layer thickness, and u_i, v_i, w_i arbitrary constants which are determined from the impulse and continuity equations and from the boundary conditions at the edge of the boundary layer. If the spectral velocity components u, v, w are introduced in the impulse equations, quadratic algebraic equations (QAE) for the determination of the coefficients u_i, v_i, w_i are obtained as in [1] and [2]. If the collocation method is used, the impulse equations, written at K discrete points Q_k $(k = 1, 2, ..., K)$, are

$$\sum_{i=1}^{N}\sum_{j=1}^{N} u_i \left(A_{ijk}^{(1)} u_j + B_{ijk}^{(1)} v_j + C_{ijk}^{(1)} w_j\right) = A_{0k}^{(1)} + \sum_{i=1}^{N} \left(A_{ik}^{(1)} u_i + B_{ik}^{(1)} v_i + C_{ik}^{(1)} w_i\right),$$

$$\sum_{i=1}^{N}\sum_{j=1}^{N} v_i \left(A_{ijk}^{(2)} u_j + B_{ijk}^{(2)} v_j + C_{ijk}^{(2)} w_j\right) = B_{0k}^{(2)} + \sum_{i=1}^{N} \left(A_{ik}^{(2)} u_i + B_{ik}^{(2)} v_i + C_{ik}^{(2)} w_i\right),$$

$$\sum_{i=1}^{N}\sum_{j=1}^{N} w_i \left(A_{ijk}^{(3)} u_j + B_{ijk}^{(3)} v_j + C_{ijk}^{(3)} w_j\right) = \sum_{i=1}^{N} C_{ik}^{(3)} w_i. \tag{2a, b, c}$$

The boundary conditions and the continuity equation, written at P discrete points Q_p $(p = 1, 2, \ldots, P$ with $P = 3N - 3K - 7$ for three-dimensional and $P = 2N - 2K - 4$ for two-dimensional flow), are linear forms in u_i, v_i, w_i, that is,

$$\sum_{i=1}^{N} u_i = 1, \quad \sum_{i=1}^{N} i u_i = 0, \quad \sum_{i=1}^{N} i(i-1) u_i = 0, \tag{3a, b, c}$$

$$\sum_{i=1}^{N} v_i = 1, \quad \sum_{i=1}^{N} i v_i = 0, \quad \sum_{i=1}^{N} i(i-1) v_i = 0, \qquad (4a, b, c)$$

$$\sum_{i=1}^{N} w_i = 1, \quad \sum_{i=1}^{N} (D_{ip} u_i + E_{ip} v_i + F_{ip} w_i) = 0, \qquad (5a, b)$$

and can be written explicitly with respect to $P + 7$ coefficients for the three-dimensional flow (and $P+4$ for the two-dimensional flow). These coefficients are eliminated from the QAE (2a,b,c). An equivalent quadratic algebraic system (QAS) with $M = 3K$ variables for the three-dimensional flow and $M = 2K$ for the two-dimensional flow, is obtained in original and in canonical form:

$$\sum_{i=1}^{M} \left[\sum_{j=1}^{M} a_{ij}^{(k)} X_i X_j + 2a_{i,M+1}^{(k)} X_i \right] + a_k = 0, \quad \sum_{i=1}^{M} \lambda_i^{(k)} X_{ik}^{''2} + a_k^{''} = 0. \qquad (6a, b)$$

Henceforth, $a_k = a_{M+1,M+1}^{(k)}$, $k = 1, 2, .., M$, and

$$X_i = \begin{matrix} u_i \\ v_i \\ w_i \end{matrix}, \; a_{ijk}^{(1)} = \begin{matrix} A_{ijk}^{(1)} \\ B_{ijk}^{(1)} \\ C_{ijk}^{(1)} \end{matrix}, \; a_{ijk}^{(2)} = \begin{matrix} A_{ijk}^{(2)} \\ B_{ijk}^{(2)} \\ C_{ijk}^{(2)} \end{matrix}, \; a_{ijk}^{(3)} = \begin{matrix} A_{ijk}^{(3)} \\ B_{ijk}^{(3)} \\ C_{ijk}^{(3)} \end{matrix}, \; a_{ij}^{(k)} = \begin{matrix} a_{ijk}^{(1)} & 1 < i < M \\ a_{ijk}^{(2)}, & M+1 < i < 2M. \\ a_{ijk}^{(3)} & 2M+1 < i < 3M \end{matrix}$$

Solving the partial differential equations of the three-dimensional boundary layer reduces to solving the equivalent quadratic algebraic system (6a). This method is useful for the determination of the friction drag coefficient and for the velocity field in the CBL, as in [1]–[3], for the viscous aerodynamic design as in [1] and [3], and for the qualitative analysis of the partial differential equations of the CBL, which we discuss here.

2. Qualitative analysis of the elliptic partial differential equations

The canonical form of the k th QAE (6b) is used for the qualitative analysis and the visualisation of the asymptotic behaviour of each partial differential equation of the CBL in the vicinity of the critical points. In the k th canonical QAE (6b), $\lambda_i^{(k)}$ are the eigenvalues, $a_k^{''} = \Delta_k / \delta_k$ is the free term, Δ_k and δ_k are the great determinants and the discriminants of the k th QAE (6a), and $X_{ik}^{''}$ is the canonical system of coordinates of this equation. In the k th QAE, the free term a_k, which is proportional to the pressure at the edge of the boundary layer, is systematically varied and all the other coefficients are maintained constant. The character of this QAE remains unchanged during this variation. If this QAE is of elliptic or hyperbolic type, the equation reaches its critical point, which is located in its centre. The critical value $a_k = a_c$ is obtained from the linear equation (with respect to a_c), $\Delta_k = 0$. The asymptotic behaviour of the elliptic QAE in the vicinity of its singular point is visualised in the form of coaxial hyperellipsoids whose sizes decrease as the free term a_k increases. For the critical value $a_k = a_c$, they degenerate into a point. For $a > a_c$, the coaxial hyperellipsoids collapse. The visualisation of this asymptotic behaviour up to $M = 4$ can be done only in cuts. Fig. 1a–f shows this visualisation for $M = 4$, in the six cuts made by the coordinate hyperplanes for the elliptic equation (used as an example)

$$3x^2 + 5y^2 + 4z^2 + 2t^2 + 4xy - 6x - 3y + a = 0. \tag{7}$$

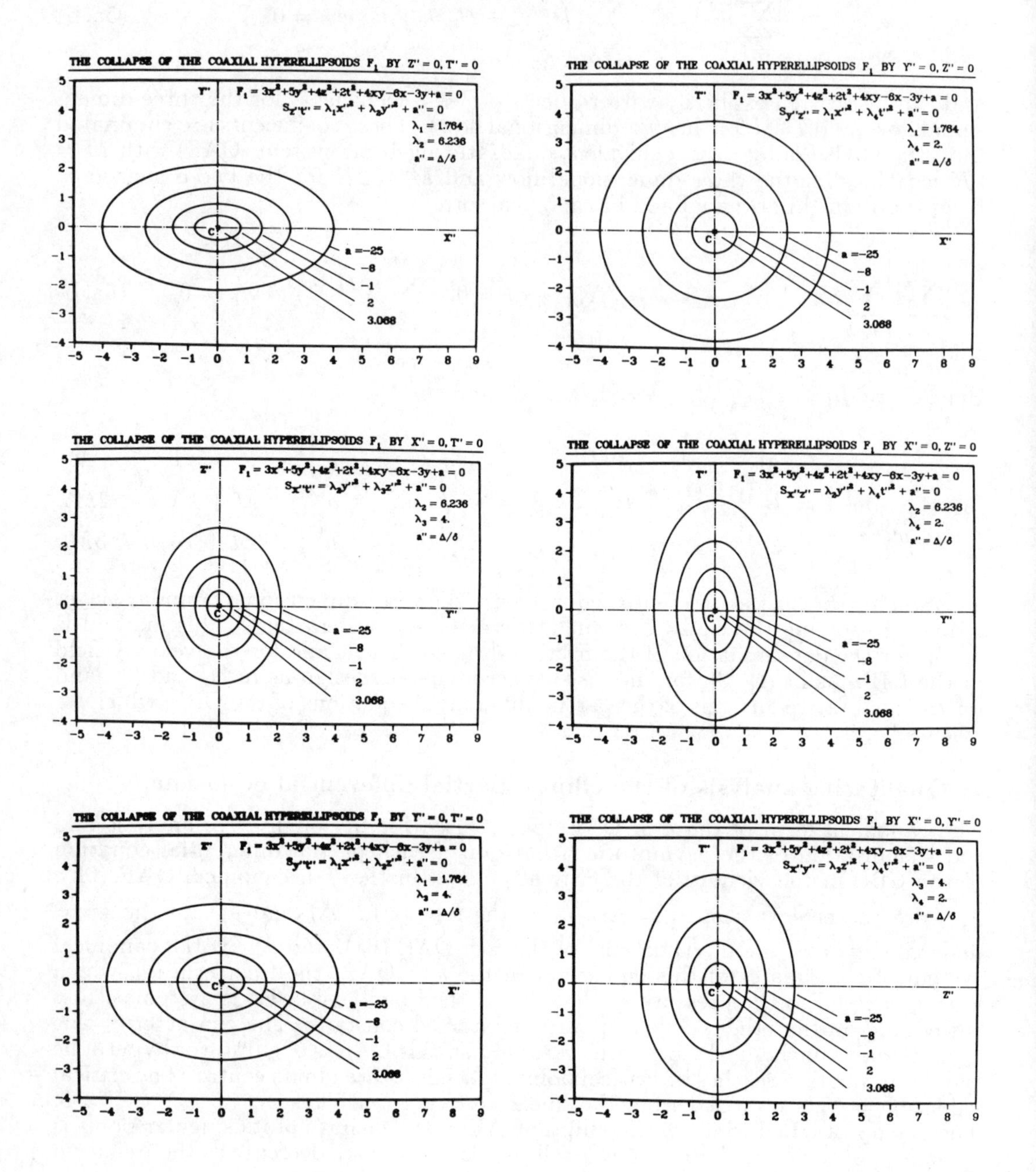

Fig. 1a–f. Cuts in the hyperellipsoids for $M = 4$.

The canonical form of this QAE, after translation and rotation, is (6b) for $M = 4$, the eigenvalues λ_i are given in Fig. 1a–f and the free term a_k'' is given by $a_k'' = a'' = \Delta/\delta$ with $\delta = 88$ and $\Delta = 8(11a - 135/4)$. The critical value of a is obtained by cancellation of Δ and is equal to $a = a_c = 3.0682$. The position of the hypercentre with respect to the initial coordinates is $C(x = 1.091,\ y = -0.136,\ z = 0,\ t = 0)$.

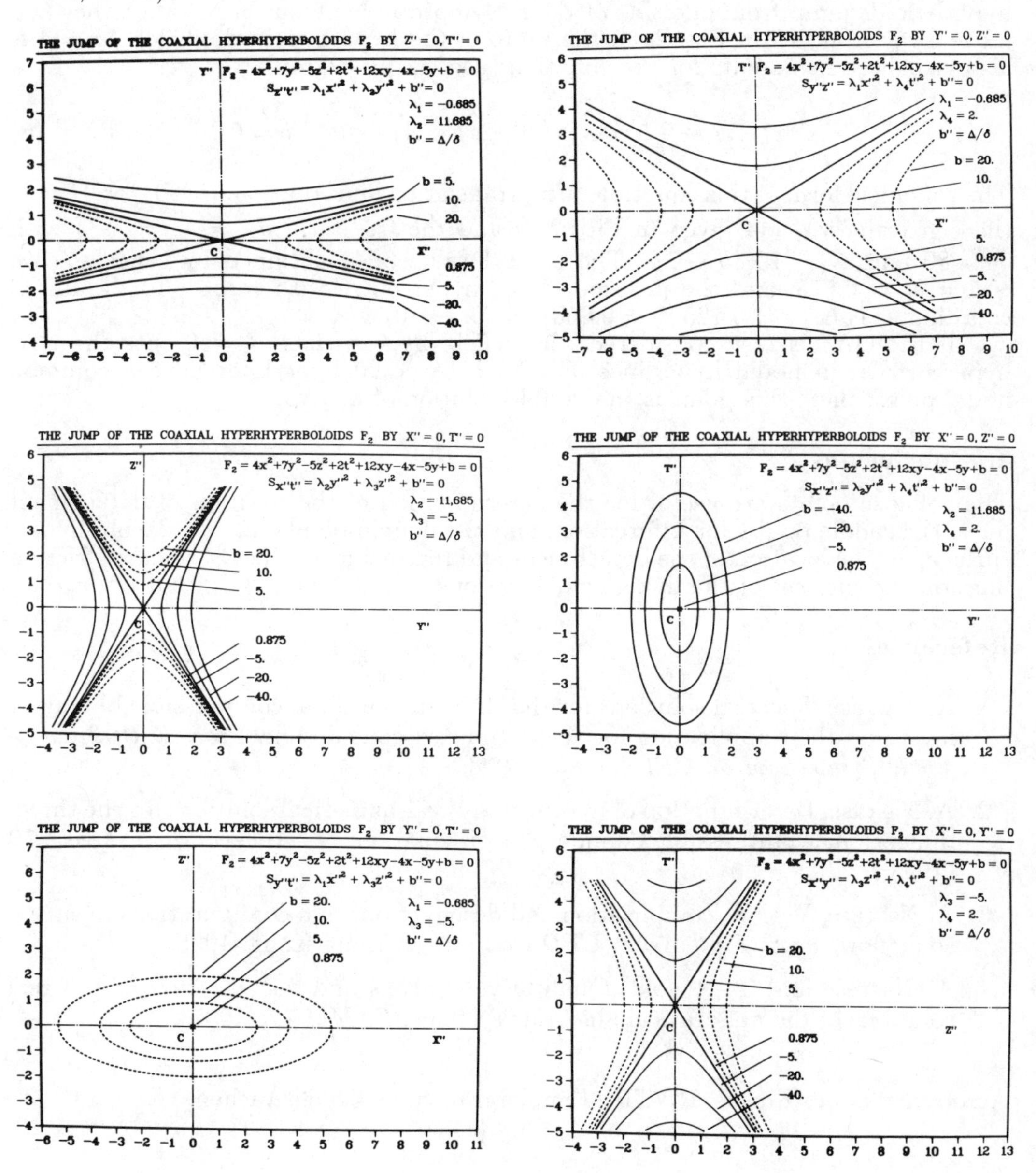

Fig. 2a–f. Cuts in the hyperhyperboloids for $M = 4$.

3. Qualitative analysis of the hyperbolic partial differential equations

If the free term $a_k = b$ is systematically varied, the hyperbolic equations are represented in the form of coaxial hyperhyperboloids. For $b = b_c$ these hyperhyperboloids present a saddle point. If b increases and $b < b_c$, the coaxial hyperhyperboloids are approaching their asymptotic hypersurfaces, for $b = b_c$ they degenerate into the asymptotic hypersurface and, by passing from $b < b_c$ to $b > b_c$, the coaxial hyperhyperboloids jump from one side of their asymptotic hypersurface to the other and move away from it. In Fig. 2a–f, the jump of the hyperhyperboloids for $M = 4$ is also visualised in six cuts for the equation (used as an example)

$$4x^2 + 7y^2 - 5z^2 + 2t^2 + 12xy - 4x - 5y + b = 0. \tag{8}$$

The canonical form of this equation, after translation and rotation, is (6b) for $M = 4$, the eigenvalues λ_i are given in Fig. 2a–f and the free term $a_k'' = b'' = \Delta/\delta$ with $\delta = 80$ and $\Delta = 10(8b - 7)$. The critical value of b, obtained by cancelling Δ, is $b = b_c = 0.875$, and the position of the hypercentre with respect to the initial coordinates is C ($x = 0.125$, $y = 0.250$, $z = 0$, $t = 0$).

If the QAS is now considered, the Jacobi hypersurface $J = 0$ cuts the QA-hypersurfaces in nodal hyperlines. If all these nodal hyperlines have a common nodal point, then, this point is the double solution of a QAS.

4. Conclusions

The collapse points are useful for the determination of the position of detachment lines, the saddle points for bifurcation, and the double points, on the Jacobi hypersurface, for the location of the detachment and reattachment lines of the skin friction lines on the surfaces of the flying configurations.

References

1. A. Nastase, Spectral solutions for the three-dimensional compressible boundary layer and their application for the optimal viscous design, *Proc. Fifth International Symposium on CFD*, Sendai, 1993.

2. A. Nastase, Determination of hybrid analytical-numerical solutions for the three-dimensional compressible boundary layer equations, *Z. Angew. Math. Mech.* **73** (1993).

3. A. Nastase, Viscous computation and design of optimal configurations in supersonic flow. *Proc. First Asian CFD Conference*, Hong Kong, 1995.

4. A. Nastase and T. Arnold, Qualitative analysis and visualisation of the real solutions of the quadratical algebraic systems, *ZAMM* **77** (1997).

Aerodynamik des Fluges, RWTH, Templergraben 55, 52062 Aachen, Germany

G.A. NESENENKO

Effect of focusing on catacaustics of irregular heat fields in composites with moving boundaries

1. Introduction

In this report, we propose and describe a method for investigating the effect of focusing of irregular heat fields on the space-time catacaustics. The method is based on the asymptotic analysis of irregular heat fields, namely, on the integral representation of solutions of the given problems.

During the investigation of irregular heat fields near the catacaustics, the author introduced new types of sample integrals for two types of catacaustics [1], which differ from those used to describe the effect of focusing the wave fields near catacaustics [2].

As in the theory of wave propagation, we have the following situation for irregular heat fields: if the "point of observation" belongs to the time-space catacaustics, then the corresponding "rays" and stationary points merge. In particular, it follows from the theory presented in this report that curvilinear time-spacing cracks in optical polymers accelerate their damage by a pulsed laser beam [3].

2. Statement of the problem

In this paper we discuss an algorithm for obtaining asymptotic expansions (A.E.) of the integral representation of Green's functions for corresponding singularly perturbed heat BVP in domains with moving boundaries.

Thus, let $\Gamma(x,t;y,s)$ be the Green's function of the singularly perturbed Dirichlet problem [4] in a semi-bounded domain with a moving boundary [5]

$$\frac{\partial \Gamma}{\partial t} = \varepsilon \frac{\partial^2 \Gamma}{\partial x^2}, \quad (x,t) \in M_t = \{(x,t)\colon -\infty < x < N(t),\ 0 < t \le 1\}, \tag{2.1}$$

$$\Gamma(x,t;y,s) \to \delta(x-y) \ \text{ as } \ t \to s+0, \tag{2.2}$$

$$\Gamma(x,t;y,s) = 0, \quad x = N(t), \tag{2.3}$$

$$\Gamma(x,t;y,s) \to 0 \ \text{ as } \ x \to -\infty,\ 0 < t \le 1,\ 0 \le s < t \le 1,\ -\infty < y < N(s). \tag{2.4}$$

The need to consider the possibility of the presence of catacaustics as a result of a moving boundary $x = N(t)$ is a distinguishing feature of a singularly perturbed BVP. For example, if the moving boundary is a circle, that is, $N(t) = \sqrt{R^2 - t^2}$, then for such a boundary there exist moving catacaustics lying in a boundary layer of this boundary [2]. Our problem is as follows: assuming that a catacaustic emerges, we need to devise an algorithm that includes its influence on the Green's function asymptotics, and, as a consequence, also to develop an algorithm that takes into account its influence on the solutions of the singularly perturbed BVP.

Hence, following the general approach of the heat potential method [5], we find the Green's function (2.1)–(2.4) in the form of a sum of two functions:

$$\Gamma(x,t;y,s) = G(x,t;y,s) + \Gamma_1(x,t;y,s), \tag{2.5}$$

where

$$G(x,t;y,s) = \frac{1}{2\sqrt{\pi\varepsilon(t-s)}} \exp\left\{-\frac{(x-y)^2}{4\varepsilon(t-s)}\right\} \tag{2.6}$$

is a fundamental solution of the differential equation (2.1). To obtain a mathematically correct A.E. of the Green's function $\Gamma(x,t;y,s)$, we write out the function $\Gamma_1(x,t;y,s)$ as a combination of heat potentials of simple and double layers that takes into account the shape of the moving boundary [5]:

$$\Gamma_1(x,t;y,s) = \frac{1}{4\pi\varepsilon} \int_s^t \Psi(t,s_1) \exp\left\{-\frac{H(t,s_1)}{4\varepsilon}\right\} p(s_1)ds_1, \tag{2.7}$$

where

$$\begin{aligned} \Psi(t,s_1) &= \frac{N'(s_1)(t-s_1)+N(s_1)-x}{(t-s_1)^{3/2}(s_1-s)^{1/2}}, \\ H(t,s_1) &= \frac{[x-N(s_1)]^2}{t-s_1} + \frac{[N(s_1)-y]^2}{s_1-s}, \end{aligned} \tag{2.8}$$

and $p(t)$ is the solution of the Volterra equation of the second kind

$$-1 = p(t) + \int_s^t \frac{\psi(t,s_1)}{\sqrt{\varepsilon}} \exp\left\{-\frac{h(t,s_1)}{4\varepsilon}\right\} p(s_1)ds_1 \tag{2.9}$$

with

$$\begin{aligned} \psi(t,s_1) &= \frac{\sqrt{t-s}}{2\sqrt{\pi}} \frac{N'(s_1)(t-s_1)+N(s_1)-N(t)}{(t-s_1)^{3/2}(s_1-s)^{1/2}}, \\ h(t,s_1) &= \frac{[N(t)-N(s_1)]^2}{t-s_1} + \frac{[N(s_1)-y]^2}{s_1-s} - \frac{[N(t)-y]^2}{t-s}. \end{aligned} \tag{2.10}$$

In [5], [7] and [8] it has been proved that if $N(t)$ satisfies certain conditions (which happens, for example, when $N(t) = \sqrt{R^2-t^2}$), then the solution $p(t)$ of the integral equation (2.9)–(2.10) has a uniformly (with respect to $t \in [s,1]$) convergent A.E.

$$p(t) \underset{\varepsilon\to 0}{\sim} -1 + \sum_{i=1}^{\infty} c_i(t)\varepsilon^i, \tag{2.11}$$

whose coefficients c_i can be found explicitly by means of the Laplace method [6]. We emphasise that an A.E. of the form (2.11) could be found since the function $\Gamma_1(x,t;y,s)$ was chosen in a specific manner that took into account the shape of the moving boundary $x = N(t)$.

Since the function $p(t)$ has the uniformly convergent A.E. (2.11), we can substitute it into (2.7).

3. The main result

We introduce some notation. Let $A(y,s)$ be the "source point" and $B(x,t)$—the "observation point". The fact that the "observation point" $B(x,t)$ belongs to the "light zone" (A) is evident: $B(x,t) \in (A)$ if the line segment AB—the so-called "forward ray"—does not intersect the graph of the function $x = N(t)$.

Suppose that the "observation point" $B(x,t)$ belongs to the boundary layer of the moving boundary, and that it is in the "light zone", for example,

$$N(t) - x = O(\varepsilon^q) \quad \text{as } \varepsilon \to 0, \quad q > 1, \quad (x,t) \in (A). \tag{3.1}$$

Theorem. *Let $N(t) \in C^{\infty}_{[0,1]}$, let $B(x,t)$ be in the boundary layer but outside $N(t)$, and suppose that $B(x,t) \in (A)$ and that the curve $x = N(t)$ is such that*

(a) *in the boundary layer of the curve $x = N(t)$ there are no catacaustics.*

Then the A.E. of the Green's function $\Gamma(x,t;y,s)$ of the singularly perturbed BVP (2.1)–(2.4) *has the form*

$$\Gamma(x,t;y,s) \underset{\varepsilon\to 0}{\sim} \frac{1}{2\sqrt{\pi\varepsilon(t-s)}}\Bigg[\exp\left\{-\frac{(x-y)^2}{4\varepsilon(t-s)}\right\} + \frac{1}{\sqrt{\pi d_{-1}}}\exp\left\{-\frac{d_0}{4\varepsilon}\right\}\sum_{i=0}^{\infty} a_i(\varepsilon)\left(\frac{d_{-1}}{\varepsilon^2}\right)^i I^{(b)}_{-(i-1/2)}\left[\frac{d_{-1}}{\varepsilon^2}\frac{d_1}{4}\right]\Bigg], \tag{3.2}$$

where d_{-1} and d_0 are the coefficients of the Laurent series of $H(t,s_1)$ in the vicinity of the point $s_1 = t$, $a_i(\varepsilon)$ are polynomials with $\deg a_i(\varepsilon) \le i$, *and $I^{(b)}_{-(i-1/2)}(z)$ are sample integrals of the form*

$$I^{(b)}_{-(i-1/2)}(z) = \int_0^{\infty} w^{-(i-1/2)-1}\exp\left\{-w-\frac{z}{w}\right\}dw, \quad z > 0; \tag{3.3}$$

(b) *in the boundary layer of the curve $x = N(t)$ there is a smooth catacaustic.*
In this case, if

(b_0) *the "observation point" $B(x,t)$ does not belong to the smooth catacaustic but is in its vicinity, then the A.E. of the Green's function $\Gamma(x,t;y,s)$ is determined by the formula* (3.2), *but with different coefficients;*

(b_s) *the "observation point" $B(x,t)$ belongs to the smooth catacaustic, then the A.E. of the Green's function $\Gamma(x,t;y,s)$ is determined by the formula*

$$\Gamma(x,t;y,s) \underset{\varepsilon\to 0}{\sim} \frac{1}{2\sqrt{\pi\varepsilon(t-s)}}\Bigg[\exp\left\{-\frac{(x-y)^2}{4\varepsilon(t-s)}\right\} + \frac{1}{\sqrt{\pi d_{-1}}}\exp\left\{-\frac{d_0}{4\varepsilon}\right\}\sum_{i=0}^{\infty} a_i^{(s)}(\varepsilon)\left(\frac{d_{-1}}{\varepsilon^2}\right)^i I^{(sc)}_{-(i-1/2)}\left[\varepsilon^2\left(\frac{d_{-1}}{\varepsilon^2}\right)^3\frac{d_3}{4^4}\right]\Bigg], \tag{3.4}$$

where d_{-1} and d_0 are similar to those in (3.2), d_3 is the coefficient in the expansion of $H(t, s_1)$, $a_i^{(s)}(\varepsilon)$ are polynomials of degree not exceeding i, and $I^{(sc)}_{-(i-1/2)}(z)$ are sample integrals of the form

$$I^{(sc)}_{-(i-1/2)}(z) = \int_0^\infty w^{-(i-1/2)-1} \exp\left\{-w - \frac{z}{w^3}\right\} dw, \quad z > 0; \tag{3.5}$$

(c) *in the boundary layer of the curve $x = N(t)$ there is a cuspoid catacaustic (that is, a catacaustic with a "beak").*

In this case, if

(c_0) *the "observation point" $B(x,t)$ does not belong to the cuspoid catacaustic, but is in its vicinity, then the A.E. of the Green's function $\Gamma(x,t;y,s)$ is determined by (3.2) but with different coefficients $a_i(\varepsilon)$;*

(c_s) *the "observation point" $B(x,t)$ belongs to the smooth part of the cuspoid catacaustic (that is, outside its "beak"), then the A.E. of the Green's function $\Gamma(x,t;y,s)$ is determined by (3.4) but with different coefficients $a_i^{(s)}(\varepsilon)$;*

(c_c) *the "observation point" $B(x,t)$ coincides with the cusp ("beak") of the cuspoid catacaustic, then the A.E. of the Green's function $\Gamma(x,t;y,s)$ is determined by the formula*

$$\Gamma(x,t;y,s) \underset{\varepsilon\to 0}{\sim} \frac{1}{2\sqrt{\pi\varepsilon(t-s)}}\left[\exp\left\{-\frac{(x-y)^2}{4\varepsilon(t-s)}\right\}\right.$$
$$\left. + \frac{1}{\sqrt{\pi d_{-1}}} \exp\left\{-\frac{d_0}{4\varepsilon}\right\} \sum_{i=0}^{\infty} a_i^{(cc)}(\varepsilon) \left(\frac{d_{-1}}{\varepsilon^2}\right)^i I^{(cc)}_{-(i-1/2)}\left[\varepsilon^3 \left(\frac{d_{-1}}{\varepsilon^2}\right)^4 \frac{d_4}{4^5}\right]\right], \tag{3.6}$$

where d_{-1} and d_0 are similar to those in (3.2), d_4 is the coefficient in the expansion of $H(t, s_1)$, $a_i^{(cc)}(\varepsilon)$ are polynomials of degree not exceeding i, and $I^{(cc)}_{-(i-1/2)}(z)$ are sample integrals of the form

$$I^{(cc)}_{-(i-1/2)}(z) = \int_0^\infty w^{-(i-1/2)-1} \exp\left\{-w - \frac{z}{w^4}\right\} dw, \quad z > 0. \tag{3.7}$$

References

1. G.A. Nesenenko, Taking into account the influence of moving catacaustics over the structure of irregular heat fields: application of catastrophe theory and "geometric-optical" asymptotic method to finding the asymptotic expansion of Green's functions in domains with moving boundaries, in *Methods and algorithms of parametric analysis of linear and non-linear transfer models*, vol. 11, Moscow, 1995, 3–29 (Russian).

2. T. Poston and I. Stewart, *Catastrophe theory and its applications*, Pitman, London, 1978.

3. K.M. Dyumaev, A.A. Manenkov et al., *Interaction of laser radiation with optical polymers*, Nauka, Moscow, 1991 (Russian).

4. H.S. Carslaw and J.C. Jeager, *Conduction of heat in solids*, Oxford University Press, Oxford, 1959.

5. G.A. Nesenenko, *Boundary layer in unsteady temperature fields of solids*, Moscow State Correspondence Pedagogic Institute Press, Moscow, 1991 (Russian).

6. M.V. Fedoryuk, *Method of saddle point*, Nauka, Moscow, 1977 (Russian).

7. G.A. Nesenenko, *Bibliographic handbook on the geometric-optical asymptotic method of solving non-linear singularly perturbed problems of heat and mass transfer in multilayer media with cracks*, Moscow State Open Pedagogic Institute Press, Moscow, 1994 (Russian).

8. G.A. Nesenenko, *Bibliographic survey, statements of theorems and algorithms on the geometric-optical asymptotic method for solving singularly perturbed boundary value problems of non-linear heat and mass transfer in domains with moving boundaries*, Moscow State Open Pedagogic University Press, Moscow, 1995 (Russian).

Department of Applied Mathematics, Bauman Moscow State Technical University, 2nd Baumanskaya Street 5, 107005 Moscow, Russia

S. NOMURA

Thermoelastic analysis of functionally graded materials

1. Introduction

Functionally graded materials (FGMs) distribute the material functions throughout the material body to achieve the maximum heat resistance and mechanical properties [1] ideal for space planes where one side is exposed to extremely high temperature and the other side is exposed to extremly low temperature. Typically, ceramics can be used on one side exposed to high temperature while metals can be used on the other side, where mechanical strength is in demand (see Fig. 1).

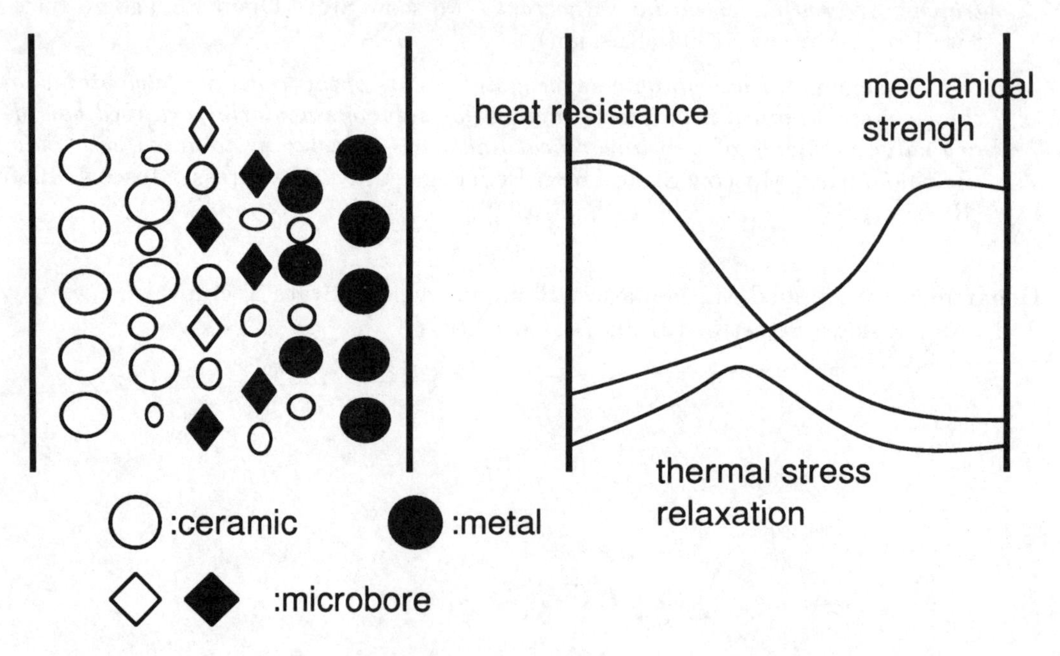

Fig. 1. A typical FGM.

Continuous mixtures of ceramics and metals are distributed in the intermediate range, thus eliminating distinct phase interface that often causes cracks due to different thermal expansion at the interface. Because of continuously varying material properties, the analysis of FGM is complicated and many analytic methods developed for composites with distinct phases may not be directly applicable to FGMs.

In this paper, a micromechanical approach is used to obtain thermoelastic fields in FGMs semi-analytically by the eigenfunction expansion method. The physical

This material is based in part on work supported by the Texas Advanced Technology Program under Grant No. 003656-044.

fields in FGM can be expanded in eigenfunctions defined for the associated eigenvalue problem, and the coefficient of each eigenfunction can be determined by the Galerkin method. The eigenfunctions obtained can also be used to construct a numerical Green's function for FGMs that can handle the change in body force and boundary conditions [2]. Although this method is well known, the implementation of the method for multi-dimensional and heterogeneous bodies is the first attempt, to the author's best knowledge. As an example, thermal stress distribution in a 2D rectangular elastic body under steady-state heat conduction subject to the first kind of displacement boundary condition is shown for demonstration. As the obtained expressions retain all the relevant parameters, the procedure is suited to sensitivity analysis.

2. Analysis

The equilibrium equations of linear elasticity for general heterogeneous materials with temperature effects is expressed as

$$(C_{ijkl}(\mathbf{r})u_k(\mathbf{r})_{,l})_j - (C_{ijkl}(\mathbf{r})\alpha_{kl}(\mathbf{r})\Delta T(\mathbf{r}))_{,j} = 0 \tag{1}$$

with a prescribed boundary condition, where $C_{ijkl}(\mathbf{r})$, $u_k(\mathbf{r})$, $\alpha_{kl}(\mathbf{r})$ and $\Delta T(\mathbf{r})$ are the elastic modulus, displacement, thermal conductivity and temperature rise, respectively. The summation convention is used where applicable.

The displacement field $u_k(\mathbf{r})$ can be expressed as [3]

$$u_k(\mathbf{r}) = \sum_A c^A \psi_k^A(\mathbf{r}), \tag{2}$$

$$c^A = \frac{\int_\Omega b_i(\mathbf{r})\psi_i^A(\mathbf{r})d\mathbf{r}}{\lambda^A}, \tag{3}$$

where, for brevity, $b_i(\mathbf{r})$ represents the second term on the right-hand side of equation (1) and $\psi_i^A(\mathbf{r})$ and λ_i^A are the Ath eigenfunction and eigenvalue, respectively, defined for the eigenvalue problem

$$(C_{ijkl}(\mathbf{r})\psi_{k,l}^A(\mathbf{r}))_{,j} + \lambda^A\psi_i^A(\mathbf{r}) = 0, \tag{4}$$

where no summation is taken over A and the eigenfunction $\psi_k^A(\mathbf{r})$ must satisfy the homogeneous boundary condition. The symbol Ω in (3) denotes the domain of all material points. The eigenfunctions are mutually orthonormal, that is,

$$\int_\Omega \psi_i^A(\mathbf{r})\psi_i^B(\mathbf{r})d\mathbf{r} = \delta_{AB}, \tag{5}$$

where δ_{AB} is the Kronecker delta. In order to approximate the eigenfunction $\psi_i^A(\mathbf{r})$, the Galerkin method is adopted, in which the eigenfunction $\psi_i^A(\mathbf{r})$ is expanded in a series of admissible functions $f^\beta(\mathbf{r})$, each of which satisfies the accompanying homogeneous boundary condition:

$$\psi_k^A(\mathbf{r}) = \sum_\beta c_k^{A\beta} f^\beta(\mathbf{r}), \tag{6}$$

where $c_k^{A\beta}$ are unknown coefficients to be determined and the $f^\beta(\mathbf{r})$ are admissible functions chosen from a complete set of functions that satisfy the homogeneous boundary condition. For example, if the boundary shape is 2D circular, then the admissible function can be chosen as $f(x, y) = (x^2 + y^2 - a^2)x^i y^j$, where a is the radius of the circle and i and j are ordered integers.

By substituting (6) into (4), multiplying by $f^\gamma(\mathbf{r})$ on both sides and integrating over the entire material domain, the eigenvalue problem of equation (4) in partial differential equation form is converted into an algebraic eigenvalue problem as

$$A\mathbf{c} + \lambda B\mathbf{c} = 0, \tag{7}$$

where

$$A_{ik}^{\beta\gamma} = \int_\Omega (C_{ijkl}(\mathbf{r}) f_{,l}^\beta(\mathbf{r}))_{,j} f^\gamma(\mathbf{r}) d\mathbf{r} \tag{8}$$

and

$$B^{\beta\gamma} = \int_\Omega f^\beta(\mathbf{r}) f^\gamma(\mathbf{r}) d\mathbf{r}. \tag{9}$$

In 2D, equation (9) can be written as

$$\begin{pmatrix} A_{11} & A_{12} \\ A_{21} & A_{22} \end{pmatrix} \begin{pmatrix} c_1^A \\ c_2^A \end{pmatrix} + \lambda^A \begin{pmatrix} B & 0 \\ 0 & B \end{pmatrix} = 0, \tag{10}$$

where each component is a matrix or vector. The components of equations (8) and (9) can be evaluated exactly using a computer algebra system.

We remark that the Green's function [2] for the FGM can be constructed from the eigenfunctions as

$$g_{km}(\mathbf{r}, \mathbf{r}') = \sum_A \frac{\psi_m^A(\mathbf{r}) \psi_k^A(\mathbf{r}')}{\lambda^A}, \tag{11}$$

where $g_{km}(\mathbf{r}, \mathbf{r}')$ is defined as

$$(C_{ijkl}(\mathbf{r}) g_{km}(\mathbf{r}, \mathbf{r}')_{,l})_j + \delta_{im} \delta(\mathbf{r} - \mathbf{r}') = 0 \tag{12}$$

with the homogeneous boundary condition, and $\delta(\mathbf{r}-\mathbf{r}')$ is the Dirac delta function. If the boundary condition is given by other than the Dirichlet condition, the admissible functions need to be modified accordingly.

The steady-state heat conduction equation for heterogeneous materials is expressed as

$$(K_{ij}(\mathbf{r}) T(\mathbf{r})_{,j})_{,i} + g(\mathbf{r}) = 0, \tag{12}$$

where $K_{ij}(\mathbf{r})$ is the anisotropic thermal conductivity, $T(\mathbf{r})$ is the temperature distribution and $g(\mathbf{r})$ is the heat generation term. The solution procedure developed for thermal stresses can be used to solve for the temperature field. It is noted that the use of the computer algebra system is essential in evaluating $A_{ik}^{\beta\gamma}$ and $B^{\beta\gamma}$ [4].

3. Examples

Fig. 2 is an example of thermal stresses under steady-state heat conduction in a rectangular body defined over $\{(x, y),\ 0 < x < a,\ 0 < y < b\}$ subject to the Dirichlet type homogeneous boundary condition ($T = 0$ and $u_i = 0$) on the boundary with a uniform internal heat generation [5]. The thermal conductivity, elastic modulus and thermal expansion coefficient are assumed to vary in the form of $k_0 + k_1 x + k_2 y$, where the k_i are constants. Fig. 2 shows the (non-parametrised) stress component σ_{11} over the same region. As the method presented can retain all the relevant material parameters, it is suitable for sensitivity analysis for the optimum composition.

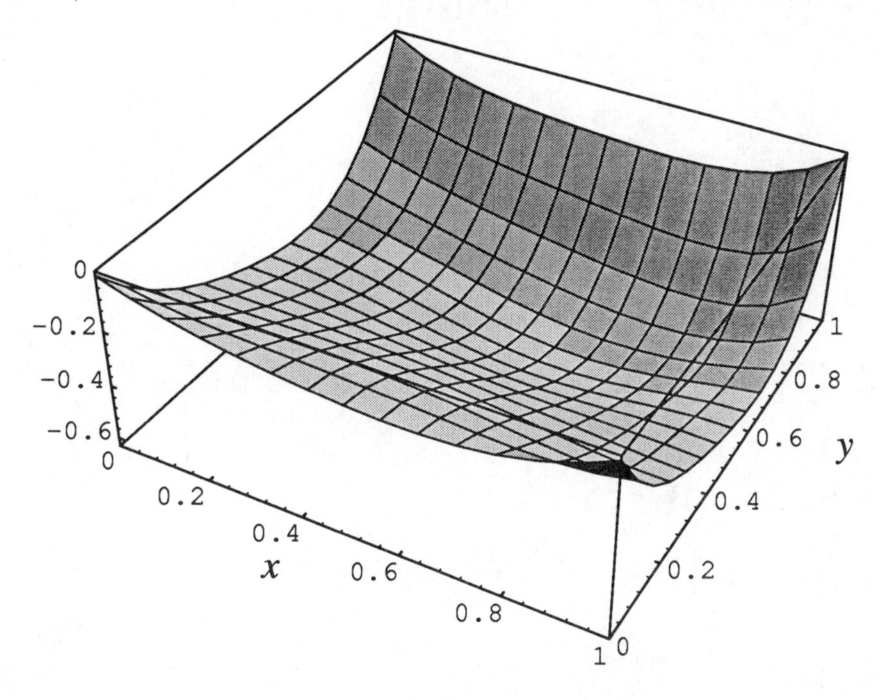

Fig. 2. Stress distribution (σ_{xx}) in FGMs.

References

1. M. Yamanouchi, M. Koizumi, T. Hirai and I. Shiota, eds., *Proc. First Internat. Symp. on Functionally Graded Materials, Sendai, 1990*, Functionally Gradient Materials Forum and the Society of Non-traditional Technology, Tokyo, 1990.

2. T. Mura, *Micromechanics of defects in solids*, M. Nijhoff, The Hague, 1982.

3. R. Diaz and R. Nomura, Green's function applied to solution of Mindlin plate, *Comput. & Structures* **60** (1996), 41–48.

4. S. Nomura, Symbolic algebra approach to micromechanics of composite materials analysis, in *The IUTAM Symp. on Local Mechanics Concepts for Composite Material Systems, 1992*, Springer-Verlag, Heidelberg-New York, 383–393.

5. S. Nomura and D.M. Sheahen, Galerkin function approach to thermomechanical analysis of functionally graded materials, in *Proc. Third Internat. Conf. on Composite Engineering, New Orleans, 1996*, 635–636.

Department of Mechanical and Aerospace Engineering, University of Texas at Arlington, Arlington, TX 76019-0023, USA

Email: nomura@uta.edu

D. POLJAK and V. ROJE

The integral equation method for ground wire input impedance

1. Introduction

The study of the transient behaviour of a wire buried in a lossy half-space, as part of a grounding system, is of great practical importance in electromagnetic pulse protection design. The most important parameter leading to such an analysis is the transient response of the ground electrode or its counterpart in the frequency domain—the input impedance spectrum. The most conventional approach for obtaining the input impedance of the grounding system is usually referred to as the transmission line method (TLM) [1]. However, the TLM approach is valid for long horizontal conductors, but it is not convenient for vertical or interconnected conductors. A more accurate approach, based on integral equations arising from wire antenna theory, was offered by Grčev (see [2] and [3]). Grčev solved the corresponding electric field integral equation (EFIE) by using the point-matching technique. Unfortunately, the method is numerically inefficient, because it requires too long a computational time for the evaluation of the broadband frequency spectrum. In this paper a new finite element procedure for the input impedance calculation is proposed.

2. Governing equations in the frequency domain

We consider a straight end-fed wire antenna of length $2L$ and radius $2a$, immersed in a lossy half-space at depth h (Fig. 1). The electric field in the vicinity of the wire can be obtained by applying the thin wire approximation and Sommerfeld's theory of radiation in the presence of a lossy half-space. The normal electric field component is given by

$$E_z(x,z) = \frac{1}{4\pi\omega j\varepsilon_{\text{eff}}}\left\{\int_0^L \left[\frac{\partial^2}{\partial z^2} + k_1^2\right]\frac{\partial W_{11}}{\partial x} I(x')\,dx'.\right.$$
$$\left. + \int_0^L \frac{\partial^2}{\partial x \partial z}\left[g_0(x,x') - g_i(x,x') + U_{11}\right] I(x')\,dx'\right\}. \quad (1)$$

If we assume that the wire is perfectly conducting, then the total tangential electric field E_x vanishes along the antenna surface, and the electric field integral equation can be written as

$$\frac{1}{4\pi\omega j\varepsilon_{\text{eff}}}\left\{\int_0^L \frac{\partial^2}{\partial x^2}\left[g_0(x,x') - g_i(x,x') + k_1^2 V_{11}\right] I(x')\,dx'\right.$$
$$\left. + k_1^2 \int_0^L \left[g_0(x,x') - g_i(x,x') + U_{11}\right] I(x')\,dx'\right\} = -E_x^{\text{inc}}, \quad (2)$$

where E_x^{inc} is the incident field (the excitation function) and $I(x')$ is the unknown current distribution, $g_0(x,x') = \exp(-\gamma R_1)/R_1$ denotes the free space Green's function, while $g_i(x,x') = \exp(-\gamma R_2)/R_2$ comes from image theory. The influence of the lossy ground is taken into account by means of "attenuation terms" in the form of Sommerfeld integrals U_{11}, V_{11} and W_{11} defined by [4] $U_{11} = k_1^2 V_{11} - \mathrm{d}W_{11}/\mathrm{d}z$, where

$$V_{11} = 2\int_0^\infty \frac{e^{-\mu_1(h-z)}}{k_2^2\mu_1 + k_1^2\mu_2} J_0(\lambda\rho)\lambda\, d\lambda, \quad W_{11} = 2\int_0^\infty \frac{(\mu_1 - \mu_2)e^{-\mu_1(h-z)}}{k_2^2\mu_1 + k_1^2\mu_2} J_0(\lambda\rho)\lambda\, d\lambda, \quad (3)$$

where $k_1 = -\gamma^2 = -\omega^2\mu\varepsilon_{\text{eff}}$ and $k_2 = \omega^2\mu\varepsilon_0$ are the phase constants of the lossy ground and free space, respectively, and R_1 and R_2 are the distances from the antenna and its image to the observation point. Also, $\varepsilon_{\text{eff}} = \varepsilon_r\varepsilon_0 - j\sigma/\omega$ is the complex permittivity of the lossy ground, $J_0(\lambda\rho)$ is the zero-order Bessel function, h is the distance from the interface to the antenna, $\mu_{1,2} = (\lambda^2 - k_{1,2}^2)^{1/2}$ and $\rho = |x - x'|$.

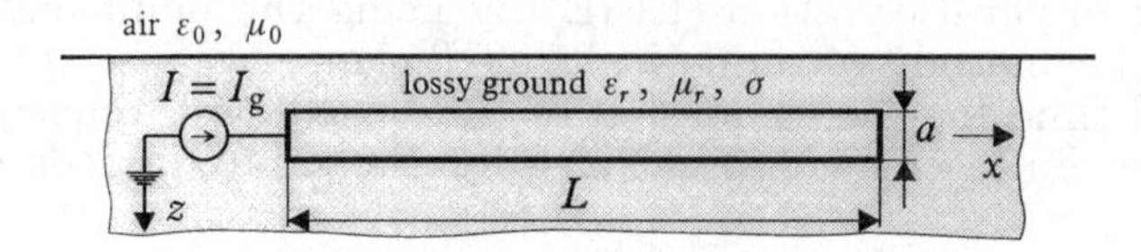

Fig 1. The geometry of ground wire.

Solving the integral equation (2), we obtain the equivalent antenna current. The input impedance is given by the ratio of the voltage V_g and current I_g at the monopole antenna input; that is, $Z_{\text{in}} = V_g/I_g$.

The feed-point voltage V_g can be obtained by integrating the normal electric field component from the electrode surface to infinity, as done in [1]. The spectrum can be obtained by repeating this procedure in a wide frequency band.

3. Solution of the integral equation and input impedance evaluation

The integral equation (2) is often modelled by a point-matching technique, while the differential operator, the inside integral equation kernel, is handled by finite difference approximation in order to avoid the appearence of a quasi-singularity (see [1] and [2]). However, it is well known that this approach suffers from a poor convergence rate. On the other hand, finite differences applied to the calculation of Sommerfeld integrals increase the complexity of the algorithm.

In order to avoid these disadvantages, in this paper we use the weak formulation of the finite element integral equation method (FEIEM). A similar approach was promoted in [4], where the FEIEM was successfully applied to solving the half-space antenna problem. The FEIEM is outlined in a few steps. The unknown current $I(x')$ is expanded into a linear combination of the independent basis functions $N_i(x')$ with unknown complex coefficients α_i: $I(x') = \{N\}^T\{\alpha\}$. According to the weighted residual approach, choosing the set of test functions $W_j = N_j$ and exploiting the integral equation kernels properties $dG_1/dx = -dG_1/dx'$ and $dG_2/dx = -dG_2/dx'$, where $G_1(x,x') = g_0(x,x') - g_i(x,x') + k_1^2 V_{11}$ and $G_2(x,x') = g_0(x,x') - g_i(x,x') + U_{11}$, after integrating by parts we find that

$$\sum_{i=1}^{n} \alpha_i \frac{1}{4\pi\omega j\varepsilon_{\text{eff}}} \Bigg\{ - \int_0^L \frac{dN_j(x)}{dx} \int_0^L \frac{dN_i(x')}{dx'} G_1(x,x')\, dx'\, dx$$
$$+ k_1^2 \int_0^L N_j(x) \int_0^L N_i(x') G_2(x,x')\, dx'\, dx \Bigg\} = \int_0^L E_x^{\text{inc}}(x) N_j(x)\, dx\,, \quad j = 1,\ldots,n. \tag{4}$$

Equation (4) represents the weak Galerkin-Bubnov formulation of (2). The right-hand side contains the excitation function in the form of an incident field. However, the ground wire considered in this paper is neither excited by a voltage generator, nor illuminated by a plane wave, so the right-hand side vector is zero. Applying the finite element algorithm, from (4) we obtain the homogeneous system of equations

$$\sum_{i=1}^{M} [a]_{ji} \{\alpha\}_i = 0\,, \quad j = 1, 2, \ldots, M, \tag{5}$$

where $[a]_{ji}$ is the local matrix (mutual impedance matrix) connecting the ith source finite element to the jth observation finite element:

$$[a]_{ji} = -\frac{1}{4\pi\omega j\varepsilon_{\text{eff}}} \Bigg(\int_{\Delta l_j} \int_{\Delta l_i} \{D\}_j \{D'\}_i^T G_1(x,x')\, dx'\, dx$$
$$+ k_1^2 \int_{\Delta l_j} \int_{\Delta l_i} \{N\}_j \{N'\}_i^T G_2(x,x')\, dx'\, dx \Bigg). \tag{6}$$

The matrices $\{N\}$ and $\{N'\}$ contain the shape functions $N_k(x)$ and $N_k(x')$, while $\{D\}$ and $\{D'\}$ contain their derivatives. M is the total number of boundary elements and Δl_i, Δl_j are the widths of the ith and jth boundary elements, respectively. Since the wire is end-driven by a current generator, the source current is taken into account as a forced condition in the first node of the solution vector: $I_1 = I_g$, while the current at the free end of the wire is assumed to be zero. Consequently, the linear equation system can now be properly solved.

This approach avoidsthe quasi-singularity since the second-order differential operator is removed from the kernel. In addition, in order to avoid numerical integration, the Sommerfeld integrals are solved by means of the exponential approximation technique, fully documented in [4]. According to this technique, certain identities are used to transform the Sommerfeld integral into the sum with unknown coefficients, which can be written symbolically in the form

$$\int_0^\infty f(\lambda) J_0(\lambda\rho) e^{-\psi(\lambda)} \lambda\, d\lambda = \sum_{k=1}^{N} A_k \frac{1}{(\xi_k^2 + \rho^2)^{3/2}}, \tag{7}$$

where $f(\lambda)$ and $\psi(\lambda)$ are the corresponding parts of the integrands of V_{11} and W_{11}, respectively. The coefficients ξ_k must be conveniently chosen, and A_k can be determined from the linear equation system.

However, this approach requires a time-consuming repeated matrix inversion for several frequencies. In order to simplify this procedure, the rigorous Sommerfeld approach is used only in the near field zone, while in the rest of the integration path ($h > \lambda/2$ and $k_1 R_2 \gg 1$) the Fresnel reflection coefficient (RC) approximation is preferred. The use of the RC R_M implies the following approximation in kernels: $G_1 = g_0(x, x', z') - R_M g_i(x, x', z')$ and $G_2 = 0$. The corresponding R_M is given by

$$R_M = \frac{\cos\theta - \sqrt{\varepsilon_{\text{eff}}(1 - \varepsilon_{\text{eff}} \sin^2\theta}}{\cos\theta - \sqrt{\varepsilon_{\text{eff}}(1 - \varepsilon_{\text{eff}} sin^2\theta)}}. \tag{8}$$

To evaluate the input impedance, we calculate the feed voltage by integrating the E_z component (1) from infinity to the electrode surface. Using integration by parts in (1), we find that the E_z component of the electric field is

$$E_z(x, z) = -\frac{1}{4\pi\omega j\varepsilon_{\text{eff}}} \frac{d}{dz}\left[I(x')G_1(x, x', z)\big|_{x'=0}^{x'=L} - \int_0^L \frac{\partial I(x')}{\partial x'} G_1(x, x', z)\, dx' \right]. \tag{9}$$

It is obvious that differentiation with respect to z is performed outside the integral sign. Moreover, integrating the field from the electrode surface to infinity yields

$$\begin{aligned} V_g = \int_a^\infty E_z\, dz &= -\frac{1}{4\pi\omega j\varepsilon_{\text{eff}}} \int_a^\infty \frac{d}{dz} I(x')G_1(x, x', z)\bigg|_{x'=0}^{x'=L} \\ &\qquad - \int_a^\infty \frac{d}{dz} \int_0^L \frac{\partial I(x')}{\partial x'} G_1(x, x', z)\, dx'\, dz \\ &= \frac{1}{4\pi\omega j\varepsilon_{\text{eff}}}\left[I(0)G_1(x, 0, z) - \int_0^L \frac{\partial I(x')}{\partial x'} G_1(x, x', z)\, dx' \right]\Bigg|_{z=a}^{z=\infty}, \end{aligned} \tag{10}$$

and the tedious numerical integration over the infinite domain is avoided. Finally, the input impedance of the ground wire is defined by the ratio V_g/I_g:

$$Z_{\text{in}} = -\frac{1}{4\pi\omega j\varepsilon_{\text{eff}}} \frac{1}{I_g}\left[I_g G_1(x, 0, a) + \sum_{i=1}^{M} \int_{\Delta l_i} \{D'\}_i^T G_i(x, x', a)\, dx' \{\alpha\}_i \right]. \tag{11}$$

4. Numerical results and concluding remarks

All the results refer to the Fourier transform of the time domain delta-function, that is, to the unit current excitation $I_g = 1e^{j0}$. In accordance with the previous application of the FEIEM, the optimal number of 31 elements over the entire wire length is chosen for satisfactory convergence and computational time [4]. Figs. 2 and 3 show the spectrum of the input impedance magnitude for the wire radiating in the presence of a lossy half-space. The results obtained in this paper seem to be in good agreement with the results in [2].

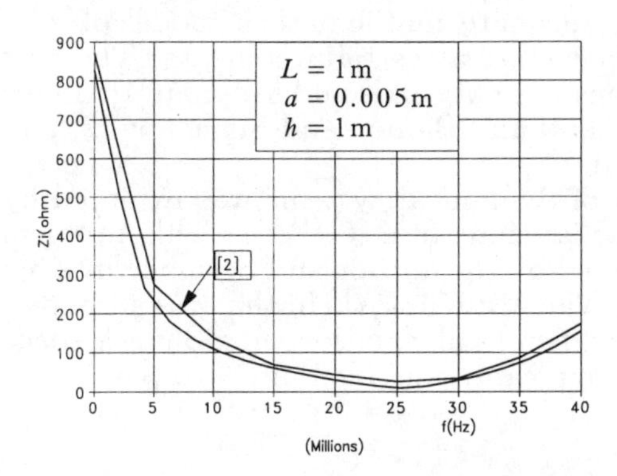

Fig. 2. The input impedance for $\sigma = 0.01\,\mathrm{S/m}$, $\varepsilon_r = 10\varepsilon_0$.

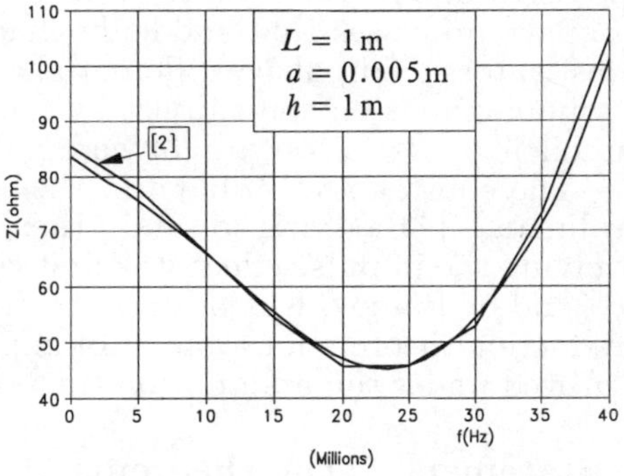

Fig. 3. The input impedance for $\sigma = 0.001\,\mathrm{S/m}$, $\varepsilon_r = 10\varepsilon_0$.

The impulse behaviour of the ground wire immersed in a lossy half-space is analysed in the frequency domain. The corresponding EFIE is solved via the weak finite element formulation adjusted to the integro-differential operator. The effect of the lossy half-space is taken into account through a combination of the Sommerfeld rigorous approach and an RC approximation. After solving the EFIE and obtaining the corresponding current distribution along the electrode, we obtain the input impedance by analytically performing the integration from the electrode surface to the remote point. This procedure shows some advantages over the usual point-matching techniques, especially in computational efficiency and accuracy.

References

1. R. Velazquez and D. Muhkedo, Analytical modelling of grounding electrodes transient behaviour, *IEEE Trans. Power Appar. Systems* **103** (1984), 1314–1322.
2. L. Grčev and F. Dawalibi, An electromagnetic model for transients in grounding systems, *IEEE Trans. Power Delivery* **4** (1990), 1773–1781.
3. L. Grčev, Calculation of the transient impedance of grounding systems, D.Sc. Thesis, University of Zagreb, 1986 (Croatian).
4. D. Poljak, New numerical approach in the analysis of a thin wire radiating over a lossy half-space, *Internat. J. Numer. Methods Engrg.* **38** (1995), 3803–3816.

Department of Electronics, University of Split, R. Boskovica bb, Split, Croatia

A. REINFELDS

Decoupling of impulsive differential equations in a Banach space

1. Introduction

Impulsive differential equations provide an adequate mathematical model of evolutionary processes that suddenly change their state at certain moments. The first investigators of impulsive differential equations were Myshkis and Mil'man [1]. Later on, monographs on the subject by Lakshmikantham, Bainov and Simeonov [2] and Samoilenko and Perestyuk [3] were published.

The equivalence problem in the theory of dynamical systems was explored by Grobman [4], Hartman [5] and other mathematicians (see [6]–[12]). This problem involving the impulse effect was first considered by the author and Sermone in [13]–[16], and by Bainov, Kostadinov and Nguyen Van Minh in [17]. In the present paper, a reduction theorem for systems of impulsive differential equations in a Banach space is proved under the assumption that the system can be divided into two parts.

2. Statement of the theorems

Let $\mathbf{X}$ and $\mathbf{Y}$ be Banach spaces, and let $\mathcal{L}(\mathbf{X})$ and $\mathcal{L}(\mathbf{Y})$ be the corresponding Banach spaces of bounded linear operators. Consider the system of impulsive differential equations

$$\begin{aligned} dx/dt &= A(t)x + f(t,x,y), \\ dy/dt &= B(t)y + g(t,x,y), \\ \Delta x\big|_{t=\tau_i} &= x(\tau_i+0) - x(\tau_i-0) \\ &= C_i x(\tau_i-0) + p_i(x(\tau_i-0), y(\tau_i-0)), \\ \Delta y\big|_{t=\tau_i} &= y(\tau_i+0) - y(\tau_i-0) \\ &= D_i y(\tau_i-0) + q_i(x(\tau_i-0), y(\tau_i-0)), \end{aligned} \tag{1}$$

and its linear truncation

$$\begin{aligned} dx/dt &= A(t)x, \\ dy/dt &= B(t)y, \\ \Delta x\big|_{t=\tau_i} &= x(\tau_i+0) - x(\tau_i-0) = C_i x(\tau_i-0), \\ \Delta y\big|_{t=\tau_i} &= y(\tau_i+0) - y(\tau_i-0) = D_i y(\tau_i-0), \end{aligned} \tag{2}$$

where

(i) the maps $A{:}\,\mathbb{R} \to \mathcal{L}(\mathbf{X})$ and $B{:}\,\mathbb{R} \to \mathcal{L}(\mathbf{Y})$ are locally integrable in the Bochner sense;

This work was partly supported by the Latvian Council of Science under Grant 93.809.

(ii) the maps $f: \mathbb{R} \times \mathbf{X} \times \mathbf{Y} \to \mathbf{X}$ and $g: \mathbb{R} \times \mathbf{X} \times \mathbf{Y} \to \mathbf{Y}$ are locally integrable in the Bochner sense with respect to t for fixed x and y, and, in addition, they satisfy the estimates

$$|f(t,x,y) - f(t,x',y')| \le \varepsilon(|x-x'| + |y-y'|),$$
$$|g(t,x,y) - g(t,x',y')| \le \varepsilon(|x-x'| + |y-y'|),$$
$$\sup_{t,x} |g(t,x,0)| < +\infty;$$

(iii) for $i \in \mathbb{Z}$, $C_i \in \mathcal{L}(\mathbf{X})$, $D_i \in \mathcal{L}(\mathbf{Y})$, the maps $p_i: \mathbf{X} \times \mathbf{Y} \to \mathbf{X}$, $q_i: \mathbf{X} \times \mathbf{Y} \to \mathbf{Y}$ satisfy the estimates

$$|p_i(x,y) - p_i(x',y')| \le \varepsilon(|x-x'| + |y-y'|),$$
$$|q_i(x,y) - q_i(x',y')| \le \varepsilon(|x-x'| + |y-y'|),$$
$$\sup_{i,x} |q_i(x,0)| < +\infty;$$

(iv) the maps $(x,y) \mapsto (x + C_i x + p_i(x,y), y + D_i y + q_i(x,y))$ and $(x,y) \mapsto (x + C_i x, y + D_i y)$ are homeomorphisms;

(v) the moments τ_i of impulse effect form a strictly increasing sequence

$$\cdots < \tau_{-2} < \tau_{-1} < \tau_0 < \tau_1 < \tau_2 < \cdots,$$

where the limit points may be only $\pm\infty$.

Definition. By a *solution* to a system of impulsive differential equations we mean a piecewise absolutely continuous map with discontinuities of the first kind at the points $t = \tau_i$, which for almost all t satisfies system (1) and for $t = \tau_i$ satisfies the conditions of a "jump".

Let $U(t,\tau)$ and $V(t,\tau)$ be Cauchy evolution operators of the system of impulsive linear differential equations (2). We assume that the evolution operators satisfy the estimates

$$\nu_1 = \sup_t \int_{-\infty}^{t} |V(t,\tau)||U(\tau,t)|\, d\tau + \sup_t \sum_{\tau_i \le t} |V(t,\tau_i)||U(\tau_i - 0, t)| < +\infty,$$

$$\nu_2 = \sup_t \int_{t}^{+\infty} |V(\tau,t)||U(t,\tau)|\, d\tau + \sup_t \sum_{t < \tau_i} |V(\tau_i - 0, t)||U(t,\tau_i)| < +\infty,$$

$$\mu = \sup_t \left(\int_{-\infty}^{t} |V(t,\tau)|\, d\tau + \sum_{\tau_i \le t} |V(t,\tau_i)| \right) < +\infty.$$

Let $\Phi(\cdot, t_0, x_0, y_0) = (x(\cdot, t_0, x_0, y_0), y(\cdot, t_0, x_0, y_0)): \mathbb{R} \to \mathbf{X} \times \mathbf{Y}$ be the solution of system (1), where $\Phi(t_0 + 0, t_0, x_0, y_0) = (x_0, y_0)$. At the break points τ_i, the values of all solutions are taken at $\tau_i + 0$ unless otherwise specified. For simplicity, we use the notation $\Phi(t) = (x(t), y(t))$.

Theorem 1. *If* $4\varepsilon \max\{\nu_1, \nu_2\} \le 1$ *and* $2\varepsilon \max\{\sup_i |(id_x + C_i)^{-1}|, \mu\} < 1 + \sqrt{1 - 4\varepsilon\nu_1}$, *then there is a unique piecewise continuous bounded map* $h: \mathbb{R} \times \mathbf{X} \to \mathbf{Y}$ *such that*

(i) $h(t, x(t, t_0, x_0, h(t_0, x_0))) = y(t, t_0, x_0, h(t_0, x_0))$ *for all* $t \in \mathbb{R}$;

(ii) $|h(t, x_0) - h(t, x_0')| \le \lambda |x_0 - x_0'|$;

(iii) *there holds the inequality*

$$\int_{t_0}^{+\infty} |U(t_0, t)||y(t, t_0, x_0, y_0) - h(t, x(t, t_0, x_0, y_0))|\, dt$$
$$+ \sum_{t_0 < \tau_i} |U(t_0, \tau_i)||y(\tau_i - 0, t_0, x_0, y_0) - h(\tau_i - 0, x(\tau_i - 0, t_0, x_0, y_0))|$$
$$\le \nu_2 (1 - \varepsilon(1 + \lambda)\nu_2)^{-1} |y_0 - h(t_0, x_0)|,$$

where $\lambda = (2\varepsilon\nu_1)^{-1}(1 - 2\varepsilon\nu_1 - \sqrt{1 - 4\varepsilon\nu_1})$.

The properties (i) and (ii) for impulsive systems with additional assumptions were proved in [18]. The inequality in (iii) characterises the integral distance between an arbitrary solution and the integral manifold. We remark that $\lambda < 1$ when $4\varepsilon \max\{\nu_1, \nu_2\} < 1$.

Let $\mathbf{U}$ be a Banach space. Consider two impulsive differential equations

$$du/dt = P(t, u), \quad \Delta u\big|_{t=\tau_i} = S_i(u(\tau_i - 0)) \tag{3}$$

and

$$du/dt = Q(t, u), \quad \Delta u\big|_{t=\tau_i} = T_i(u(\tau_i - 0)) \tag{4}$$

that satisfy the conditions of the existence and uniqueness theorem. We assume that the maximum interval of the existence of the solutions is $\mathbb{R}$. Let $\phi(\cdot, t_0, u_0): \mathbb{R} \to \mathbf{U}$ and $\psi(\cdot, t_0, u_0): \mathbb{R} \to \mathbf{U}$ be the solutions of the above systems, respectively, and suppose that there is a function $e: \mathbf{U} \to \mathbb{R}$ such that

$$\max\{|P(t, u) - Q(t, u)|, \sup_i |S_i(u) - T_i(u)|\} \le e(u).$$

Definition. The two impulsive differential equations (3) and (4) are *dynamically equivalent in the large* if there exists a map $H: \mathbb{R} \times \mathbf{U} \to \mathbf{U}$ and a positive constant c such that

(i) $H(t, \cdot): \mathbf{U} \to \mathbf{U}$ is a homeomorphism;

(ii) $H(t, \phi(t, t_0, u_0)) = \psi(t, t_0, H(t_0, u_0))$ for all $t \in \mathbb{R}$;

(iii) $\max\{|H(t, u) - u|, |H^{-1}(t, u) - u|\} \le ce(u)$;

(iv) if the systems of differential equations are autonomous and have no impulse effect, then the map H does not depend on t.

We remark that without (iii) and (iv) the concept of dynamical equivalence would be trivial, since in this case the equality $H(t_0, u_0) = \psi(t_0, 0, \phi(0, t_0, u_0))$ gives a dynamical equivalence. It is significant that in the case of the classical global Grobman-Hartman theorem (see [4] and [5]) for autonomous differential equations, the corresponding function $e(x) = a > 0$ and appropriate constant c depend only on the linear truncation.

Suppose there exist maps $g_0: \mathbb{R} \times \mathbf{Y} \to \mathbf{Y}$ and $q_{i0}: \mathbf{Y} \to \mathbf{Y}$ locally integrable in the Bochner sense with respect to t for fixed y and such that

$$|g_0(t,y) - g_0(t,y')| \le \varepsilon |y - y'|, \quad |q_{i0}(y) - q_{i0}(y')| \le \varepsilon |y - y'|,$$
$$\sup_{t,x,y} |g(t,x,y) - g_0(t,y)| < +\infty, \quad \sup_{i,x,y} |q_i(x,y) - q_{i0}(y)| < +\infty;$$

suppose also that the maps $y \mapsto y + D_i y + q_{i0}(y)$ are homeomorphisms.

Next, consider the reduced system of impulsive differential equations

$$\begin{aligned} dx/dt &= A(t)x + f(t,x,h(t,x)), \\ dy/dt &= B(t)y + g_0(t,y), \\ \Delta x\big|_{t=\tau_i} &= x(\tau_i + 0) - x(\tau_i - 0) \\ &= C_i x(\tau_i - 0) + p_i(x(\tau_i - 0), h(\tau_i - 0, x(\tau_i - 0))), \\ \Delta y\big|_{t=\tau_i} &= y(\tau_i + 0) - y(\tau_i - 0) \\ &= D_i y(\tau_i - 0) + q_{i0}(y(\tau_i - 0)). \end{aligned} \tag{5}$$

This system can be divided into two parts, of which the first does not contain the variable y, while the second is independent of x.

Let $\Psi(\cdot, t_0, x_0, y_0) = (x_0(\cdot, t_0, x_0), y_0(\cdot, t_0, y_0)): \mathbb{R} \to \mathbf{X} \times \mathbf{Y}$ be a solution of system (5), where $\Psi(t_0 + 0, t_0, x_0, y_0) = (x_0, y_0)$. For simplicity, we use the notation $\Psi(t) = (x_0(t), y_0(t))$.

Theorem 2. *If* $4\varepsilon \max\{\nu_1, \nu_2\} < 1$ *and* $2\varepsilon \max\{\mu, \sup_i |(id_x + C_i)^{-1}|\} < 1 + \sqrt{1 - 4\varepsilon\nu_1}$, *then systems* (1) *and* (5) *are dynamically equivalent in the large.*

References

1. V.D. Mil'man and A.D. Myshkis, On the stability of motion in the presence of impulses, *Sibirsk. Mat. Zh.* **1** (1960), 233–237 (Russian).
2. V. Lakshmikantham, D.D. Bainov and P.S. Simeonov, *Theory of impulsive differential equations*, World Scientific, River Edge, N.J., 1989.
3. A.M. Samoilenko and N.A. Perestyuk, *Differential equations with impulse effect*, Vishcha Shkola, Kiev, 1987 (Russian).
4. D.M. Grobman, Topological classification of neighbourhoods of a singularity in n-space, *Mat. Sb.* **56** (1962), 77–94 (Russian).
5. P. Hartman, *Ordinary differential equations*, Birkhäuser, Boston, 1982.
6. U. Kirchgraber and K.J. Palmer, *Geometry in the neighbourhood of invariant manifolds of maps and flows and linearization*, Pitman Res. Notes Math. Ser. **233**, Longman, New York, 1991.

7. A. Reinfelds, A reduction theorem, *Differential Equations* **10** (1974), 645–649.

8. A. Reinfelds, The reduction theorem for closed trajectories, *Differential Equations* **11** (1975), 1353–1358.

9. A. Reinfelds, A reduction theorem for extensions of dynamical systems, *Latv. Mat. Ezhegodnik* **33** (1989), 67–75 (Russian).

10. A. Reinfelds, The reduction principle for discrete dynamical and semidynamical systems in metric spaces, *Z. Angew. Math. Phys.* **45** (1994), 933–955.

11. A. Reinfelds, The reduction of discrete dynamical and semidynamical systems in metric spaces, in B. Aulbach and F. Colonius (eds.), *Six lectures on dynamical systems*, World Scientific, River Edge, New Jersey, 1996, 267–312.

12. A.N. Shoshitaishvili, The bifurcation of the topological type of singular points of vector fields that depend on parameters, *Trudy Sem. Petrovsk.* **1** (1975), 279–309 (Russian).

13. A. Reinfelds and L. Sermone, Equivalence of differential equations with impulse action, *Latv. Univ. Zinātn. Raksti* **553** (1990), 124–130 (Russian).

14. A. Reinfelds and L. Sermone, Equivalence of nonlinear differential equations with impulse effect in Banach space, *Latv. Univ. Zinātn. Raksti* **577** (1992), 68–73.

15. A. Reinfelds, A reduction theorem for systems of differential equations with impulse effect in a Banach space, *J. Math. Anal. Appl.* **203** (1996), 187–210.

16. L. Sermone, Equivalence of linear differential equations with impulse effect, *Proc. Latvian Acad. Sci. B* **1994**, no. 2(559), 78–80.

17. D.D. Bainov, S.I. Kostadinov and Nguyen Van Minh, *Dichotomies and integral manifolds of impulsive differential equations*, World Scientific, River Edge, N.J., 1994.

18. D.D. Bainov, S.I. Kostadinov, Nguyen Van Minh, Nguyen Hong Thai and P.P. Zabreiko, Integral manifolds of impulsive differential equations, *J. Appl. Math. Stochastic Anal.* **5** (1992), 99–110.

Institute of Mathematics of Latvian Academy of Sciences and University of Latvia, Akadēmijas Laukums 1, LV–1524 Rīga, Latvia

E-mail: reinf@cclu.lv

R. RIKARDS, E. BARKANOV, A. CHATE and A. KORJAKIN

Modelling, damping analysis and optimisation of sandwich and laminated composite structures

1. Introduction

Sandwich and laminated composite structures are widely used in structural applications requiring high stiffness-to-weight and strength-to-weight ratios. For this reason they are applied in various fields of engineering: aerospace technology, shipbuilding, automotive industry, mechanical engineering etc. The ability to absorb the vibration energy is also a very well-known quality of sandwich and laminated composite structures. Damping in the viscoelastic layers is taken into account using the complex modulus of elasticity [1]. Due to the increase in the applications of high damped structures, significant progress has recently been achieved in the analysis and optimisation of the damping properties of such structures. The present investigations are devoted to the damping analysis and optimisation of sandwich and laminated composite structures by various finite elements and methods of dynamic analysis.

2. Finite element models

In the present investigation, the finite element method has been used. For the numerical analysis of the examined structures, the finite element models of sandwich and laminated composite beams (see [2] and [3]), plates ([4] and [5]) and shells ([6] and [7]) have been developed. The distinctive feature of these sandwich finite element models is the use of the Mindlin-Reissner hypothesis for each layer of sandwich beam, plate and shell (Fig. 1).

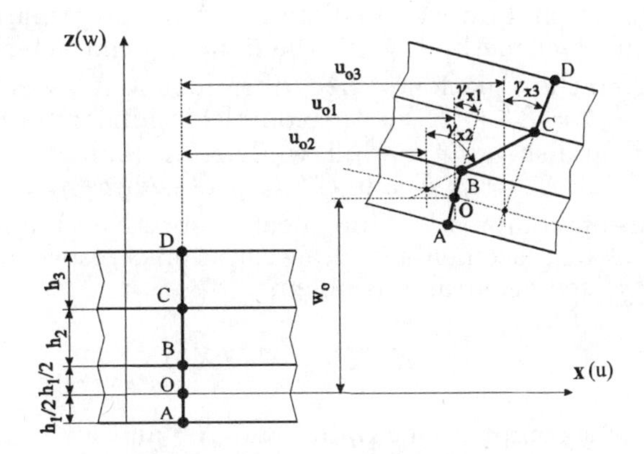

Fig. 1. Kinematic assumptions for the sandwich plate in the ZX-plane.

This work was supported by grant N 93415 from the Latvian Council of Science. The presentation at the Conference IMSE96 was supported by the Soros Foundation-Latvia and by the Division of Mathematics and the Department of Mathematical Sciences (University of Oulu).

For laminated composite models this hypothesis is already used for the entire laminate (Fig. 2). In this case, transverse shear stiffness is obtained by means of shear correction factors.

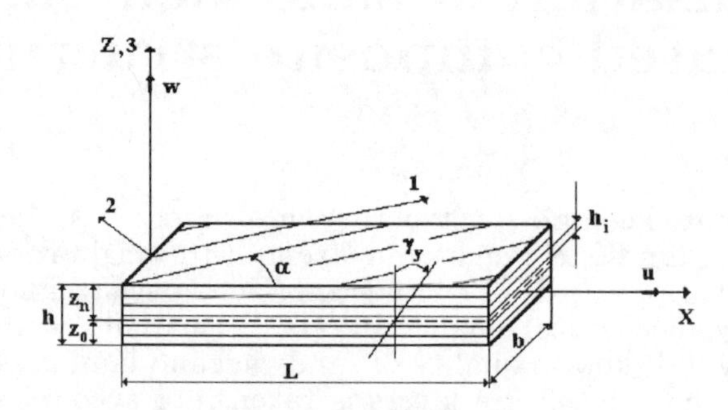

Fig. 2. The laminated composite beam.

3. Methods of damping analysis

The dynamic analysis of sandwich and laminated composite structures with viscoelastic damping is carried out in respect of free vibration analysis and frequency response analysis. As a result, the damped eigenfrequencies, corresponding loss factors and displacements have been obtained.

3.1. Free vibration analysis

The main object of this analysis is the calculation of damping in harmonic oscillations at frequencies close to the natural vibrations of the structure. Depending on the form of the discretised equations of motion, the free vibration analysis of damped structures can be performed by one of three methods: complex eigenvalues, potential energy of eigenmodes, and direct calculation of frequency characteristics.

The first is an exact method, when the damping characteristics of the structure are determined from the equation of free vibrations $M\ddot{X}^* + K^*X^* = 0$, where M is the mass matrix, $K^* = K' + iK''$ is the complex stiffness matrix (K' is determined from the elastic E' and shear G' moduli, while K'' is found from the imaginary parts of the complex moduli $E'' = \eta_E E'$ and $G'' = \eta_G G'$, where η_E and η_G are the material loss factors in tension-compression and shear, respectively), and X^* and $\ddot{X}^*$ are the complex vectors of displacements and accelerations, respectively. Transformation leads to the generalised eigenvalue problem

$$K^*\bar{X}^* = \lambda^* M\bar{X}^*, \tag{1}$$

where $\lambda^* = \omega^{*2}$ is a complex eigenvalue and $\bar{X}^*$ and $\omega^* = \omega' + i\omega''$ are the corresponding complex eigenvector and complex eigenfrequency, respectively. The real part ω' represents the damped frequency of the natural vibration of the structure and the imaginary part ω'' determines the rate of decay of the process in time. The solution of the generalised eigenvalue problem (1) is complicated since K^* is a complex symmetric matrix where the permutation law on the Hermitian components does not hold. The matrix M is real and symmetric; it is necessary to determine only the lowest eigenvalues, that is, to solve a partial eigenvalue problem; the generalised eigenvalue problem has a large dimension. These difficulties can be surmounted by the Lanczos algorithm proposed in [8]. The Lanczos method is programmed in a truncated version, where the generalised eigenvalue problem is transformed into a standard eigenvalue problem with a reduced-order symmetric tridiagonal matrix. Orthogonal projection operations are employed with greater economy and elegance using elementary reflection matrices. The program is flexible enough to allow the user to define any storage model that particularly suits the sparsity pattern of each problem. As a result, the complex eigenvalues and eigenfrequencies for the damped structure are obtained in this case. The loss factor for each mode can be determined from the relation $\lambda = \lambda''/\lambda' = 2\tan^{-1}(\omega''/\omega')$; for structures with small damping it is possible to re-write this expression as $\lambda \approx 2\omega''/\omega'$. This approach makes it possible to preserve the exact mathematical formulation for the examined damping model and to calculate structures with high damping.

The second method is approximate, since in this approach it is assumed that the damping characteristics of the structure can be calculated by means of the equation of natural vibrations of the corresponding undamped structure, namely

$$M\ddot{X} + KX = 0.$$

Transformation yields the generalised eigenvalue problem

$$K\bar{X} = \lambda M\bar{X}. \tag{2}$$

The eigenvalue problem (2) is solved by the subspace iteration method [9]. The loss factors of the structure in this case are calculated as the ratio of dissipated and elastic strain energy for one cycle of steady state vibrations, that is, $\eta = (\bar{X}^t K''\bar{X})/(\bar{X}^t K'\bar{X})$. It is necessary to note that this approach can be used only for structures made from viscoelastic materials with small loss factors since in this case the difference between the natural modes of damped and undamped structures is negligible.

The third method allows the frequency-dependent damping model to be taken into account. This is very important since the dynamic modulus and loss factors of real viscoelastic materials depend, to a significant extent, on frequency. The damping characteristics of the structure are determined in this case from the equation of free vibrations

$$M\ddot{X}^* + K^*(\omega)X^* = 0.$$

Transformation leads to the non-linear generalised eigenvalue problem

$$K^*(\omega)\bar{X}^* = \lambda^* M\bar{X}^*. \tag{3}$$

The calculations start with a constant frequency ($\omega = \text{const}$). Then at each step the linear ($K^*(\omega) = \text{const}$) generalised eigenvalue problem (3) is solved by the Lanczos method. The process terminates when the condition

$$\frac{|\omega'_{i+1} - \omega'_i|}{\omega'_i} * 100\% \leq \xi$$

is satisfied; here ω'_{i+1} is the eigenfrequency of the structure, ω'_i is the frequency from the equation describing the loss factor of the viscoelastic material, and ξ is the desired precision. Only a few iterations are needed to obtain the final results.

3.2. Frequency response analysis

The frequency response is obtained by means of the Gauss algorithm [9] from the system of complex linear equations

$$[K^* - \omega^2 M]\bar{X}^* = \bar{F},$$

where $\bar{F}$ is the amplitude of applied force. This method also allows the frequency-dependent damping model to be taken into account. However, in this case it is necessary to recalculate the complex stiffness matrix $K^*(\omega)$ for each of the frequencies. This does not recommend it for use in the design of complicated structures with frequency-dependent viscoelastic damping in a wide frequency range.

It is well known that the dynamic characteristics of the structure (eigenfrequencies and corresponding loss factors) can easily be obtained using the results of the frequency response analysis. The eigenfrequencies ω_i of the structure represent the points of the real part of the response spectrum where the displacements are zero, but the corresponding loss factors can be determined from the analysis of the resonant peaks for a particular mode [10].

This method involves a considerable amount of computing time, since the general solution requires that the dynamic stiffness matrix $[K^* - \omega^2 M]$ be re-calculated, decomposed and stored at each of the many frequency steps. For this reason, for a system with a large number of degrees of freedom and for a large number of calculated dynamic characteristics it is more effective to use the results of the free vibration analysis. The frequency response analysis can be successfully applied in the case when it is necessary to determine only a small number of desired dynamic characteristics or the eigenfrequency of the undamped structure is already known and it is necessary to specify it for the damped structure and to calculate the corresponding loss factor.

4. Numerical optimisation

The direct optimisation methods used at present are ineffective, as they require a large volume of calculations to obtain the dynamic characteristics of structures with damping. Therefore, the present investigation is devoted to the optimal design procedure based on the planning of experiments (see [11] and [12]). This engineering approach of optimisation consists of three fundamental stages: the carrying out of numerical calculations by an informative plan of experiments, the approximation of the results obtained, and the solution of the optimisation problem by the method of non-linear programming. For each of these stages, it is possible to solve a problem by various methods.

The constrained non-linear optimisation problem can be written in the form

$$\min F(x); \quad H_i(x) \geq 0; \quad G_j(x) = 0; \quad i = 1, 2, \ldots, I; \; j = 1, 2, \ldots, J,$$

where I and J are the numbers of inequality and equality constraints, respectively. This problem is replaced by an unconstrained minimisation problem in which the constraints are taken into account with penalty functions. A new version of the random search method is used to solve the optimisation problem.

5. Numerical examples

For sandwich and laminated composite beams, plates and shells, the damping analysis for a wide range of material and geometric parameters was carried out in [2]–[8].

Numerical optimisation results for maximum damping and minimum weight of a sandwich and laminated composite structure, can be found in [11] and [12].

As an example, the effect of the ply angle on the frequency and loss factor is examined for a cylindrical shell under clamped end conditions with the material properties

$$E_1 = 37.88\,\text{GPa}, \quad E_2 = E_3 = 10.9\,\text{Gpa},$$
$$G_{12} = G_{13} = G_{23} = 4.91\,\text{GPa}, \quad \nu_{12} = \nu_{13} = 0.3,$$
$$\rho = 1813.9\,\text{kg/m}^3, \quad \psi_1 = 0.87\%, \quad \psi_2 = 5.05\%, \quad \psi_{12} = \psi_{13} = \psi_{23} = 6.91\%.$$

The shell radius is $R = 0.22\,\text{m}$, its thickness is $t = 0.00205\,\text{m}$ and its length is $L = 0.227\,\text{m}$. The stacking sequence of layers is assumed to be $[\beta, -\beta, \beta, -\beta]_S$. The variation of the eigenfrequencies and the corresponding loss factors with respect to the ply angle are shown in Fig. 3.

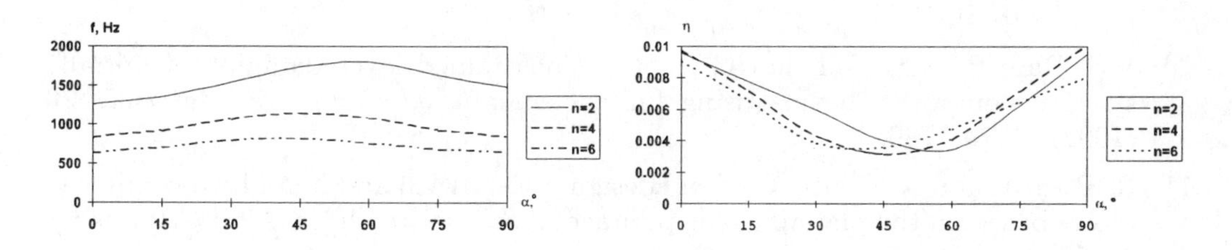

Fig. 3. Dependence of the eigenfrequencies and corresponding loss factors of the cylindrical laminated composite shell from the angle of layer.

6. Conclusion

The same results have been obtained by means of different finite element models and methods of damping analysis for various types of sandwich and laminated composite structures. Recommendations for a more effective use of the different methods of dynamic analysis of structures with viscoelastic damping have been made.

References

1. R.M. Christensen, *Theory of viscoelasticity. An introduction*, Academic Press, New York, 1971.
2. R. Rikards, A. Chate and E. Barkanov, Finite element analysis of damping the vibrations of laminated composites, *Comput. & Structures* **47** (1993), 1005–1015.
3. E. Barkanov and J. Gassan, Frequency response analysis of laminated composite beams, *Mech. Composite Materials* **30** (1994), 664–674.

4. A. Chate, R. Rikards, K. Mäkinen and K.-A. Olsson, Free vibration analysis of sandwich plates on flexible supports, *Mech. Composite Materials Structures* **2** (1995), 1–18.

5. R. Rikards, A. Chate and A. Korjakin, Vibration and damping analysis of laminated composite plates by the finite element method, *Engrg. Comput.* **12** (1995), 61–74.

6. A. Chate, R. Rikards and A. Korjakin, Damped free vibrations of sandwich cylindrical shells, *Proc. Internat. Conf. on Education, Practice and Promotion of Computational Methods in Engineering Using Small Computers, Macao, 1995*, vol. 2, 1185–1190.

7. A. Chate, R. Rikards and A. Korjakin, Analysis of free damped vibrations of laminated composite cylindrical shells, *Mech. Composite Materials* **31** (1995), 646–659.

8. E.N. Barkanov, Method of complex eigenvalues for studying the damping properties of sandwich type structures, *Mech. Composite Materials* **29** (1993), 90–94.

9. K.-J. Bathe and E.L. Wilson, *Numerical methods in finite element analysis*, Prentice-Hall, 1976.

10. V.S. Rao, B.V. Sankar and C.T. Sun, Constrained layer damping of initially stressed composite beams using finite elements, *J. Composite Materials* **26** (1992), 1752–1766.

11. R. Rikards and A. Chate, Optimal design of sandwich and laminated composite plates based on the planning of experiments. *Structural Optimisation* **10** (1995), 46–53.

12. E. Barkanov, R. Rikards and A. Chate, Numerical optimisation of sandwich and laminated composite structures, *Proc. Fourth Internat. Conf. on Computer Aided Optimum Design of Structures, Miami, 1995*, 311–318.

Institute of Computer Analysis of Structures, Riga Technical University, Kalku St. 1, 1658 Riga, Latvia

O. ROMAN

On the maximal symmetry group and exact solutions of non-linear higher order wave equations

A scalar uncharged particle in quantum field theory is known to be described by the wave equation

$$\Box u = F(u), \tag{1}$$

where $\Box$ is the d'Alembertian in 4-dimensional space-time, u is a real function, and $F(u)$ is an arbitrary smooth function. Recently a number of works (see, for example, [1]–[3]) have appeared concerning some generalisations of (1), specifically, higher order non-linear equations. The most appropriate technique for investigating these equations is the symmetry method (see [1], [4] and [5]). Here we analyse the higher order wave equation

$$\Box^l u = F(u), \quad l \in \mathbb{N},\ l > 1, \quad \Box^l = \Box(\Box^{l-1}) \tag{2}$$

($\Box = \partial^2_{x_0} - \partial^2_{x_1} - \ldots - \partial^2_{x_n}$ is the d'Alembertian in the $(n+1)$-dimensional pseudo-Euclidean space $\mathbb{R}(1,n)$ with metric tensor $g_{\mu\nu} = \mathrm{diag}(1,-1,\ldots,-1)$, $\mu,\nu = \overline{0,n}$) and show that the symmetry properties of (2) are virtually the same as those of the wave equation (1). But in special cases (2) has a more remarkable symmetry than (1). For example, unlike (2), the wave equation (1) with $F(u) = \lambda e^u$ is not invariant under the conformal group (see [1]).

A symmetry classification of (2) has been carried out in [6]. Here we mention some important statements.

Theorem 1. *The maximal symmetry group of* (2) *with an arbitrary smooth function $F(u)$ is the Poincaré group $P(1,n)$ generated by the operators*

$$P_\mu = \partial_{x_\mu}, \quad J_{\mu\nu} = x^\mu \partial_{x_\nu} - x^\nu \partial_{x_\mu}, \tag{3}$$

where rising or lowering of the vector indices is performed by means of the metric tensor $g_{\mu\nu}$ (for example, $x^\mu = g^{\mu\nu} x_\nu$) and covariant summation from 0 to n over repeated indices is understood.

Theorem 2. *Equation* (2) *admits the extended Poincaré group $\widetilde{P}(1,n)$ generated by the operators $P_\mu, J_{\mu\nu}$ of the form* (3) *and the dilation operator D if and only if it is equivalent to either*

$$\Box^l u = \lambda u^k, \quad k \neq 1, \quad D = x_\mu \partial_{x_\mu} + 2l(1-k)^{-1} u \partial_u, \tag{4}$$

This work was supported in part by the grant PSU 061088 from the International Soros Science Education Programme.

or

$$\Box^l u = \lambda e^u, \quad D = x_\mu \partial_{x_\mu} - 2l\partial_u. \tag{5}$$

$\widetilde{P}(1,n)$ is the maximal symmetry group of the non-linear equations (4) and (5).

Theorem 3. *Equation (2) admits the conformal group $C(1,n)$ generated by the operators P_μ and $J_{\mu\nu}$ of the form (3), the dilation operator D and the conformal ones if and only if it is equivalent to either*

$$\begin{gathered} \Box^l u = \lambda u^{(n+1+2l)/(n+1-2l)}, \quad n+1 \neq 2l, \\ D = x_\mu \partial_{x_\mu} + \tfrac{1}{2}(2l-n-1)u\partial_u, \quad K_\mu = 2x^\mu D - (x_\nu x^\nu)\partial_{x_\mu}, \end{gathered} \tag{6}$$

or

$$\Box^l u = \lambda e^u, \quad n+1 = 2l, \quad D = x_\mu \partial_{x_\mu} - 2l\partial_u, \quad K_\mu = 2x^\mu D - (x_\nu x^\nu)\partial_{x_\mu}. \tag{7}$$

$C(1,n)$ is the maximal symmetry group of the non-linear equations (6), (7).

The proofs of Theorems 1–3 are based on the Lie infinitesimal algorithm (see [1], [4] and [5]). They require rather cumbersome computations and for this reason they are omitted. The proofs of these theorems for the biwave equations (when $l = 2$) are given in [7].

It should be noted that the symmetry of (4)–(6) has been established in [1] by means of the Baker-Campbell-Hausdorff formula, and a symmetry classification of wave equation (1) has been carried out in [8].

It follows from Theorem 3 that, if $n+1 = 2l$, then the linear wave equation

$$\Box^l u = 0$$

admits two non-equivalent representations of the conformal algebra $C(1,n)$, namely $\langle P_\mu^{(1)}, J_{\mu\nu}^{(1)}, D^{(1)}, K_\mu^{(1)} \rangle$, where

$$\begin{aligned} P_\mu^{(1)} &= P_\mu = \partial_{x_\mu}, & J_{\mu\nu}^{(1)} &= J_{\mu\nu} = x^\mu \partial_{x_\nu} - x^\nu \partial_{x_\mu}, \\ D^{(1)} &= x_\mu \partial_{x_\mu}, & K_\mu^{(1)} &= 2x^\mu D^{(1)} - (x_\nu x^\nu)\partial_{x_\mu}, \end{aligned}$$

and $\langle P_\mu^{(2)}, J_{\mu\nu}^{(2)}, D^{(2)}, K_\mu^{(2)} \rangle$, where

$$\begin{aligned} P_\mu^{(2)} &= P_\mu = \partial_{x_\mu}, & J_{\mu\nu}^{(2)} &= J_{\mu\nu} = x^\mu \partial_{x_\nu} - x^\nu \partial_{x_\mu}, \\ D^{(2)} &= x_\mu \partial_{x_\mu} + \partial_u, & K_\mu^{(2)} &= 2x^\mu D^{(2)} - (x_\nu x^\nu)\partial_{x_\mu}. \end{aligned}$$

One of the important applications of the Lie groups in mathematical physics is finding the exact solution of differential equations. Here we restrict our attention to the non-linear multi-dimensional biwave equation

$$\Box^2 u = \lambda u^k, \quad n > 1. \tag{8}$$

It follows from Theorem 1 that (8) is invariant under the Poincaré group $P(1,n)$ generated by the operators (3). Therefore, its exact solutions can be found in the form [1]

$$u = \varphi(\omega_1, \omega_2), \tag{9}$$

where $\omega_1 = \omega_1(x)$, $\omega_2 = \omega_2(x)$. If ω_1 and ω_2 are invariants of some sub-algebra of the Poincaré algebra $P(1,n)$, then (9) reduces (8) to an equation with two independent variables. We consider only two cases.

1. Let $\omega_1 = \alpha x \equiv \alpha_\mu x^\mu$ and $\omega_2 = \beta x$. If

$$\alpha^2 = 1, \quad \beta^2 = -1, \quad \alpha\beta = 0, \tag{10}$$

then (9) reduces (8) to the one-dimensional biwave equation

$$\left(\frac{\partial^2}{\partial\omega_1^2} - \frac{\partial^2}{\partial\omega_2^2}\right)^2 \varphi = \lambda\varphi^k. \tag{11}$$

Making use of the exact solutions of (11) given in [7], we obtain for (8) the solutions

$$u = \left(\frac{64\,(k+1)^2}{\lambda\,(k-1)^4}\right)^{1/(k-1)} [(\alpha x + \beta x + c_1)(\alpha x - \beta x + c_2)]^{-2/(k-1)}, \quad k \neq \pm 1,$$

$$u = \left(\frac{8}{\lambda}\frac{(k+1)(k+3)(3k+1)}{(k-1)^4}\right)^{1/(k-1)} (\alpha x + c_3)^{4/(1-k)}, \quad k \neq \pm 1, -3, -\tfrac{1}{3},$$

$$u = \left(\frac{8}{\lambda}\frac{(k+1)(k+3)(3k+1)}{(k-1)^4}\right)^{1/(k-1)} (\beta x + c_4)^{4/(1-k)}, \quad k \neq \pm 1, -3, -\tfrac{1}{3},$$

where c_1, c_2, c_3, c_4 are arbitrary constants and the parameters α and β satisfy (10).

2. Let $\omega_1 = \alpha x$ and $\omega_2 = \beta x + (\alpha x)(\gamma x)$. If

$$\alpha^2 = \alpha\beta = \alpha\gamma = \beta\gamma = 0, \quad \beta^2 = -1, \quad \gamma^2 = -1, \tag{12}$$

then (9) reduces (8) to the ordinary differential equation

$$\frac{\partial^4\varphi}{\partial\omega_2^4}(1+\omega_1^2)^2 = \lambda\varphi^k. \tag{13}$$

The solutions of (13) can be found in the form

$$\varphi = a(\omega_1)[\omega_2 + c(\omega_1)]^m, \tag{14}$$

where the functions $a(\omega_1)$, $c(\omega_1)$ and m are determined after substitution of (14) into (13). In this case, we obtain for (8) the exact solutions

$$u = \left(\frac{8(k+1)(k+3)(3k+1)}{(1-k)^4}\right)^{1/(k-1)} [1+(\alpha x)^2]^{2/(k-1)} \times [\beta x + (\gamma x)(\alpha x) + c(\alpha x)]^{4/(1-k)},$$

where $c(\cdot)$ is an arbitrary function and parameters α, β and γ satisfy (12).

To generalise the solutions obtained, we consider for (8) the two solution forms

$$u = [f(\omega_1) + g(\omega_2)]^m, \quad \omega_1 = \alpha x, \; \omega_2 = \beta x, \tag{15}$$

and

$$u = [f(\omega, \omega_1) + g(\omega, \omega_2)]^m, \quad \omega = \alpha x, \quad \omega_1 = \beta x, \quad \omega_2 = \gamma x. \tag{16}$$

Substitution of (15) or (16) into (8) gives us some conditions on α, β, γ, m and the functions f and g. As they are complicated, we do not write them here. We only state that if

$$\alpha^2 = \alpha\beta = 0, \quad \beta^2 = -1, \tag{17}$$

then (15) leads us to the exact solution of (8)

$$u = \left(\frac{8}{\lambda}\frac{(k+1)(k+3)(3k+1)}{(k-1)^4}\right)^{1/(k-1)} [f(\alpha x) + \beta x]^{4/(1-k)}, \quad k \neq \pm 1, -3, -\tfrac{1}{3},$$

where the parameters α and β satisfy (17) and $f(\cdot)$ is an arbitrary function; on the other hand, if

$$\alpha^2 = \alpha\beta = \alpha\gamma = \beta\gamma = 0, \quad \beta^2 = -1, \quad \gamma^2 = -1, \tag{18}$$

then (16) leads us to the solutions of (8)

$$u = [\beta x a(\alpha x) + \gamma x b(\alpha x)]^{4/(1-k)},$$

where α, β and γ satisfy (18) and $a(\cdot)$ and $b(\cdot)$ satisfy

$$a^2(\omega) + b^2(\omega) = \left(\frac{\lambda(k-1)^4}{8(k+1)(k+3)(3k+1)}\right)^{1/2}.$$

In conclusion, we remark that wave equations have analogous solutions ([1], [8]).

References

1. W.I. Fushchych, W.M. Shtelen and N.I. Serov, *Symmetry analysis and exact solutions of equations of nonlinear mathematical physics*, Kluwer, Dordrecht, 1993.

2. W.I. Fushchych, Symmetry in problems of mathematical physics, in *Algebraic–Theoretic Studies in Mathematical Physics*, Kiev Inst. Math., 1981, 6–28 (Russian).

3. C.G. Bollini and J.J. Giambia, Arbitrary powers of d'Alembertians and the Huygens' principle, *J. Math. Phys.* **34** (1993), 610–621.

4. P. Olver, *Applications of Lie groups to differential equations*, Springer, New York, 1986.

5. L.V. Ovsyannikov, *Group analysis of differential equations*, Nauka, Moscow, 1978 (Russian).

6. W.I. Fushchych, O.V. Roman, and R.Z. Zhdanov, Symmetry and some exact solutions of nonlinear polywave equations, *Europhys. Lett.* **31** (1995), no. 2, 75–79.

7. W.I. Fushchych, O.V. Roman and R.Z. Zhdanov, Symmetry reduction and some exact solutions of nonlinear biwave equations, *Rep. Math. Phys.* **37** (1996), 267–281.

8. W.I. Fushchych and N.I. Serov, The symmetry and some exact solutions of the nonlinear many–dimensional Liouville, d'Alembert and eikonal equations, *J. Phys. A* **16** (1983), 3645–3658.

Institute of Mathematics, 3 Tereshchenkivska St., 252004 Kyiv, Ukraine

E-mail: olena@apmat.freenet.kiev.ua

V.Al. SAVA

An initial boundary value problem for the equations of plane flow of a micropolar fluid in a time-dependent domain

1. Introduction

In this note we consider an initial boundary value problem for the equations of plane flow of a micropolar fluid, assuming that the domain in which the motion of the fluid takes place depends on the time t.

A theory of micropolar fluids was introduced by ERINGEN in [1]. These fluids differ from non-Newtonian fluids in that they exhibit microinertial effects and can support couple stresses and body couples.

To study the mixed problem for the flow equations, we consider it in weak form.

2. Basic equations

Throughout this note a rectangular coordinate system (x_1, x_2) is employed. The indices denoted by small Greek letters take the values 1, 2.

Let Ω_t be an open bounded set in the (x_1, x_2)-plane depending on t, and let $\partial\Omega_t$ be its boundary, which we assume to consist of the lines

$$\begin{aligned} \partial\Omega_{t1} &= \{(x_1, x_2),\ x_1 = 0,\ k_1' \leq x_2 \leq k_2'\}, \\ \partial\Omega_{t2} &= \{(x_1, x_2),\ x_1 = L,\ k_1'' \leq x_2 \leq k_2''\}, \\ \partial\Omega_{t3} &= \{(x_1, x_2),\ 0 \leq x_1 \leq L,\ x_2 = \eta_\alpha(x_1, t)\}, \\ \eta_\alpha(0, t) &= k_\alpha', \quad \eta_\alpha(L, T) = k_\alpha''. \end{aligned}$$

The equations of plane motion in the region Ω_t of an incompressible micropolar fluid are

$$\begin{aligned} &v_{\alpha,\alpha} = 0, \\ &\dot{v}_\alpha - (\mu + k)\nabla^2 v_\alpha + k\varepsilon_{\beta\alpha}\varphi_{,\beta} + v_\beta v_{\alpha,\beta} + p_{,\alpha} = f_\alpha, \\ &j\dot{\varphi} - \gamma\nabla^2\varphi + k(2\varphi + \varepsilon_{\beta\alpha}v_{\beta,\alpha}) + jv_\alpha\varphi_{,\alpha} = l, \end{aligned} \tag{1}$$

where $\mathbf{v}(\mathbf{x}, t) = (v_1(\mathbf{x}, t), v_2(\mathbf{x}, t))$ $(\mathbf{x} = (x_1, x_2))$, $\varphi(\mathbf{x}, t)$ denote respectively the velocity vector and the microrotation or spin, $p(\mathbf{x}, t)$ is the pressure, $\mathbf{f}(\mathbf{x}, t) = (f_1(\mathbf{x}, t), f_2(\mathbf{x}, t))$ the body force, $l(\mathbf{x}, t)$ the couple and j the micro-inertia, μ, k are the viscosity coefficients, γ is the gyroviscosity, $\varepsilon_{\alpha\beta}$ is the alternating symbol, a comma denotes partial derivation with respect to the variables x_α and a dot denotes differentiation with respect to time t. The Clausius-Duhem inequality implies that k, μ, γ are non-negative [1].

To (1) we append the initial conditions

$$\mathbf{v}(\mathbf{x}, 0) = \mathbf{v}_0(\mathbf{x}), \quad \varphi(\mathbf{x}, 0) = \varphi_0, \quad \mathbf{x} \in \Omega_0, \tag{2}$$

and the boundary conditions

$$\begin{aligned}
\tfrac{1}{2}[v_1^2(\mathbf{x},t)+j\varphi^2(\mathbf{x},t)]+p(\mathbf{x},t)=h_\alpha(\mathbf{x},t), \quad \mathbf{x}\in\partial\Omega_{t\alpha}, \quad t\in(0,T),\\
p(\mathbf{x},t)=r(\mathbf{x},t)\mid \mathbf{v}(\mathbf{x},t)\cdot\mathbf{n}(\mathbf{x},t)\mid^2, \quad \mathbf{x}\in\partial\Omega_{t3}, \quad t\in(0,T),\\
\varphi_n(\mathbf{x},t)\equiv\partial\varphi/\partial n=0, \quad \mathbf{x}\in\partial\Omega_t, \quad t\in(0,T),
\end{aligned} \tag{3}$$

where $\mathbf{n}$ is the unit outward normal to $\partial\Omega_t$.

To study the problem (1)–(3), we consider it in weak form.

Let Ω be an open set of the $\mathbf{x}=(x_1,x_2)$-plane satisfying the cone property. We denote by $C^\infty(\Omega)$ the space of functions $\mathbf{u}(\mathbf{x})$ infinitely differentiable in Ω. We also denote by $\mathbf{K}(\Omega)$ the vector functions $\mathbf{v}(\mathbf{x})=(v_1(\mathbf{x}),v_2(\mathbf{x}))$ such that $v_\alpha(\mathbf{x})\in C^\infty(\Omega)$, $\mid\mathbf{v}(x)\mid=\mid\mathbf{v}(\mathbf{x})\cdot\mathbf{n}(\mathbf{x})\mid$ when $\mathbf{x}\in\partial\Omega$ and with null divergence in Ω. Next, $C_0^\infty(\Omega)$ is the space of functions $\varphi(\mathbf{x})\in C^\infty(\Omega)$ such that $\varphi_n(\mathbf{x})=0$ when $\mathbf{x}\in\partial\Omega$, and by $\mathbf{K}^\sigma(\Omega)$ and $H_0^\sigma(\Omega)$ are the closures of $\mathbf{K}(\Omega)$ and $C_0^\infty(\Omega)$, respectively, in the Sobolev space $H^\sigma(\Omega)$.

Let $D(A)$ be the set of elements $\mathbf{u}\in\mathbf{K}^1(\Omega)$ such that the linear form $\mathbf{u}\to\langle\mathbf{v},\mathbf{u}\rangle_{\mathbf{K}^1(\Omega)}$ is continuous in the topology of $\mathbf{K}^0(\Omega)$; then it is possible to define a linear, self-adjoint, positive operator A from $D(A)$ to $\mathbf{K}^0(\Omega)$ such that

$$\langle\mathbf{v},\mathbf{u}\rangle_{\mathbf{K}^1(\Omega)}=\langle A\mathbf{v},\mathbf{u}\rangle_{\mathbf{K}^0(\Omega)} \quad \forall\mathbf{v}\in D(A), \quad \mathbf{u}\in\mathbf{K}^1(\Omega).$$

In a similar way it is possible to define an operator B from $D(B)\subset H_0^1(\Omega)$ to $H_0^0(\Omega)$ such that

$$\langle\varphi,\psi\rangle_{H_0^1(\Omega)}=\langle B\varphi,\,\varphi\rangle_{H_0^0(\Omega)}, \quad \forall\varphi\in D(B), \quad \psi\in H_0^1(\Omega).$$

We denote by A^σ and B^σ $(\sigma\geq 0)$ the power σ of A and B, respectively, and by $V_\sigma(\Omega)=D(A^{\sigma/2})$ and $W_\sigma(\Omega)=D(B^{\sigma/2})$ the domain of $A^{\sigma/2}$ and $B^{\sigma/2}$. $V_\sigma(\Omega)$ and $W_\sigma(\Omega)$ are Hilbert spaces with inner products

$$\begin{aligned}
\langle\mathbf{v},\mathbf{u}\rangle_{V_\sigma(\Omega)}&=\langle A^{\sigma/2}\mathbf{v},A^{\sigma/2}\mathbf{u}\rangle_{\mathbf{K}^0(\Omega)},\\
\langle\varphi,\psi\rangle_{W_\sigma(\Omega)}&=\langle B^{\sigma/2}\varphi,B^{\sigma/2}\psi\rangle_{\mathbf{H}_0^0(\Omega)}.
\end{aligned}$$

We set

$$a(t,\psi,\mathbf{v})=\int_{\Omega_t}\psi(\mathbf{x})\varepsilon_{\alpha,\beta}v_{\alpha,\beta}(\mathbf{x})d\mathbf{x},$$

$$b(t,\mathbf{v},\mathbf{u},\mathbf{w})=\int_{\Omega_t}v_\alpha(\mathbf{x})u_{\beta,\alpha}(\mathbf{x})w_\beta(\mathbf{x})d\mathbf{x},$$

and assume that $\mathbf{f}(t)=\{\mathbf{f}(\mathbf{x},t),\ \mathbf{x}\in\Omega_t\}\in L^2(0,T;V_{\sigma-1}(\Omega_t))$,

$$\begin{aligned}
l(t)=\{l(\mathbf{x},t),\ \mathbf{x}\in\Omega_t\}\in L^2(0,T;W_{\sigma-1}(\Omega_t)),\\
h_a(t)=\{h_\alpha(\mathbf{x},t),\ \mathbf{x}\in\partial\Omega_{t\alpha}\}\in L^2(0,T;L^2(\partial\Omega_{t\alpha})), \quad \alpha=1,2,\\
g(t)=\{g(\mathbf{x},t),\ \mathbf{x}\in\partial\Omega_{t3}\}\in L^\infty(0,T;L^\infty(\partial\Omega_{t3})), \quad \eta_\alpha\in C^1.
\end{aligned}$$

Definition. *A pair* $(\mathbf{v}(t), \varphi(t)) = \{(\mathbf{v}(\mathbf{x},t),\ \varphi(\mathbf{x},t)),\ \mathbf{x} \in \Omega_t\}$ *is a solution of equation* (1) *satisfying the boundary conditions* (3) *if*

a) $(\mathbf{v}(t),\ \varphi(t)) \in L^2(0,T; V_{\sigma+1}(\Omega_t) \times W_{\sigma+1}(\Omega_t)) \cap L^\infty(0,T; V_\sigma(\Omega_t) \times W_\sigma(\Omega_t)) \cap H^1(0,T; V_{\sigma-1}(\Omega_t) \times W_{\sigma-1}(\Omega_t))$;

b) *for every pair* $(\mathbf{u}(t), \psi(t)) \in L^2(0,T; V_{1-\sigma}(\Omega_t) \times W_{1-\sigma}(\Omega_t))$

$$\begin{aligned}
&\int_0^T \langle(\mathbf{v}(t) + (\mu + k)A\mathbf{v}(t) - \mathbf{f}(t)), \mathbf{u}(t)\rangle dt = -\int_0^T \mathbf{M}(t, \mathbf{v}(t), \mathbf{u}(t), \varphi(t))dt, \\
&\int_0^T \langle(j\varphi'(t) + \gamma B\varphi(t) + 2k\varphi(t) - l(t)), \psi(t)\rangle dt = -\int_0^T \mathbf{N}(t, \mathbf{v}(t), \varphi(t), \psi(t))dt,
\end{aligned} \tag{4}$$

where $\langle\cdot\,,\cdot\rangle$ *denotes the duality of* $V_{\sigma-1}(\Omega_t)$ *and* $V_{1-\sigma}(\Omega_t)$ *and*

$$\begin{aligned}
\mathbf{M}(t, \mathbf{v}(t), \mathbf{u}(t), \varphi(t)) &= b(t, \mathbf{v}(t), \mathbf{v}(t), \mathbf{u}(t)) + ka(t, \varphi(t), \mathbf{u}(t)) \\
&+ \sum_{\alpha=1}^{2} \int_{\partial\Omega_{t\alpha}} [h_\alpha(\mathbf{x},t) - \tfrac{1}{2}v_1^2(\mathbf{x},t) - \tfrac{1}{2}j\varphi^2(\mathbf{x},t)](\mathbf{u}(\mathbf{x},t)\cdot\mathbf{n}(\mathbf{x},t))d\sigma_\alpha \\
&\qquad + \int_{\partial\Omega_{t3}} r(\mathbf{x},t) \mid \mathbf{v}(\mathbf{x},t)\cdot\mathbf{n}(\mathbf{x},t) \mid^2 (\mathbf{u}(\mathbf{x},t)\cdot\mathbf{n}(\mathbf{x},t)d\sigma_3, \\
\mathbf{N}(t, \mathbf{v}(t), \varphi(t), \psi(t)) &= ka(t, \psi(t), \mathbf{v}(t)) + j\langle v_\alpha(t)\varphi_{,\alpha}(t), \psi(t)\rangle.
\end{aligned} \tag{5}$$

3. Statement of the main results

In what follows we give some theorems that yield the existence of a solution (in the sense of the above definition) of (1) and (3) satisfying the initial condition

$$\mathbf{v}(0) = \mathbf{v}_0, \quad \varphi(0) = \varphi_0. \tag{6}$$

Theorem 1. *Suppose that* $(\mathbf{g}(t),\ m(t)) \in L^2(0,T; V_{\sigma-1}(\Omega_t)\times W_{\sigma-1}(\Omega_t))$, $(\mathbf{v}_0, \varphi_0) \in V_0(\Omega_0) \times W_0(\Omega_0)$, *and that the functions* $\eta_\alpha(x_1,t)$, $\alpha = 1,2$, *are continuous together with their derivatives* $\partial\eta_\alpha/\partial t$ *and* $\partial\eta_\alpha/\partial x_1$ *for* $0 \le x_1 \le L,\ \ 0 \le t \le T$. *Then there exists a pair*

$$(\mathbf{v}(t), \varphi(t)) \in L^2(0,T; V_{\sigma+1}(\Omega_t) \times W_{\sigma+1}(\Omega_t)) \cap L^\infty(0,T; V_\sigma(\Omega_t) \times W_\sigma(\Omega_t)) \cap H^1(0,T; V_{\sigma-1}(\Omega_t) \times W_{\sigma-1}(\Omega_t))$$

satisfying the equations

$$\begin{aligned}
&\int_0^T \langle(\mathbf{v}'(t) + (\mu + k)A\mathbf{v}(t) - \mathbf{g}(t)), \mathbf{u}(t)\rangle dt = 0, \\
&\int_0^T \langle(j\varphi'(t) + \gamma B\varphi(t) + 2k\varphi(t) - m(t)),\ \psi(t)\rangle dt = 0
\end{aligned} \tag{7}$$

for all $(\mathbf{u}(t), \varphi(t)) \in L^2(0,T; V_{1-\sigma}(\Omega_t) \times W_{1-\sigma}(\Omega_t))$, $0 \le \sigma < \frac{1}{2}$, *and the initial condition* (6).

The proof is based on embedding [2] and trace [3] theorems, and a result in [4].

We write

$$W_{T,\sigma} = L^2(0,T; V_{\sigma+1}(\Omega_t) \times W_{\sigma+1}(\Omega_t)) \cap L^\infty(0,T; V_\sigma(\Omega_t) \times W_\sigma(\Omega)) \\ \cap H^1(0,T; V_{\sigma-1}(\Omega_t) \times W_{\sigma-1}(\Omega_t)), \quad (8)$$

and define an operator $G : W_{T,\sigma} \to W_{T,\sigma}$ in the following way. Given any pair $(\mathbf{w}(t), \theta(t)) \in W_{T,\sigma}$, we call $(\mathbf{v}(t), \varphi(t)) = G((\mathbf{w}(t), \theta(t)))$ the pair of functions (which, as can easily be seen, exists, by Theorem 1) satisfying (6) and the equations

$$\begin{aligned} &\int_0^T \langle(\mathbf{v}'(t) + (\mu+k)A\mathbf{v}(t) - \mathbf{f}(t)), \mathbf{u}(t)\rangle dt = -\int_0^T \mathbf{M}(t, \mathbf{w}(t), \mathbf{u}(t), \theta(t))dt, \\ &\int_0^T \langle(j\varphi'(t) + \gamma B\varphi(t) + 2k\varphi(t) - l(t)), \psi(t)\rangle dt = -\int_0^T \mathbf{N}(t, \mathbf{w}(t), \theta(t), \psi(t))dt \end{aligned} \quad (9)$$

for all $(\mathbf{u}(t), \psi(t)) \in L^2(0,T; V_{1-\sigma}(\Omega_t) \times W_{1-\sigma}(\Omega_t))$.

Theorem 2. *If* $(\mathbf{v}_0, \varphi_0) \in V_\sigma(\Omega_0) \times W_\sigma(\Omega_0)$, $(\mathbf{f}(t), l(t)) \in L^2(0,T; V_{\sigma-1}(\Omega_t) \times W_{\sigma-1}(\Omega_t))$, $h_\alpha(t) \in L^2(0,T; L^2(\partial\Omega_{t\alpha}))$, $r(t) \in L^\infty(0,T; L^\infty(\partial\Omega_{t3}))$ *and* $\partial\eta_\alpha/\partial x_1$, $\partial\eta_\alpha/\partial t$ *are continuous, then for* $\frac{1}{4} < \sigma < \frac{1}{2}$ *the operator* G *is completely continuous from* $W_{T,\sigma}$ *into itself.*

Consider now the transformation $(\mathbf{v}, \varphi) = G((\mathbf{w}, \theta), \lambda)$, depending continuously on the parameter $\lambda \in [0,1]$, defined in the following way. Given any pair $(\mathbf{w}(t), \theta(t)) \in W_{T,\sigma}$, we call $(\mathbf{v}(t), \varphi(t)) = G((\mathbf{w}(t), \theta(t)), \lambda)$ a pair belonging to $W_{T,\sigma}$ and satisfying

$$\mathbf{v}(0) = \lambda\mathbf{v}_0, \quad \varphi(0) = \lambda\varphi_0 \quad (10)$$

and

$$\begin{aligned} &\int_0^T \langle(\mathbf{v}'(t) + (\mu+k)A\mathbf{v}(t) - \lambda\mathbf{f}(t)), \mathbf{u}(t)\rangle dt = -\lambda\int_0^T \mathbf{M}(t, \mathbf{w}(t), \mathbf{u}(t), \theta(t))dt, \\ &\int_0^T \langle(j\varphi'(t) + \gamma B\varphi(t) + 2k\varphi(t) - \lambda l(t)), \psi(t)\rangle dt = -\lambda\int_0^T \mathbf{N}(t, \mathbf{w}(t), \theta(t), \psi(t))dt \end{aligned} \quad (11)$$

for all $(\mathbf{u}(t), \psi(t)) \in L^2(0,T; V_{1-\sigma}(\Omega_t) \times W_{1-\sigma}(\Omega_t))$.

By the definition of G, it is obvious that $G((\mathbf{w}, \theta), 1) = G((\mathbf{w}, \theta))$.

The following result holds.

Theorem 3. *Suppose that all the conditions in Theorem* 2 *are satisfied, and let* $(\mathbf{v}(t), \varphi(t))$ *be any solution of the equation*

$$(\mathbf{v}(t), \varphi(t)) = G((\mathbf{v}(t), \varphi(t)), \lambda). \tag{12}$$

If T *is sufficiently small, then* $(\mathbf{v}(t), \varphi(t))$ *is bounded on* $[0, T]$ *uniformly with respect to* λ, *that is, there exist constants* C_1, C_2 *independent of* $(\mathbf{v}, \varphi)$ *and* λ *such that for any* $\lambda \in [0, 1]$

$$\begin{aligned}
&\int_0^T \{\|\mathbf{v}(t)\|^2_{V_{\sigma+1}(\Omega_t)} + \|\mathbf{v}'(t)\|^2_{V_{\sigma-1}(\Omega_t)}\}dt + \sup_{0\le t\le T} \|\mathbf{v}(t)\|^2_{V_\sigma(\Omega_t)} \le C_1, \\
&\int_0^T \{|\varphi(t)\|^2_{W_{\sigma+1}(\Omega_t)} + \|\varphi'(t)\|^2_{W_{\sigma-1}(\Omega_t)}\}dt + \sup_{0\le t\le T} \|\varphi(t)\|^2_{W_\sigma(\Omega_t)} \le C_2.
\end{aligned} \tag{13}$$

It is obvious that, by Theorems 2 and 3, the transformation G defined above satisfies the assumptions of the Leray-Schauder principle. Hence, the functional equation $(\mathbf{v}(t), \varphi(t)) = G((\mathbf{v}(t), \varphi(t))$, which evidently corresponds to (4), (6), admits a solution $(\mathbf{v}(t), \varphi(t)) \in W_{T,\sigma}$, and we arrive at the following assertion.

Theorem 4. *Under the assumptions in Theorem* 2, *there exists in* $[0, T]$, *for* T *sufficiently small, a solution of equations* (1) *satisfying the initial condition* (2) *and the boundary conditions* (3), *with* $1/4 < \sigma < 1/2$.

References

1. A.C. Eringen, Theory of micropolar fluids, *J. Math. Mech.* **16** (1966), 1–18.
2. S.M. Nikol'skii, On embedding, extensions and approximation of differentiable functions, *Uspeki Mat. Nauk* **16** (1961), no. 5, 63–114.
3. G. Prodi, Trace de funzioni con derivate di ordine s a quadrato sommabile su varietà di dimensione arbitraria, *Rend. Sem. Mat. Univ. Padova* **28** (1958), 402–432.
4. J.-L. Lions, Espaces intermédiaires entre espaces hilbertiens et applications, *Bull. Soc. Math. Phys. Roumanie* **2** (1958), 419–432.

Department of Mathematics, Technical University "Gh. Asachi", Iasi–6600, Romania

I. SHIRINSKAYA and M. BELYI

Discrete analogues of boundary integral equations for boundary value problems in solid mechanics

1. Introduction

The aim of this paper is to develop a technique for solving discrete boundary value problems arising in finite element and finite difference analysis analogous to the boundary element method [1]. The approach being developed is based on the elimination of the unknowns corresponding to the interior part of the domain. Such elimination results in a reduced system of equations with respect to the boundary unknowns. The basic idea of the method is not new and is used in many computational techniques. Thus, the superelement method extensively used in the finite element analysis of large and complex structural systems is based on the elimination of the interior degrees of freedom of substructures by means of static condensation [2]. A method closely related to the approach under investigation was introduced by Ryabenkii and is referred to as the difference potential method [3]. In this method a system of discrete boundary equations is formulated by using boundary projectors; this system can be considered as a discrete analogue of the boundary operator equations introduced by Calderon [4] and then investigated by Seeley [5]. In contrast to Ryabenkii's method, in this paper we investigate constructive formulations of the discrete boundary equations related to the direct and indirect formulations of the boundary integral equations in the boundary element method. In the context of the developed approach we assume that the inverse of the discrete operator with constant coefficients L^{-1} can be computed as a convolution with a discrete fundamental solution. The algorithms for calculation of a discrete fundamental solution are discussed in [6]. The self-adjoint formulations of the discrete boundary equations presented in this paper are similar to the hypersingular boundary integral equations in the continuous case [7]. But in the discrete case we do not face problems of regularisation and interpretation of hypersingular boundary integral operators. The obtained symmetric formulations of the discrete boundary equations are very suitable for the development of effective algorithms and for using discrete boundary equations in combination with the finite element method.

2. Discrete analogues of Green's formulae

Let L be a self-adjoint discrete operator with constant coefficients. We consider this operator as the finite difference or finite element approximation of an elliptic self-adjoint differential operator using a regular grid. However, all subsequent results can be applied to discrete problems of regular lattice structure analysis.

Let $\mathcal{L}_0$ be the discrete operator of the second boundary value problem for the system of equations with the operator L for some grid domain. The boundary of the domain may coincide with the grid lines or not. Let θ be the characteristic function for the set of all nodes on which the operator $\mathcal{L}_0$ is defined, and θ_0 the characteristic function for the set of internal nodes where the operator $\mathcal{L}_0$ is locally identical to the operator L; that is,

$$\theta_0 \mathcal{L}_0 = \theta_0 L.$$

The characteristic functions θ and θ_0 and all the other characteristic functions used below may be considered as diagonal discrete operators with only zero or identity diagonal elements.

We define a characteristic function of the boundary nodes as

$$\theta_\Gamma = \theta - \theta_0, \quad \theta_\Gamma \theta = \theta_\Gamma, \quad \theta_\Gamma \theta_0 = 0.$$

Then the operator $\mathcal{L}_0$ may be represented in the form

$$\mathcal{L}_0 = \theta_0 L + \theta_\Gamma \Gamma. \tag{1}$$

Here the operator Γ is the discrete analogue of the operator of intrinsic boundary conditions. Formula (1) can be considered as the discrete analogue of the first Green's formula. Using the self-adjointness of the operators L and $\mathcal{L}_0$, we obtain from (1) the equation

$$\mathcal{L}_0 = L\theta_0 + \Gamma^* \theta_\Gamma. \tag{2}$$

Combining equations (1) and (2) yields

$$L\theta_0 u = \theta_0 L u + \theta_\Gamma \Gamma u - \Gamma^* \theta_\Gamma u, \tag{3}$$

where u is an arbitrary grid vector-function. Equation (3) is the second Green's formula for the discrete operator L.

Pre-multiplying equation (1) and post-multiplying equation (2) by the characteristic function θ_Γ, we obtain the formulae

$$\theta_\Gamma \Gamma = \theta_\Gamma \mathcal{L}_0, \quad \Gamma^* \theta_\Gamma = \mathcal{L}_0 \theta_\Gamma,$$

which define the action of the operators Γ and Γ^* on the grid functions.

3. Direct formulations of the discrete boundary equations

3.1. Non-self-adjoint direct formulation. Consider a mixed discrete boundary value problem defined on a grid domain with the characteristic function θ:

$$\begin{aligned} \theta_0 L u &= \theta_0 F, \\ æ_\Gamma \Gamma u &= æ_\Gamma f, \\ \overline{æ_\Gamma} u &= \overline{æ_\Gamma} g. \end{aligned} \tag{4}$$

Here $æ_\Gamma$ and $\overline{æ_\Gamma}$ are the characteristic functions of the intrinsic and the main boundary conditions respectively, defined by

$$æ_\Gamma + \overline{æ_\Gamma} = \theta_\Gamma, \quad æ_\Gamma \overline{æ_\Gamma} = 0.$$

Substituting (4) in formula (3) yields

$$L\theta_0 u = \mathcal{F} + \overline{æ_\Gamma} v_\Gamma - \Gamma^* æ_\Gamma u_\Gamma, \tag{5}$$

where

$$\mathcal{F} = \theta_0 F + æ_\Gamma f - \Gamma^* \overline{æ_\Gamma} g, \quad v_\Gamma = \overline{æ_\Gamma} \Gamma u, \quad u_\Gamma = æ_\Gamma u \,.$$

Solving equation (5) for $\theta_0 u$, we obtain

$$\theta_0 u = L^{-1}(\mathcal{F} + \overline{æ_\Gamma} v_\Gamma - \Gamma^* æ_\Gamma u_\Gamma).$$

Pre-multiplying this equation by θ_Γ, we obtain the discrete boundary equation

$$\theta_\Gamma L^{-1}(\Gamma^* æ_\Gamma - \overline{æ_\Gamma}) w_\Gamma = \theta_\Gamma L^{-1} \mathcal{F}, \tag{6}$$

where $w_\Gamma = æ_\Gamma u_\Gamma + \overline{æ_\Gamma} v_\Gamma$. Equation (6) can be considered as the discrete analogue of the boundary integral equation corresponding to the direct formulation of the boundary element method [1]. Solving the system of equations (6) with respect to the elements of the boundary vector w_Γ, we can calculate the solution of the problem (4) by means of the formula

$$u = L^{-1}[\mathcal{F} - (\Gamma^* æ_\Gamma - \overline{æ_\Gamma}) w_\Gamma] + æ_\Gamma w_\Gamma + \overline{æ_\Gamma} g. \tag{7}$$

In the general case the operator of the boundary equation (6) is not self-adjoint.

Consider particular cases. For the Dirichlet boundary value problem we have $æ_\Gamma = 0, \overline{æ_\Gamma} = \theta_\Gamma, w_\Gamma = v_\Gamma$. In this case equation (6) has the form

$$-\theta_\Gamma L^{-1} \theta_\Gamma v_\Gamma = \theta_\Gamma L^{-1}(\theta_0 F - \Gamma^* \theta_\Gamma g).$$

We note that the operator of the discrete boundary equation is self-adjoint in this case.

For the Neumann boundary value problem we have $æ_\Gamma = \theta_\Gamma, \overline{æ_\Gamma} = 0, w_\Gamma = u_\Gamma$. Now equation (6) has the form

$$\theta_\Gamma L^{-1} \Gamma^* \theta_\Gamma u_\Gamma = \theta_\Gamma L^{-1}(\theta_0 F + \theta_\Gamma f).$$

3.2. Self-adjoint direct formulation. Substituting the expression (7) for u into the boundary conditions of the problem (4) yields

$$\begin{aligned} æ_\Gamma \Gamma L^{-1}[\mathcal{F} - (\Gamma^* æ_\Gamma - \overline{æ_\Gamma}) w_\Gamma] + æ_\Gamma \Gamma æ_\Gamma w_\Gamma + æ_\Gamma \Gamma \overline{æ_\Gamma} g = æ_\Gamma f, \\ \overline{æ_\Gamma} L^{-1}[\mathcal{F} - (\Gamma^* æ_\Gamma - \overline{æ_\Gamma}) w_\Gamma] = 0. \end{aligned} \tag{8}$$

Taking into account that $æ_\Gamma \overline{æ_\Gamma} = 0$ and combining with the system (8), we obtain the following discrete boundary equation with respect to w:

$$æ_\Gamma \Gamma æ_\Gamma w_\Gamma - (æ_\Gamma \Gamma - \overline{æ_\Gamma}) L^{-1}(\Gamma^* æ_\Gamma - \overline{æ_\Gamma}) w_\Gamma = æ_\Gamma (f - \Gamma \overline{æ_\Gamma} g) - (æ_\Gamma \Gamma - \overline{æ_\Gamma}) L^{-1} \mathcal{F}. \tag{9}$$

The operator of equation (9) is self-adjoint. This is obvious for the second term on the left-hand side of equation (9); for the first term it follows from formula (1) and the fact that $\mathcal{L}_0^* = \mathcal{L}_0$.

In the particular cases of the first (Dirichlet) and the second (Neumann) boundary value problem equation (9) is simplified as follows:

for the Dirichlet problem,

$$-\theta_\Gamma L^{-1}\theta_\Gamma v_\Gamma = \theta_\Gamma L^{-1}(\theta_0 F - \Gamma^*\theta_\Gamma g);$$

for the Neumann problem,

$$\theta_\Gamma(\Gamma - \Gamma L^{-1}\Gamma^*)\theta_\Gamma u_\Gamma = \theta_\Gamma f - \theta_\Gamma \Gamma L^{-1}(\theta_0 F + \theta_\Gamma f).$$

4. Conclusion

Along with the direct formulations of the discrete boundary equations, indirect formulations can be obtained in a similar manner. To construct the indirect formulations, we introduce the jumps of the solution (Δu_Γ) and of the intrinsic boundary condition (Δv_Γ) at the boundary, and express the solution of the problem through these jumps. Substituting this representation of the solution into the boundary conditions, we obtain a system of equations with respect to Δu_Γ and Δv_Γ. We note that the solution is expressed through two functions Δu_Γ and Δv_Γ concentrated on the boundary set of points. Therefore, at each point of the boundary set the value of Δu_Γ or Δv_Γ can be assigned arbitrarily. Using this arbitrariness, we obtain different versions of indirect formulations for the reduced system of discrete boundary equations.

References

1. P.K. Banerjee and R. Butterfield, *Boundary element methods in engineering science*, McGraw-Hill, 1981.
2. J. Argyris and H.P. Mlejnek, *Die Methode der finiten Elemente in der elementaren Struktur-Mechanik*, I–III, Vieweg, Braunschweig/Wiesbaden, 1986–1988.
3. V.S. Ryabenkii, *Difference schemes (introduction to the theory)*, Moscow, 1977 (Russian).
4. A.P. Calderon, Boundary value problems for elliptic equations, in *Proc. Soviet-American Symp. on Partial Difference Equations*, Moscow, 1963, 303–304.
5. R.T. Seeley, Singular integrals and boundary value problems. *Amer. J. Math.* **88** (1966), 781–809.
6. M.V. Belyi and I.V. Shirinskaya, Algorithms for the calculation of fundamental solutions of finite element and finite difference equations, in *Integral Methods in Science and Engineering 1996*, Pitman Res. Notes Math. Ser., Addison Wesley Longman, Harlow, 1997.
7. J.C. Nédélec, Integral equations with non-integrable kernels, *Integral Equations Operator Theory* **5** (1982), 562–572.

Department of Applied Mathematics, Moscow State University of Civil Engineering, Yaroslavskoe Shosse 26, 129337 Moscow, Russia

A.S. SILBERGLEIT and G.M. KEISER

On eddy currents from moving point sources of magnetic field in the Gravity Probe B experiment

1. Introduction

In the Gravity Probe B experiment [1], a superconducting rotating ball will be orbiting the Earth for over a year, and the drift of its axis resulting from two General Relativity effects will be measured. The drift is expected to be very small (0.042 arc-sec/year and 6.6 arc-sec/year in the East-West and North-South directions, respectively) and the measurement accuracy should be very high, so the classical (non-relativistic) torques causing the drift must be either eliminated or carefully accounted for. In particular, there will be quantum-size sources of magnetic field (fluxons) on the surface of the superconducting rotor which induce eddy currents and thus energy dissipation in surrounding normal metals. Consequently, differential damping torques are produced which must be estimated.

In this paper we give such estimates by solving explicitly two corresponding model boundary value problems in plane geometry (they may also be of interest for other applications). We are able to avoid complications of the spherical case, imminent for the GP–B experiment, since the gap between the rotor surface and normal metals around it is extremely small compared to the rotor's radius.

2. The problem for a conducting half-space

Suppose that a metal with electrical conductivity σ and magnetic constant $\mu = \mu_0$ occupies the half-space $z' > d$ of a Cartesian coordinate system $\{x', y', z'\}$. The plane $z = 0$ is a superconductor's surface in which a fluxon and antifluxon move with a constant velocity $v \mathcal{L} c$ along the x'-axis; the layer $0 < z' < d$ is a dielectric gap. According to [2, **10.00**], the set of governing equations and boundary conditions for the *quasi-stationary* magnetic induction $\mathbf{B}$ in the dimensionless coordinates $x = (x' - x'_f(t))/d$, $y = (y' - y'_f)/d$, $z = z'/d$ co-moving with the sources is

$$\nabla \cdot \mathbf{B} = 0, \quad \nabla \times \mathbf{B} = 0, \quad 0 < z < 1, \quad |x|, |y| < \infty, \tag{1}$$

$$B_z|_{z=0} = \frac{\Phi_0}{d^2}\left[\delta(x)\delta(y) - \delta(x - x_0)\delta(y - y_0)\right], \tag{2}$$

$$\nabla \cdot \mathbf{B} = 0, \quad \Delta \mathbf{B} = -\kappa \frac{\partial \mathbf{B}}{\partial x}, \quad z > 1, \quad |x|, |y| < \infty, \quad \mathbf{B}|_{z=1+0} = \mathbf{B}|_{z=1-0}. \tag{3}$$

Here Φ_0 is the flux of a point source (for GP–B, $\Phi_0 = h/2e$ is the magnetic flux quantum), $x'_{f,a}(t) = x'_{f,a} + vt$ and $y'_{f,a}$ are the coordinates of the fluxon/antifluxon, $x_0 = (x_a - x_f)/d$, $y_0 = (y_a - y_f)/d$, the rest of the notations are standard, and we have omitted the obvious decay conditions at infinity. The only dimensionless parameter of the problem is $\kappa = \sigma \mu_0 v d$.

This work was supported by the NASA grant NAS8-39225 to Gravity Probe B.

Problem (1)–(3) may be solved by means of the Fourier integral transform in the x and y variables, which leads to rather cumbersome double Fourier integrals preventing further analytic calculations and hiding the physical meaning of the result. Instead, we look for a perturbative solution as a power series in the small parameter κ, $\kappa \lesssim 0.02$, for the GP–B conditions, that is,

$$\mathbf{B}(x,y,z) = \mathbf{B}^{(0)}(x,y,z) + \kappa\mathbf{B}^{(1)}(x,y,z) + \kappa^2\mathbf{B}^{(2)}(x,y,z) + \cdots, \tag{4}$$

where the $\mathbf{B}^{(k)}$, $k = 0, 1, 2, \ldots$, are to be determined successively from the sequence of problems implied by (1)–(4), namely

$$\begin{gathered}\nabla \cdot \mathbf{B}^{(k)} = 0, \quad \nabla \times \mathbf{B}^{(k)} = 0, \quad z < 1, \\ B_z^{(k)}|_{z=0} = \frac{\Phi_0}{d^2}\left[\delta(x)\delta(y) - \delta(x - x_0)\delta(y - y_0)\right]\delta_{k0},\end{gathered} \tag{5}$$

$$\nabla \cdot \mathbf{B}^{(k)} = 0,\ \Delta\mathbf{B}^{(k)} = -(1 - \delta_{k0})\frac{\partial \mathbf{B}^{(k-1)}}{\partial x},\ z > 1,\ \mathbf{B}^{(k)}|_{z=1+0} = \mathbf{B}^{(k)}|_{z=1-0}. \tag{6}$$

The corresponding eddy current density in the conductor is ([2, **10.00**])

$$\mathbf{j} = \frac{1}{\mu_0 d}\nabla \times \mathbf{B} = (\kappa\mathbf{j}^{(1)} + \kappa^2\mathbf{j}^{(2)} + \cdots), \quad \mathbf{j}^{(k)} = \frac{1}{\mu_0 d}\nabla \times \mathbf{B}^{(k)}, \quad z > 1, \tag{7}$$

where we have taken into account the fact that $\nabla \times \mathbf{B}^{(0)} = 0$ in the whole half-space $z > 0$. Naturally, the current density is of the first order in κ, since without a conductor ($\kappa = 0$) there is no current at all. Our aim is to determine $\mathbf{j}^{(1)}$ and then to calculate the dissipated power to the first non-vanishing order.

The 'unperturbed' field $\mathbf{B}^{(0)}$ is that of the two point sources in a half-space without a conductor. It satisfies the Laplace equation in the whole half-space $z > 0$ with the boundary condition (2) and therefore is given by

$$\mathbf{B}^{(0)} = \nabla\psi^{(0)}, \quad \psi^{(0)}(x,y,z) = -\frac{\Phi_0}{2\pi d^2}\left(\frac{1}{R} - \frac{1}{R_0}\right), \quad z > 0, \tag{8}$$

where $R = \sqrt{x^2 + y^2 + z^2}$ and $R_0 = \sqrt{(x - x_0)^2 + (y - y_0)^2 + (z - z_0)^2}$.

To find $\mathbf{j}^{(1)}$ using (7), we do not need $\mathbf{B}^{(1)}$ itself (from (6) with $k = 1$), but rather its curl in the conductor. Surprisingly, the latter may be determined without completely solving the boundary value problem: from (5) and (6) we can derive a representation of the form

$$\mathbf{B}^{(1)} = \nabla\left(\psi^{(1)} - \frac{x}{2}\psi^{(0)}\right) + \nabla \times \mathbf{A}^{(1)} + \psi^{(0)}\mathbf{e}_x, \quad z > 1, \tag{9}$$

where $\psi^{(1)}$ and $\mathbf{A}^{(1)}$ are harmonic functions in $z > 1$, and $\mathbf{A}^{(1)}$ is divergenceless there. Therefore, the first two terms in (9) are curl-less, so that we have (see (7))

$$\nabla \times \mathbf{B}^{(1)} = \nabla\psi^{(0)} \times \mathbf{e}_x = \mathbf{B}^{(0)} \times \mathbf{e}_x, \quad \mathbf{j} = \kappa\mathbf{j}^{(1)} + O(\kappa^2) = \sigma v\,\mathbf{B}^{(0)} \times \mathbf{e}_x + O(\kappa^2). \tag{10}$$

The last result has a clear physical meaning: when a conductor moves with velocity $\mathbf{v}^{(0)}$ in an external magnetic field $\mathbf{B}^{(0)}$, the induced current density in it is given by the Lorentz formula $\mathbf{j} = \sigma \mathbf{v}^{(0)} \times \mathbf{B}^{(0)}$. In our case $\mathbf{v}^{(0)} = -v\,\mathbf{e}_x$, and, as the conductivity is small, the induced magnetic field may be neglected compared to the 'external' field $\mathbf{B}^{(0)}$, which amounts exactly to (10).

We now evaluate the power dissipated by a fluxon–antifluxon pair as

$$P_{fa} = \frac{d^3}{\sigma} \int\limits_{z>1} j^2\, dV \le 4P_f = \frac{d^3}{\sigma} \int\limits_{z>1} j_f^2\, dV \simeq \sigma v^2 d^3 \int\limits_{z>1} \left[\left(B_{f,y}^{(0)}\right)^2 + \left(B_{f,z}^{(0)}\right)^2 \right] dV,$$

where P_f is the power dissipated by a single fluxon, $\mathbf{B}_f^{(0)} \sim \nabla(1/R)$ (see (8)), and $\simeq$ means that we calculate to the first non-vanishing order in κ. By (8), this leads to the simplified and rather conservative estimate

$$P_{fa} \le 4P_f = \frac{3}{4\pi} \frac{\sigma v^2 \Phi_0^2}{d}. \tag{11}$$

Typically, the value of P_{fa} is at least twice smaller than this upper bound.

3. The problem for a strongly conducting thin layer

We now turn to the second problem, in which the conducting half-space is replaced by a conducting layer of thickness d_0 in the domain $d < z' < d + d_0$. The layer is relatively thin, $d_0 \mathcal{L} d$, and its conductivity is high, so that κ is no longer small (for the GP–B, $d_0 \sim 0.1d$ and $\kappa \sim 30$).

Consequently, outside the layer the magnetic field satisfies (1) with the boundary condition (2), while in the layer equations (3) are valid; the appropriate matching conditions are imposed at both surfaces of the layer (from here on we again use the unprimed dimensionless co-moving coordinates). The solution of this problem is very complicated, so another physically sound simplification of the model is required. The proper one is to replace the thin highly conducting layer of a *finite* thickness by an *infinitely thin* sheet carrying current of some surface density $\mathbf{j}^s = \mathbf{j}^s(x, y)$, and to formulate the matching conditions for the magnetic field at $z = 1$. The latter are known to be ([2], **7.21**)

$$B_z|_{z=1+0} = B_z|_{z=1-0}, \quad \mathbf{e}_z \times \left(\mathbf{B}|_{z=1+0} - \mathbf{B}|_{z=1-0}\right) = \mu_0 \mathbf{j}^s; \tag{12}$$

thus, to solve the problem, we only need to relate $\mathbf{j}^s$ to $\mathbf{B}$. For this we introduce the vector potential $\mathbf{A}$ defined by $\mathbf{B} = \nabla \times \mathbf{A}$, $\nabla \cdot \mathbf{A} = 0$, and note that the induction equation, which is valid in a real three-dimensional layer, implies that $\mathbf{j} = \sigma v\, \partial \mathbf{A}/\partial x$. Hence, for a *surface* current density we can write

$$\mathbf{j}^s(x, y) = \sigma v d_0 \left. \frac{\partial \mathbf{A}}{\partial x} \right|_{z=1}. \tag{13}$$

The verification of the matching condition (12) and (13) is provided by the asymptotic integration of the complete three-dimensional problem from the asymptotic theory of 'thin' bodies (plates, shells; cf. [3]).

In terms of **A** we still have a very cumbersome vector boundary value problem, which we simplify by setting $A_z \equiv 0$ so that $\nabla \cdot \mathbf{A} = \partial A_x/\partial x + \partial A_y/\partial y = 0,\ z > 0$. The latter is satisfied if $A_x = -\partial\Pi/\partial y$ and $A_y = \partial\Pi/\partial x$, where $\Pi = \Pi(x, y, z)$ is a new function to be determined. By this and (13),

$$j_x^s = -\sigma v d_0 \left.\frac{\partial^2 \Pi}{\partial x \partial y}\right|_{z=1}, \quad j_y^s = \sigma v d_0 \left.\frac{\partial^2 \Pi}{\partial x^2}\right|_{z=1}. \tag{14}$$

Now it is not difficult to check that all the relevant equations are valid provided that $\Pi(x, y, z)$ solves the problem

$$\Delta\Pi = 0,\ z > 0,\ z \neq 1,\ -\left.\frac{\partial^2 \Pi}{\partial z^2}\right|_{z=0} = \frac{\Phi_0}{d^2}\left[\delta(x)\delta(y) - \delta(x - x_0)\delta(y - y_0)\right], \tag{15}$$

$$\left.\left\langle \Pi \right\rangle\right|_{z=1} = 0, \quad \left.\left\langle \frac{\partial \Pi}{\partial z} \right\rangle\right|_{z=1} = -\kappa_0 \left.\frac{\partial \Pi}{\partial x}\right|_{z=1},$$

where the angle brackets stand for the jump of the quantity inside them at $z = 1$, and $\kappa_0 = \kappa\, d_0/d = \sigma\mu_0 v d_0$. With the exception of the second derivative in the boundary condition at $z = 0$, this scalar problem is a standard one; its solution for $z \geq 1$ by means of the Fourier integral transform in x and y is ($\hat{f}(\lambda, \nu, \ldots)$, which denotes the corresponding Fourier image of $f(x, y, \ldots)$):

$$\hat{\Pi}(\lambda, \nu, z) = D(\lambda, \nu)\, e^{-\gamma z}, \quad \gamma(\lambda, \nu) = \sqrt{\lambda^2 + \nu^2} \geq 0, \tag{16}$$

$$D(\lambda, \nu) = -\frac{\Phi_0}{d^2 \gamma} \frac{1 - e^{-i(\lambda x_0 + \nu y_0)}}{\gamma - i(\kappa_0/2)\lambda(1 - e^{-\gamma})}, \quad z \geq 1.$$

From (16) and (14) we obtain

$$\frac{\hat{j}_x^s(\lambda, \nu)}{\sigma v d_0} = \lambda\nu D(\lambda, \nu) e^{-\gamma}, \quad \frac{\hat{j}_y^s(\lambda, \nu)}{\sigma v d_0} = -\lambda^2 D(\lambda, \nu) e^{-\gamma}. \tag{17}$$

To compute the dissipation rate P_{fa}, we use the Parseval identity and (17), which yield

$$P_{fa} = \frac{d^2}{\sigma d_0} \int_{-\infty}^{\infty}\int_{-\infty}^{\infty} \left[\left(\hat{j}_x^s(\lambda, \nu)\right)^2 + \left(\hat{j}_y^s(\lambda, \nu)\right)^2\right] d\lambda d\nu$$

$$= \sigma v^2 d_0 d^2 \int_{-\infty}^{\infty}\int_{-\infty}^{\infty} \lambda^2 \gamma^2 e^{-2\gamma} |D(\lambda, \nu)|^2\, d\lambda d\nu.$$

Substituting (16) into this and using the inequality $\left|1 - \exp(-iQ)\right|^2 = 4\sin^2(0.5Q) \leq 4$ for simplification, we arrive at the desired estimate (cf. (11))

$$P_{fa} \leq 4P_f = \frac{4\sigma v^2 d_0 \Phi_0^2}{d^2} C(\kappa_0), \tag{18}$$

where

$$C(\kappa_0) = \int_{-\infty}^{\infty}\int_{-\infty}^{\infty} \frac{\lambda^2 e^{-2\gamma}}{\gamma^2 + (\kappa_0/2)^2\lambda^2(1-e^{-2\gamma})^2}\, d\lambda d\nu, \quad \gamma(\lambda,\nu) = \sqrt{\lambda^2+\nu^2} \geq 0.$$

The double integral representing the coefficient $C(\kappa_0)$ may be calculated explicitly as a combination of elementary functions, and also allows for a universal estimate $C(\kappa_0) < C(0) = \pi/4$, which, being introduced into (18), provides a final bound on P_{fa}, given by

$$P_{fa} \leq 4P_f < \frac{\pi\sigma v^2 d_0 \Phi_0^2}{d^2}. \tag{19}$$

For the GP–B conditions, numerical estimates of the corresponding torques and drift rates based on (11) and (19) show the latter to be smaller than those from some other differential damping torques [4], and thus do not endanger the experiment accuracy.

References

1. J.P. Turneaure, et al. The gravity Probe B relativity gyroscope experiment: development of the prototype flight instrument, *Adv. Space Res.* **9** (1989), 29–38.
2. W.R. Smythe, *Static and dynamic electricity*, 3rd ed., Hemisphere, New York, 1989.
3. I.E. Zino and E.A. Tropp, *Asymptotic methods in the problems of thermal conductivity and thermoelasticity theory*, Leningrad State University, Leningrad, 1978 (Russian).
4. C.W.F. Everitt, *Report on a program to develop a gyro test of general relativity in a satellite and associated control technology*, Stanford University, Stanford, 1980.

Gravity Probe B, HEPL, Department of Physics, Stanford University, Stanford, CA 94305–4085, USA

S.L. SKOROKHODOV and V.I. VLASOV

An exact method for torsion and bending problems for beams with polygonal cross-sections

1. Statement of the problems

We consider an elastic cylindrical beam whose cross-section is a polygonal domain $\mathcal{G}$ in the w-complex plane, $w = u + iv$. We assume that $\mathcal{G}$ is a finite simply connected polygon with N vertices. It is well known (see [1] and [2]) that the problems of torsion and bending of this beam are reduced to the Dirichlet problem for the Laplace equation. Namely, the formulation of the torsional problem is

$$\Delta\psi(w) = 0, \quad w \in \mathcal{G}; \quad \psi(w) = \tfrac{1}{2}|w|^2, \quad w \in \partial\mathcal{G}, \tag{1}$$

and the formulation of the bending problem is

$$\Delta\psi(w) = 0, \quad w \in \mathcal{G};$$
$$\psi(w) = -\tfrac{1}{3}(1 - \tfrac{1}{2}\sigma)v^3 - \tfrac{1}{2}\sigma u^2 v + 2(1+\sigma)\int^{w} uv\,du, \quad w \in \partial\mathcal{G}, \tag{2}$$

where σ is Poisson's ratio. Exact solutions for these problems have not yet been obtained even for such canonical profiles as L-, Z-, T-, U-, cruciform and frame-type cross-sections. Such solutions are important both as basic results and as tests when developing numerical methods.

This paper is devoted to a method [3] which provides an exact analytic solution for the problems (1) and (2) for polygonal domains $\mathcal{G}$. We have applied the method to a number of specific cross-sections and have obtained an exact solution, in particular for the above mentioned domains of L-, Z- and other shapes (see [4]–[6]). This method is a development of Trefftz's technique [7] and its generalisation [8].

2. The essence of the method

Consider an analytic function Ψ whose imaginary part is a solution of the boundary-value problem (1) or (2), $\operatorname{Im}\Psi(w) = \psi(w)$. We introduce an analytic function $\mathcal{F}$ which is the second derivative Ψ'' of the above function for the torsional problem, and the third derivative Ψ''' for bending. The function $z = \mathcal{F}(w)$ maps conformally the initial polygonal domain $\mathcal{G}$ on to some domain $\mathcal{D}$ in the z-plane. It appears [7], [8] that the latter domain is also a polygon. This fact provides a way to construct this conformal mapping $\mathcal{F}$ by means of a Schwarz-Christoffel integral and its generalisation [9], and as a result to obtain the solution ψ of the initial problem in closed analytic form.

The method can also be extended to the Poisson equation with higher-order polynomial right-hand side on a polygon; in this case we have to operate with higher-order derivatives of the function Ψ [8]. Another generalisation of the method pertains

This work was supported by the Russian Foundation for Basic Research, Grant No. 95-01-01367a, and by Grant No. 95-0-4.3-51 from St. Petersburg State University.

to mixed boundary conditions of Dirichlet or Neumann type, or boundary conditions with an oblique derivative containing a polynomial boundary function [3]. For the sake of simplicity, we consider only the torsional problem in this paper.

3. The general scheme of Trefftz's method

The initial boundary-value problem is reduced to the determination of the mapping function $\mathcal{F}$ which transforms the initial domain $\mathcal{G}$ on to the domain $\mathcal{D}$. The latter is not known a priori, so we have to find both the function $\mathcal{F}$ and the domain $\mathcal{D}$ simultaneously. Some information on $\mathcal{D}$ is provided by the Trefftz theorem [7]: 1) the mapping $\mathcal{F}$ transforms a vertex of $\mathcal{G}$ into a vertex of $\mathcal{D}$; 2) a neighbourhood of the corner of angle $\pi\beta$ is mapped on to a neighbourhood of the corner of angle $\pi\gamma$, where $\gamma = 1 - 2\beta$. Besides, there may be some branch points in the domain $\mathcal{D}$ where $\mathcal{F}'(w) = 0$.

We express the mapping $z = \mathcal{F}(w)$ as a superposition of two auxiliary mappings $w = f_1(\zeta)$ and $z = f_2(\zeta)$, that is,

$$\mathcal{F}(w) = f_2 \circ f_1^{-1}(w), \tag{3}$$

where $w = f_1(\zeta)$ and $z = f_2(\zeta)$ transform the upper half-plane $\mathcal{H}$ $\{\operatorname{Im}\zeta > 0\}$ on to the polygons $\mathcal{G}$ and $\mathcal{D}$, respectively.

According to [9], the function f_1 can be represented as a Schwarz-Christoffel integral

$$f_1(\zeta) = \mathcal{K} \int_{\zeta_0}^{\zeta} \prod_{n=1}^{N} (t - b_n)^{\beta_n - 1} \, dt, \tag{4}$$

and the function f_2 as a generalised Schwarz-Christoffel integral

$$f_2(\zeta) = C_0 \int_{\zeta_0}^{\zeta} R_M(t) \prod_{n=1}^{N} (t - b_n)^{-2\beta_n} \, dt. \tag{5}$$

This expression contains a polynomial $R_M(t)$ as a cofactor. By the result of Trefftz [7], the polynomial is of a degree M that is related to the number of vertices N by the formula $M = 2N - 6$, and its coefficients are real numbers. These coefficients can be obtained with the help of boundary conditions for the first and second derivatives of Ψ, deduced from the boundary condition $\operatorname{Im}\Psi(w) = |w|^2/2$ following from (1):

$$\operatorname{Im}[e^{i\pi\eta_n}\Psi'(w)] = s_n(w) + \operatorname{Re}[e^{-i\pi\eta_n} W_n], \quad w \in L_n; \tag{6}$$

$$\operatorname{Im}[e^{2i\pi\eta_n}\Psi''(w)] = 1, \quad w \in L_n. \tag{7}$$

Here L_n is the side of $\mathcal{G}$ connecting the vertices B_n and B_{n+1}, $\pi\eta_n$ is the angle between the real positive axis and L_n, $\pi\eta_n = \arg(B_{n+1} - B_n)$, and $s_n(w)$ is the arc coordinate of a point $w \in L_n$ measured from $W_n \in L_n$ in the positive direction.

If the auxiliary mappings f_1 and f_2 are obtained, then the solution $\psi(w)$ of the problem can be expressed by means of the integral

$$\psi(w) = \operatorname{Im}\left\{ \int^{\zeta} \left\{ \int^{\xi} f_2(t) f_1'(t) dt \right\} f_1'(\xi) d\xi \right\}. \tag{8}$$

Here the variable ζ is related to w through the inverse function $\zeta = f_1^{-1}(w)$.

4. Difficulties of the method

In spite of the exact integral representation (8) of the solution, the method has very seldom been applied, and now it is almost forgotten. This is because it employs a hard analytic technique which gives rise to some non-trivial difficulties.

The first principal difficulty is connected with the determination of vertex originals b_n and the factor $\mathcal{K}$ in the Schwarz-Christoffel integral (4) in explicit form for a large number of vertices.

The second principal difficulty is connected with the construction of the function f_2 and lies in the fact that both f_2 and the domain $\mathcal{D}$ are unknown and must be determined jointly.

For a certain class of domains $\mathcal{G}$ we have overcome these difficulties (see [3]–[6]) by using results of the modern qualitative and the constructive theory of conformal mapping [10].

This class consists of all quadrangles and figures that can be constructed by joining the quadrangles to their mirror images relative to their sides.

Below we present in detail the results of the torsional problem solution for two specific domains: L-shaped and frame-type domains.

5. Representation of the solution and intensity coefficients

The analytic solutions obtained in the form of integral (8) enabled us to derive two types of expansions suitable for the representation of the solution in different (closed) subdomains covering the initial (closed) domain $\mathcal{G}$ completely.

The first expansion is adequate in a subset of $\mathcal{G}$ which is a sector $\{w : \arg w \in [0, \pi\beta], |w| < R_0\}$ of angle $\pi\beta$ with sides belonging to the boundary $\partial\mathcal{G}$. If $\beta \neq 1/2, 3/2$, it has the form

$$\psi(w) = \mathrm{Im}\left\{\frac{w^2}{2\cos(\pi\beta)} + \sum_{k=1}^{\infty} a_k w^{k/\beta}\right\}; \tag{9}$$

if $\beta = 1/2$ or $3/2$, then it has the form

$$\phi(w) = \mathrm{Im}\left\{\frac{w^2}{\pi\beta}\ln w + \sum_{k=1}^{\infty} a_k w^{k/\beta}\right\}. \tag{10}$$

The second expansion is adequate in a subset of $\mathcal{G}$ which is a rectangle $\{w : \mathrm{Im}\, w \in [0,1], \mathrm{Re}\, w \in (a,b)\}$ with upper and lower sides belonging to $\partial\mathcal{G}$; it has the form

$$\phi(w) = \mathrm{Im}\left\{\frac{w^2}{2} + w + \sum_{k=-\infty}^{\infty} a_k e^{\pi k w}\right\}. \tag{11}$$

These expansions converge in appropriate subsets together with all their derivatives. The expansions reflect adequately the behaviour of the solution in marked subsets, in particular near geometric singularities. The coefficients a_k in expansions (9), (10) are called intensity coefficients; they are of great importance in applications. Our method has allowed us to obtain these values explicitly for the above domains (see [3]–[6]).

6. L-shaped domain

Consider an L-shaped domain with a re-entrant corner of angle $\pi\beta$; its legs have length d measured from the vertex of the re-entrant corner, and unit width. We introduce a parameter δ by setting $\delta = e^{(-\pi d)}$. It can be verified that δ is always smaller than 1.

Naturally, the solution of the torsional problem in this domain can be represented by expansions (9)–(11). We note that all the coefficients a_k with even subscripts k are zero in series (9), (10) near the re-entrant corner; this fact follows from the symmetry of the domain about the bisector of the corner.

The coefficients in the above expansions have been obtained in explicit form. In particular, we mention a formula for the first intensity coefficient $a_1(\beta,\delta)$ in the re-entrant corner which is given by the series $a_1(\beta,\delta) = a_1(\beta,0) + \sum_{p=1}^{\infty} a_1^{(p)}\delta^p$ in powers of the parameter δ. Here the coefficient $a_1(\beta,0)$ is expressed by the explicit formula

$$a_1(3/2,0) = \frac{9G}{\pi}\left(\frac{4}{3\pi}\right)^{4/3}$$

and

$$a_1(\beta,0) = \frac{2}{\pi^2}\left(\frac{\pi\beta}{2}\right)^{1/\beta}\{\tan(\pi\beta)\left[\psi(\tfrac{1}{2}(2-\beta)) - \psi(\tfrac{1}{2}(\beta-1)\right] - \pi\}$$

for $\beta \in (1,3/2)\cup(3/2,2)$, where G is the Catalan constant and $\psi(z)$ is the logarithmic derivative of the Gamma function.

In the important case $\beta = 3/2$ these formulae become

$$a_1(\delta) = \frac{9G}{\pi}\left(\frac{4}{3\pi}\right)^{4/3}\left[1 + \sum_{p=1}^{\infty} a_1^{(p)}\delta^p\right], \quad a_1^{(1)} = -\frac{4}{G}\,e^{-\pi/2}.$$

For an L-shaped domain with a re-entrant corner of angle $\frac{3}{2}\pi$ and infinite legs (that is, for $\delta = 0$) the first intensity coefficient is $a_1(0) = (9G/\pi)/(4/(3\pi))^{4/3} = 0.83693221\ldots$ instead of the erroneous value 1.48 obtained by Trefftz [11].

7. Frame-type profile

Consider a frame-type domain $\{w : \operatorname{Im} w \in (-A--1, A+1), \operatorname{Re} w \in (-A-1, A+1)\} \setminus \{w : \operatorname{Im} w \in (-A,A), \operatorname{Re} w \in (-A,A)\}$ with walls of unit thickness and inner side length $2A$. An important mechanical characteristic for this profile is the torsional rigidity

$$C = 8R_0A^2 + 8\int_{\mathcal{G}} \left[\psi(w) - \tfrac{1}{2}|w|^2\right] du\, dv,$$

where R_0 is the Bredt constant [1] appearing in the boundary condition for the inner contour of the domain.

2A	Our results	Roark-Young	Error
5	236.05	216	9.2%
3	73.82	64	15.3%
2	33.06	27	22.4%
1	11.19	8	39.9%
0.5	5.478	3.375	62.3%

The following explicit formulae have been obtained for these quantities:

$$R_0 = \frac{8G}{\pi^2} + \left\{A^2 + \left[1 - \frac{8G}{\pi^2}\right]A + Q_1\right\}\left\{A - \frac{\pi Q_0}{4}\right\}^{-1} + \sum_{p=1}^{\infty} r_p \delta^p,$$

$$C = 8R_0A(A+1) + \tfrac{8}{3}A + \alpha_0 R_0 + \alpha_1 + \sum_{p=1}^{\infty} c_p \delta^p,$$

where the coefficients $Q_0, Q_1, \alpha_0, \alpha_1, r_p$ and c_p have been found in closed form [6].

In the table above we give some numerical results for the torsional rigidity of a frame-type profile, and compare them with those in the well-known handbook [12] on structural mechanics.

References

1. S.P. Timoshenko, *Theory of elasticity*, 1st ed., McGraw-Hill, London-New York, 1934.
2. P. Frank and R. von Mises (eds.), *Die Differential- und Integralgleichungen der Mechanik und Physik.* vols. 1 and 2, F. Vieweg & Sohn, Braunschweig, 1930, 1935.
3. V.L. Vlasov and S.L. Skorokhodov, On an improvement of Trefftz' method. *Russian Acad. Sci. Dokl. Math.* **50** (1995), 157–163.
4. V.L. Vlasov and S.L. Skorokhodov, An analytic solution of the Dirichlet problem for the Poisson equation in a class of polygonal domains, *Comm. Appl. Math.*, Comp. Centre Akad. Sci. USSR, 1988 (Russian).
5. S.L. Skorokhodov, An analytic solution of the Dirichlet problem for T-shaped and cross-shaped domains, in *Analytic and numerical methods for solving problems of mathematical physics*, Comp. Centre Acad. Sci. USSR, Moscow, 1989, 58–70 (Russian).
6. S.L. Skorokhodov, An analytic solution of the Dirichlet problem for one class of doubly connected domains, *Comm. Appl. Math.*, Comp. Centre Akad. Sci. USSR, 1990 (Russian).
7. E. Trefftz, Uber die Torsion prismatischer Stabe von polygonalem Querschnitt, *Math. Ann.* **82** (1921), 97–112.
8. P.P. Kufarev, On the question of torsion and bending of polygonal cross-section beams, *Prikl. Mat. Mekh.* **1** (1937), 43–76 (Russian).
9. W. Koppenfels and F. Stallmann, Praxis der Konformen Abbildung, Springer-Verlag, Berlin-Göttingen-Heidelberg, 1959.
10. V.I. Vlasov, On the variation of the mapping function under deformation of a domain, *Soviet Math. Dokl.* **29** (1984), 377–379.
11. E. Trefftz, Uber die Wirkung einer Abrundung auf die Torsionsspannung in der inneren Ecke eines Winkeleisens, *Z. Angew. Math. Mech.* **2** (1922), 263–267.
12. W.C. Young, *Roark's formulas for stress and strain*, 6th ed., McGraw-Hill, 1989.

Computing Centre, Russian Academy of Science, Moscow

E-mail: vlasov@ccas.ru

TADIE

Steady vortex rings in an ideal fluid: asymptotics for variational solutions

1. Introduction

Steady vortex rings (without swirl) which propagate through an ideal fluid occupying the whole space $\mathbb{R}^3$ are considered. The flow is axisymmetric. In the meridional half-plane ($\theta =$ const, $r > 0$) $\Pi = \{x = (r, z) \mid r > 0;\ z \in \mathbb{R}\}$, where (r, θ, z) denotes the system of cylindrical coordinates, the problem is formulated as follows (see [4]):

Given a flux constant $k > 0$, a uniform velocity $W > 0$ of the rings relative to the fluid at ∞, and a vorticity function $f \in C(\mathbb{R}; \mathbb{R}_+) \cap C^{0,\mu}((0, \infty))$, which is non-decreasing and has support in $[0, \infty)$, find a stream function $\Psi \equiv \Psi_k$ and a cross-section $A \equiv A_k = \{x \in \Pi \mid \Psi(x) > 0\}$ such that $\Psi \in C^{1,\mu}(\bar{A}) \cap C^{2,\mu}(A)$,

$$L\Psi := (1/r^2)\{\partial_r^2 - (1/r)\partial_r + \partial_z^2\}\Psi = -f(\Psi) \quad \text{in} \quad \Pi,$$
$$\Psi|_{r=0} = -k, \Psi_z/r \to 0, \quad \Psi_r/r \to -W \quad \text{at} \quad \infty.$$

The existence of such vortex rings with a small cross-section has been established by many authors (see [2]–[5]). In [2] a unique solution Ψ is found as a perturbation of a vortex cylinder (see (2.1a)).

In [4], given $k, W, \eta > 0$, solutions $\psi(x) := \Psi(x) + Wr^2/2 + k$ of $L\psi = -\lambda f(\Psi)$ are found as maximisers on $S_\eta := \{u \in H(\Pi) \mid ||u||_L^2 := -\int_\Pi uLu\, rdrdz = \eta\}$ of the functional $J(u) := \int_\Pi F(U) rdrdz$, where $F(t) = \int_0^t f(s)ds$ and $U(x) = u(x) - Wr^2/2 - k$; here $H(\Pi)$ denotes the completion of $C_0^\infty(\Pi)$ in the norm $||.||_L$. The parameter $\lambda = \lambda(\psi)$ is a Lagrangian multiplier to be determined a posteriori. These solutions satisfy $\psi(r, z) = \psi(r, -z)$ for any x with $\psi_z < 0$ for $z > 0$; $\psi \in C^{2,\mu}(\bar{\Pi})$ if f is continuous at 0, and $\psi \in C^{1,\mu}(\bar{\Pi}) \cap C^{2,\mu}(\Pi \setminus \partial A)$ and $\Psi^{-1}(0)$ has zero measure otherwise; A is simply connected if $f(t) \leq tf'(t)$, and has a finite or countable number of connected components if $k > 0$. These solutions will be refered to as *variational solutions.* We show that for these solutions, as $k \nearrow \infty$, $\Psi \equiv \Psi_k$ converge to "the same vortex cylinder".

In what follows c or C will denote generic constants and $d\tau := rdrdz$.

2. Some results derived from [2] and [3]

In [2] and [3], for some prescribed $l > 0$, a small $\varepsilon > 0$ and a non-negative $\Omega \in C^1([0, 1))$ with support in $[0, 1]$, the existence of vortex rings such that $\operatorname{diam} A = 2l\varepsilon$ and $\nabla\Psi(l, 0) = 0$ is obtained as a perturbation of the vortex cylinder

$$V'' + V'/s = -\Omega(s), \quad s \in [0, 1], \tag{2.1a}$$

where (s, t) denotes the system of polar coordinates with origin at $(l, 0)$ in Π. Parametrically,

$$\lambda f(\Psi) = U\Omega/(\varepsilon l)^2, \quad \Psi = Ul^2\{V(s) - V(1)\}, \tag{2.1b}$$

where U is a prescribed velocity element. The solutions obtained in this way will be referred to as *perturbation solutions.* The function Ψ is unique when ε is small enough. From the estimates given in [3] for the perturbation solution Ψ_p, say, corresponding to $\Omega \equiv 1$ in $[0,1)$ (where W and $\eta = T/(\rho\pi)$ are eliminated) we deduce the following assertion.

Lemma 2.1. *For ψ_p with $c_0 = 8\pi(2/3)^{3/2}$, as $\tilde{k} \equiv K$ becomes large, we have*
1) $\varepsilon = C\,\exp(-c_0 K^{3/2})\,\{1 + O(K^{-3/2})\}$;
2) $\tilde{U} = c\,K^{-3/2}\{1 + C\,K^{-3/2} + O(K^{-3})\}$;
3) $\tilde{l}^2 = (2/3)K\{1 + c\,K^{-3/2} + O(K^{-3})\}$.

The tilde indicates non-dimensional elements. From now on, $W = \eta = 1$ and the elements will be non-dimensional. For technical reasons, for some $f_0 > 0$ and $m_0 \geq 0$ we take

$$f(t) = f_0 t_+^{m_0} \quad \text{with} \quad t_+ := \max\{0, t\}.$$

Lemma 2.2. *Let $b \geq 0$, and let $Q \equiv Q_b$ be the unique solution of*

$$Q'' + Q'/t = -Q(t)_+^b, \quad t \geq 0; \quad Q(0) = 1, \quad Q'(0) = 0, \tag{2.2}$$

with $a \equiv a(b)$ its unique positive zero and $Q(s)_+ = \max\{Q(s)\,, 0\}$. If

$$\Omega(s) = \frac{a\{Q_b(as)_+\}^b}{2|Q_b{}'(a)|},$$

then for some $\lambda > 0$ the perturbation solutions Ψ satisfy $L\Psi = -\lambda f(\Psi)$ in Π with $f(\theta) = c\,\theta_+^b$ for some $c > 0$.

Proof. This is obtained by means of some simple identifications, where use is made of (2.1b).

Lemma 2.3. *If ψ_p is the solution in [3] and ψ a variational solution, then for large values of k*

$$\begin{aligned} \lambda(\Psi_p) &= O(k^{-5/2}\exp(2c_0 k^{3/2})), \\ |\Psi_p|_{C(A_p)} &= O(k^{-1/2}), \\ \lambda(\psi) &\leq c\,k^{(m_0-2)/2}\exp(2c_0 k^{3/2}). \end{aligned} \tag{2.3}$$

Proof. By Lemma 2.1, the estimates for λ_p and Ψ_p follow from (2.1b). As $|A_p|_\tau := \int_{A_p} d\tau \simeq l^3\varepsilon^2$, we find that $J_0(\psi_p) := \int_{A_p} \Psi_p d\tau \simeq k\exp(-2c_0 k^{3/2})$. Since ψ maximises J on S_1, we deduce that

$$\lambda = \left\{\int_A \psi f(\Psi) d\tau\right\}^{-1} \leq 1/J(\psi) \leq 1/J(\psi_p).$$

From the inequality

$$\int_{A_p} \Psi_p d\tau \le |\Psi_p|_{L^{m_1+1}(A_p)} |A_p|_\tau^{m_1/(m_1+1)}$$

we now obtain $J(\psi_p) \ge c\, k^{(2-m_1)/2} \exp(-2c_0 k^{3/2})$.

3. Some estimates related to the variational solutions

For a very small $\alpha > 0$ and $x_0 \in A$ we define

$$A_\alpha(x_0) := \{x \in A \,|\, s = |x - x_0| = [(r-r_0)^2 + (z-z_0)^2]^{1/2} < \alpha r_0\}.$$

The fundamental solution of L in Π satisfies (see [4] and [7])

$$P(x, x_0) \le r_0\{\log(6/\alpha) + \log(\alpha r_0/s)\} \quad \text{if} \quad x \in A_\alpha(x_0). \tag{3.1}$$

Let $r_1\,(r_2) := \inf(\sup)\{r > 0 \,|\, (r,0) \in A\}$, $A^0 = A \cap \{z = 0\} = \bigcup(a_i, b_i)$, $A_\theta = \{x \,|\, \Psi + \theta > 0; r_1 < r < r_2\}$ and $A_\theta^0 = \bigcup(\alpha_j, \beta_j)$.

Lemma 3.1. *If k is fixed and $r_2 \le \rho(k) < \infty$, then A_θ has a finite number of disjoint connected components for any $\theta > 0$.*

Proof. We define $u(r) = \Psi(r, 0)$, $u \in C^1([0,\infty)) \cap C^2([0,\infty) \setminus \{a_i, b_i\})$. Since $\psi_z < 0$ for $z > 0$ and $\psi(r,z) = \psi(r,-z)$, it follows that $\Psi_{zz}(r,0) \le 0$ for any $r > 0$. So u has only local minima in (b_i, a_{i+1}). As for any i there is i' such that $[a_{i'}, b_{i'}] \subset (\alpha_i, \beta_i)$, if A_θ had infinitely many disjoint connected components, then A_θ^0 would contain an infinite family of disjoint intervals (α_i, β_i) with $\alpha_i < a_{i'} < b_{i'} < \beta_i$. The sequences (α_i), (β_i), $(a_{i'})$ and $(b_{i'})$, increasing and bounded above by $\rho(k)$, would converge to a common limit. But this would mean that u had two distinct values, 0 and θ, at that point.

In the asymptotic process, only the components E such that $\lambda \int_E \psi f(\Psi) d\tau > c$ $\forall k > 0$ for some $c > 0$ will be considered.

Let $E \subset \Pi$. The *L-capacity* of E in Π (if it exists) is defined as

$$\mathrm{Cap}_L(E) := \inf\left\{ ||u||^2_{\Pi\setminus\bar{E}} = -\int_{\Pi\setminus E} uLu d\tau \,|\, u \in C_0^\infty(\Pi);\ u|_E \ge 1\right\}.$$

Theorem 3.2. *For variational solutions ψ, for large values of k we have*

$$1)\quad r_\nu = O(k^{1/2}),\ \nu = 1,2; \quad 2)\quad \lambda \ge c\,\exp(ck^{3/2}); \tag{3.2}$$

$$\text{(i)}\quad |\psi|_{C(A)} = O(k); \quad \text{(ii)}\quad |(3/2)r_\nu^2 - k| \le c\,k^{-1/2}\log k; \tag{3.3}$$

$$\mathrm{diam}\, A \le c\, k^{1/2}\exp(-c_0 k^{3/2}). \tag{3.4}$$

Proof. From [4], p. 42, we have the impulse identity

$$3J(\psi) = \int_A (r^2 - \psi/2) f(\Psi) d\tau \tag{Ip}$$

for any variational solution. So, $r_2^2 \geq r_1^2/4 + k/2$. Let $x \in A$, $A_\alpha \equiv A_\alpha(x)$. From (3.1), $\forall \alpha > 0$ we have $r^2/2 + k \leq \psi(x) \leq c\{(r/k)\log(6/\alpha) + \lambda r_2^{m'}\alpha\}$ for some $m' > 1$. Setting $\alpha = r_2^{-m'}$, we notice that λ cannot be bounded and that $r_2^2/2 + k \leq c(r_2/k)\log\lambda$ and $r_1^2/2 + k \leq c(r_1/k)\log\lambda$ for $\alpha = (\lambda r_2^{m'})^{-1}$. (3.2)2) is obtained from (2.3). Let $E \subset \Pi$ be a connected component of A, and let t_2 $(t_1) = \sup\ (\inf)\{r > 0 : (r, 0) \in E\}$. As $t_i = O(k^{1/2})$ and $\psi|_E > t_1^2/2 + k$ [6], if Λ is a line-segment of length $2a\varepsilon$ parallel to Oz or Or with its extremities in E, then

$$\begin{aligned}\mathrm{Cap}_L(\Lambda) = (2\pi/\{a\log(16e^{-2}/\varepsilon)\})\{1 + O(\varepsilon\log(1/\varepsilon))\} \\ \leq \mathrm{Cap}_L(E) \leq 1/(t_1^2/2 + k)^2),\end{aligned}$$

hence, $\varepsilon \leq c\exp(-2\pi(t_1^2/2 + k)^2/t_2$. Let $t_i = sk^{1/2}$; $y(s) := 2\pi\{s^{3/2}/2 + s^{-1/2}\}^2$ has its minimum $c_0 = 8\pi(2/3)^{3/2}$ at $s_0 = \sqrt{2/3}$, whence $\varepsilon \leq c\exp(-c_0 k^{3/2})$; (3.4) is thus obtained for E (or for A if A has a finite number of connected components [6]). Since there is $c > 0$ such that $(1/c)\lambda\int_E \psi f(\Psi)d\tau \to 1$ as $k \nearrow \infty$, for large k we have $1 \leq ck^{m'}\exp(2c_0k^{3/2})\varepsilon^2$. (3.3) (ii) is then obtained for such E or for A if A has a finite number of components. The fact that (3.3) (ii) holds for each component of A implies that $\operatorname{diam} A_\theta/a \searrow 0$ as $k \nearrow \infty$, where $a := (r_1 + r_2)/2$. So, as $A \subset A_\theta$, we see that $\varepsilon \leq c\exp\{-c_0([1 - \theta/k]k)^{3/2}\}$ for any $\theta > 0$.

4. Problem in the stretched plane $\hat{\Pi}$

Let a, ε be such that $\nabla\Psi(a,0) = 0, \Psi(a,0) = |\Psi|_{C(A)}$ and $\operatorname{diam} A = 2a\varepsilon$. Consider the transformation $x \mapsto \zeta = (\xi, \eta)$, where $r = a(1 + \varepsilon\xi)$ and $z = a\varepsilon\eta$. $\hat{\Pi} := \{\zeta \mid \xi > -1/\varepsilon,\ \eta \in (-\infty, \infty)\}$ is the image of Π and $\hat{E}$ the image of $E \subset \Pi$ under this transformation. For u defined in Π, let $u(\zeta) := u(a(1 + \varepsilon\xi), a\varepsilon\eta)$. The fundamental solution of L in $\hat{\Pi}$ satisfies (see [2])

$$P(\zeta, \zeta_0) := (a/2\pi)\{\log(8e^{-2}/(\varepsilon|\zeta - \zeta_0|)) + R_1\log(8/\varepsilon|\zeta - \zeta_0|)) + R_2\},$$

where $|D^\alpha R_i| = O(\varepsilon)\ \forall|\alpha| \in \mathbb{N}$ if $\varepsilon|\zeta|, \varepsilon|\zeta_0| < 1$.

Lemma 4.1. *If $d\hat{\tau}' = a^3\varepsilon^2(1 + \varepsilon\xi')d\eta'd\xi'$ and $\mu = a^4\varepsilon^2\lambda/2\pi$, then for large k*
 (i) $\varepsilon\lambda\int_{\hat{A}} f(\Psi)\log(1/(a\varepsilon|\zeta - \zeta'|))\, d\hat{\tau}' = O(\varepsilon\log(1/\varepsilon)), |\Psi|_{C(A)} = O(k^{-1/2})$;
 (ii) $\Psi(\zeta) = \mu\int_{\hat{A}} f(\Psi)\log(1/|\zeta - \zeta'|\,)d\xi'd\eta' + O(\varepsilon\log 1/\varepsilon), k^{1/2}|\Psi(\zeta)| \leq c|\zeta|$.

Proof. (i) follows from simple computations. Let B be the disk centered at $(a, 0)$ with radius $k\varepsilon$. For large k it contains A_p and A; so, using (Ip) and (3.3)(ii), we find that $c \leq k^{1/2}|\Psi|_{C(A)} \leq c \log k$ as ψ maximises J on S_1. From (i) we get (ii), since $d\underline{\zeta} = (1+\varepsilon\xi)d\xi d\eta$ implies that

$$\psi(\zeta) = ck + (\mu/2\pi)\int_{\hat{A}} f(\Psi)\log(\,1/|\zeta-\zeta'|\,)d\underline{\zeta'} + O(\varepsilon \log 1/\varepsilon).$$

From (Ip), $\lambda k \int_A f(\Psi)d\tau \simeq \lambda k f(|\Psi|_{C(A)})|A|_\tau = O(1)$ for large k, whence, since $|\Psi|_{C(A)} = k^{-\beta}$, the second part of (i) is derived from $a\mu f(k^{-\beta}) = \lambda a^5\varepsilon^2 f(k^{-\beta}) = O(1)$. As $\hat{A}$ is compact, the last part of (ii) follows from the inequality $|\nabla\psi(\zeta)| \leq c\,k^{-1/2}$.

5. Estimate of $\partial\hat{\mathbf{A}}$ as $\mathbf{k} \nearrow \infty$

We define $V(\zeta) = k^{1/2}\Psi(\zeta)$ and $u_k(\zeta) = V(\zeta)$ $(k \in \mathbb{N})$; by Lemma 4.1, $\nabla u_k(0) = 0$, $\operatorname{diam}\hat{A}_k = 2$ and $|u_k(\zeta)| \leq c\,|\zeta|$.

Theorem 5.1. *Let $\nu \in (0,1)$; (u_k) has a subsequence which converges in $C^{1,\nu}(\hat{\Pi})$ to u, say. Then $u \in C^{2,\nu}(\hat{\Pi})$ if f is continuous, and $u \in C^{1,\nu}(\hat{\Pi}) \cap C^{2,\nu}(\hat{\Pi} \setminus \partial\hat{A})$ otherwise.*

Proof. Let B be a bounded disk centered at O and of radius greater than 1. As $LV = -\lambda k^{1/2} f(\Psi)$ in Π, we have

$$l_\varepsilon u_k := \{\,\partial_\xi^2 - \varepsilon\partial_\xi/(1+\varepsilon\xi) + \partial_\eta^2\,\}u_k = -\lambda a^2(1+\varepsilon\xi)^2 k^{1/2} f(\Psi) \quad \text{in} \quad \hat{\Pi}.$$

For large k, l_ε is an elliptic operator with regular coefficients in B; hence,

$$||u_k||_{W_p^2(B)} \leq C(p,\hat{A}) \quad \forall p > 1.$$

This uniform bound (for fixed m) and elliptic theory complete the proof.

Corollary 5.2. *If u is as in Theorem 5.1, then for $\chi = f_0\sqrt{6}/(4|\hat{A}_u|u(0)^{m_0})$*

$$2\pi u(\zeta) = \chi\int_{\hat{A}_u} u(\zeta')_+^{m_0}\log(1/|\zeta-\zeta'|)d\xi' d\eta', \quad \Delta u = \chi(u_+)^{m_0} \quad \text{in} \quad \hat{\Pi}. \qquad (5.1)$$

Proof. (5.1) follows from Lemma 4.1 (i) and (Ip).

Theorem 5.3. (i) *The limit function u in Theorem 5.1 is radial and $\hat{A}_u = \{\zeta \in \hat{\Pi} : |\zeta| < 1\}$.*
(ii)) *For $\sigma = |\zeta|$ and $m = m_0$, any limit function u in Theorem 5.1 satisfies*

$$u(\sigma) = \{f_0\sqrt{6}/(4\pi a(m)^2)\}Q_m(a(m)\sigma), \qquad (5.2)$$

where Q_m and $a(m)$ are those in (2.2). *Consequently, the sequence $(u_k)_{k\in\mathbb{N}}$ converges to u in $C^{1,\nu}(\hat{\Pi})$.*

Proof. (i) (5.1) implies that u is radial (see Theorem 7.1 in [1] or [8]).

(ii) As u is radial, $u''+u'/\sigma = -\chi(u_+)^m$ and $u'(0) = 0$. For $t = \{\chi u(0)^{m-1}\}^{1/2}\sigma$ and $W(t) = u(\sigma)/u(0)$, W satisfies (2.2) with $b = m$; thus,

$$u(0)W(t) = u(0)Q_m(t) = u(0)Q_m(\{\chi u(0)^{m-1}\}^{1/2}\sigma) = u(\sigma).$$

As $u(1) = 0$, we have $a(m)^2 = \chi u(0)^{m-1} = f_0\sqrt{6}/4\pi u(0)$, whence

$$u(0) = f_0\sqrt{6}/(4\pi a(m)^2),$$

and (5.2) is obtained. Because $u(0)$ does not depend on the choice of the subsequence, any convergent subsequence of (u_k) (in $C^{1,\nu}(\hat{\Pi})$) converges to u.

References

1. L.A. Caffarelli and A. Friedman, Asymptotic estimates for the plasma problem, *Duke Math. J.* **47** (1980), 705–742.
2. L.E. Fraenkel, On steady vortex rings of small cross-section in an ideal fluid, *Proc. R. Soc. Lond. A* **316** (1970), 29–62.
3. L.E. Fraenkel, Examples of steady vortex rings of small cross-section in an ideal fluid, *J. Fluid Mech.* **51** (1972), 119–135.
4. L.E. Fraenkel and M.S. Berger, A global theory of steady vortex rings in an ideal fluid, *Acta Math.* **132** (1974), 13–51.
5. J. Norbury, A family of steady vortex rings, *J.Fluid Mech.* **57** (1973), 417–431.
6. Tadie, Problèmes elliptiques à frontière libre axisymétrique: estimation du diamètre de la section au moyen de la capacité, *Potential Anal.* **5** (1996), 61–72.
7. Tadie, On the bifurcation of steady vortex rings from a Green function. *Math. Proc. Cambridge Philos. Soc.* **116** (1994), 555–568.
8. Tadie, Radial functions as fixed points of some logarithmic operators, *Potential Anal.* **6** (1997) (to appear).

Matematisk Institut, Universitetsparken 5, 2100 Copenhagen, Denmark

E-mail: tad@math.ku.dk

TADIE

Solutions of semilinear elliptic problems based only on weak supersolutions

1. Introduction

We consider the problems

$$E(f,u) := \Delta u + f(r,u) = 0 \quad \text{in} \quad \mathbb{R}^n, \tag{E}$$
$$E(f,u) := \Delta u + f(r,u) = 0 \quad \text{in} \quad \Omega, \tag{E$_\Omega$}$$

where Ω is a bounded domain in $\mathbb{R}^n$. For f, with $\mathbb{R}_+ = [0,\infty)$,

$$\exists u_0 > 0 \text{ such that } f \text{ is continuous and positive in } \mathbb{R}_+ \times [0, \mathrm{u}_0] \tag{f}$$

and $f = f_1 + f_2$, where $f_1 \in C^1(\mathbb{R}_+^2)$ and $f_2 \in C^{0,\nu}(\mathbb{R}_+^2)$ is positive and non-decreasing in u. In this note we use an extension of the super-sub-solution method to obtain the existence of positive solutions of (E), based only on the existence of a continuous supersolution of (E).

Definition 1. *Let $B \subset \mathbb{R}^n$. For $\nu = 0, \alpha$, $\alpha \in (0,1]$, $v, w \in C^\nu(\bar{B})$ are said to be E_B-compatible if*

1) $w \le v$ in $\bar{B}$;

2) $E(f;w) \ge 0$ (that is, w is a weak subsolution of ($\mathrm{E_B}$)*) and $E(f;v) \le 0$ (that is, v is a weak supersolution of* ($\mathrm{E_B}$)*).*

When these super and sub-solutions are radial, they will be assumed to be C^2 except at a finite number of points.

2. Solutions in bounded domains

The following assertion is well known.

Lemma 2.1. 1) *If $\partial\Omega \in C^{2,\alpha}$, $g \in C^{2,\alpha}(\partial\Omega)$ and $h \in C^{0,\alpha}(\overline{\Omega})$, then the problem*

$$\Delta u = h \ \ in \ \ \Omega, \quad u|_{\partial\Omega} = g \tag{2.1}$$

has a classical solution (that is, in $C^{2,\alpha}(\overline{\Omega})$) (Kellogg's theorem).

2) *If $\partial\Omega \in C^{1,\alpha}$, $g \in C^{1,\alpha}(\partial\Omega)$ and $h \in C^{0,\alpha}(\overline{\Omega})$, then* (2.1) *has a solution in* $C^{1,\alpha}(\overline{\Omega})$ (see [3]).

Theorem 2.2. 1) *Suppose that $\partial\Omega \in C^{2,\alpha}$, and that there is a strictly positive supersolution $v \in C^{0,\alpha}(\overline{\Omega})$ of* (E_Ω)*. Then* (E_Ω) *has a classical solution u such that $0 \le u \le v$ in Ω.*

2) *With v as above, if $\partial\Omega \in C^{1,\alpha}$, then there is such a solution $u \in C^{1,\alpha}(\overline{\Omega})$.*

Proof. There is a radial subsolution $w \in C^1(\mathbb{R}_+)$ of (E), positive on Ω and with compact support, such that $w \leq v/4$ in Ω; it can be constructed from (26) in [7], using some suitable multiplying parameters.

As v is strictly positive and Ω is bounded, we can take $g \in C^{2,\alpha}(B)$, where $\overline{\Omega} \subset B$ is such that $w \leq g \leq v$ on $\partial\Omega$. Such g is easy to construct, for example, a radial function with B a ball.

1) Because of 1) in Lemma 2.1, the sequence $(\phi_k)_{k\in\mathbf{N}}$, where $\triangle\phi_{m+1}+f(r,\phi_m) = 0$ in Ω and $\phi_{m+1} = g$ on $\partial\Omega$, with ϕ_1 obtained from v, has all its terms in $C^{2,\alpha}(\overline{\Omega})$. Elliptic theory now completes the proof.

2) This follows from the same process, using 2) in Lemma 2.1.

Theorem 2.3. 1) *Let $b \in (0, u_0)$ and $N > 0$ be such that $b + \int_0^N rf(r,u_0)dr < u_0$. Then for $D = B_N$, $(\mathrm{E_D})$ has a radial solution $u \in C^2(\overline{D})$ such that $u(N) = b$ and $0 < u < u_0$.*

2) *Suppose that there is a strictly positive and decreasing supersolution $v \in C([0,M]) \cap C^2([0,M] \setminus \{r_1,r_2\})$ in $[0,M] \setminus \{r_1,r_2\}$. Then $\forall A \in (0, v(M)]$ and $B = B_M$, $(\mathrm{E_B})$ has a radial solution $u \in C^2(\overline{B})$ such that $u(M) = A$ and $0 < u \leq v$ in B.*

Proof. 1) For $a = n-1$, we define on $E = E_N(b,u_0) := \{u \in C([0,N]) \,|\, b \leq u \leq u_0\}$ an operator K by

$$u_1(t) \equiv Ku(t) := b + \int_t^N \int_0^r (s/r)^a f(s,u).$$

Then $u_1 \in E \cap C^2([0,N])$ and

$$|u_1|_{C^2([0,N])} \leq C_N := b + (N+1)|sf(s,u_0)|_N + (a+1)|f(s,u_0)|_N, \tag{2.3}$$

where $|\cdot|_N$ denotes the norm in $C([0,N])$. The sequence $\phi_1 = Kb$, $\phi_{i+1} = K\phi_i$, has a subsequence which converges to such a decreasing solution.

2) This time we define $E = \{u \in C([0,M]) \,|\, w \leq u \leq v \text{ in } I_M = [0,M]\}$ and K as before, with A replacing b, where w is a positive subsolution in $C^1(I_M)$ such that $w \leq v/2$ in I_M. For $V = Kv$, $I' = I_M \setminus \{r_1,r_2\}$ and $W = Vv' - V'v := V^2(v/V)'$ we have

$$t^a W(t) \leq \int_0^t s^a(v-V)f(s,v) \; \forall t \leq M.$$

So, as $V(M) = A < v(M)$, $V \leq v$ in I'. It is easy to verify that $w \leq Ku \leq V$ if $u \in E$. Moreover, $u \leq \phi$ implies that $Ku \leq K\phi$, and the sequence $v_1 = V$, $v_{k+1} = Kv_k$, has a limit which, as a fixed point of K, is such a solution.

3. Solutions in $\mathbb{R}^n$

3.1. Non-radial solutions

According to elliptic theory, if $\overline{D} \subset \Omega$ and u is a generalised solution of (E_Ω), then u is a classical solution of $(\mathrm{E_D})$.

Theorem 3.1. *Let $v \in C^{0,\alpha}([0,\infty))$ be a positive supersolution of* (E) *that decreases to* 0 *at* ∞. *Then* (E) *has a positive classical solution* u *such that* $u \leq v$ *in* $\mathbb{R}^n$.

Proof. Let $(D_k)_{k\in\mathbf{N}}$ be a family of $C^{2,\alpha}$-domains such that $\overline{D_m} \subset D_{m+1}$ and $\mathbb{R}^n = \bigcup D_m$. For any m, let u_m be the weak solution of $E(f,u) = 0$ in D_{m+1}, $u|_{\partial D_{m+1}} = v$, extended by v outside D_{m+1}. From elliptic theory [1], $u_m \in C^{2,\alpha}(\overline{D_m})$ and

$$\forall k, m \in \mathbb{N}, \quad ||u_{m+k}||_{W^2_p(D_m)} \leq C(p, D_m, v) \quad \forall p \geq 1. \tag{3.1}$$

Using this estimate, we obtain the required solution as the inductive limit of the sequence $\{D_m; v_m\}_{m\in\mathbf{N}}$, where v_1 is the limit in $C^{2,\alpha}(\overline{D_1})$ of a subsequence (u^1_m) of (u_k), v_2 is the limit in $C^{2,\alpha}(\overline{D_2})$ of a subsequence (u^2_m) of (u^1_m), and so on [7].

3.2. Radial solutions

Theorem 3.2. *If there is a positive supersolution* $v \in C(\mathbb{R}_+)$ *of* (E) *that decreases to* 0 *at* ∞, *then* (E) *has a decreasing radial solution* $u \in C^1(\mathbb{R}_+)$ *such that* $0 \leq u \leq v$.

Proof. For (D_m) defined in the proof of Theorem 3.1, the corresponding solutions as in 2) of Theorem 2.3 satisfy

$$|u_{m+k}|_{C^2(\overline{D_m})} \leq C(D_m, v) \quad \forall m, k \in \mathbb{N}, \tag{3.2}$$

and the existence of such a solution is shown similarly.

Lemma 3.3. (Pokhozhaev's identity) *If $u \in C^2(B_R)$ is a radial solution of* (E_{B_R}) *such that* $u(R) = 0$, *then, setting* $F(r,T) = \int_0^T f(r,s)ds$ *and*

$$Hf(\phi) := 2nF(r,\phi) + 2rF_r(r,\phi) - (n-2)\phi f(r,\phi), \tag{3.3}$$

we have

$$R^n u'(R)^2 = \int_0^R r^{n-1} Hf(u)dr. \tag{3.4}$$

The proof of this assertion can be found, for example, in [5].

Lemma 3.4. *Let* $g \in C^1(\mathbb{R}_+)$ *be positive and strictly decreasing. Then for* $\nu > 0$, $R > t_0 > 0$, *there is a continuous subsolution* $w \equiv w(g, t_0, \nu, R)$, *say, of* $(\mathrm{E_D})$ *such that* $g \leq w \leq g + \nu$ *in* D, *where* $D = [t_0, R]$.

Proof. The step function $h(r) = r_i$ in $[r_i, r_{i+1})$, where $r_1 = t_0$, $g(r_i) = \nu + g(r_{i+1})$, $i = 1, 2, \ldots, p$, with $r_p \geq R$, lies between g and $\nu + g$ in D. Here h is a discontinuous subsolution. For any $d, k, T > 0$, $\bar{T} := T\sqrt{n/(n+2)}$ and $y_T(r) := d + k(T^2 - r^2)^2$ we have

$$\Delta y_T = 4k\{(n+2)r^2 - nT^2\} \geq 0 \quad \text{for} \quad r \in [\bar{T}, T]. \tag{3.5}$$

For $\tau = \frac{1}{2}\min\{t>0 \,|\, g(r) = g(r+t);\ r_1 \le r \le r_p\}$ and $\alpha_0 = \frac{1}{2}r_1\{\sqrt{(n+2)/n} - 1\}$, let $\alpha = \min\{\alpha_0, \tau\}$, $\rho_i = r_i + \alpha$ for $i > 1$ and $k_i = \nu/\{\alpha(2r_i+\alpha)\}^2$. The function

$$y_i(r) = g(r_i) + k_i(\rho_i^2 - r^2)^2 \tag{3.6}$$

is used to smooth the jump of h at r_i as $y_i(\rho_i) = g(r_i)$ and $y_i(r_i) = g(r_{i-1})$.

Theorem 3.5. 1) *If $Hf(t) \le 0$ $\forall t \ge 0$, then* (E) *has a positive radial solution $u \in C^2(\mathbb{R}_+)$ such that $u \le u_0$. If, in addition, $\int_0^\infty rf(r,t) = \infty$ $\forall t > 0$, then this solution tends to 0 at ∞.*

2) *If v is a positive radial supersolution of* (E) *such that for some $t_0 > 0$*

(i) *$Hf(v) > 0$ for $0 \le r \le t_0$ and $Hf(\phi) \le 0$ for $r > t_0$ and any ϕ such that $0 \le \phi \le v$, and*

(ii) *$\exists R_0 > t_0$ such that $\int_0^{t_0} r^{n-1}Hf(v)dr + \int_{t_0}^R r^{n-1}Hf(v/2)dr < 0$ $\forall R > R_0$, then* (E) *has a classical positive radial solution u such that $u \le v$.*

Proof. 1) By (3.4), the classical solution in 1) of Theorem 2.3 is extendable to a classical radial solution positive in the whole space (see, for example, Theorem 4.5 in [5]). This solution u, say, has the representation $u(r) = b + \int_r^\infty t^{1-n}\int_0^t s^{n-1}f(s,u)$ if $u \searrow b > 0$ at ∞. This implies that $u_0 \ge u(0) > \int_0^\infty rf(r,b)dr$, and the conclusion follows.

2) Taking $g = v/2$ and $\nu = v(R_0)/4$ in Lemma 3.4, from the subsolution $w = w(v/2, v(R_0)/4, R_0)$ we have a classical radial solution u of (E_{R_0}) such that

$$\int_0^{R_0} r^{n-1}Hf(u)dr \le \int_0^{t_0} r^{n-1}Hf(v)dr + \int_{t_0}^{R_0} r^{n-1}Hf(v/2)dr \le 0;$$

u is extendable to such a solution.

4. Concluding remarks

1) If $f(r,t) \simeq (r^b t^p)/\{(1+r)^\theta\}$, $\theta - b \in [0,2)$ and $p \in ((n+2-2(\theta-b))/(n-2),\ (n+2)(n-2))$ for $t \in [0, u_0]$, then $\exists m, \alpha > 0$ such that $v(r) = (1+r^\alpha)^{-m}$ satisfies the requirements of 2) in Theorem 3.5 [7].

2) For radial solutions, in [2] f is required to be non-increasing with respect to r; the method used here does not need this condition.

3) The requirement of the supersolutions to be C^2 except at a finite number of points is a big advantage in their construction.

References

1. S. Agmon, A. Douglis and L. Nirenberg, Estimates near the boundary for solutions of elliptic differential equations satisfying general boundary conditions. I, *Comm. Pure Appl. Math.* **12** (1959), 628–727.

2. B. Gidas, W.M. Ni and L. Nirenberg, Symmetry and related properties via the maximum principle, *Comm. Math. Phys.* **68** (1979), 209–243.

3. D. Gilbarg and N.S. Trudinger, *Elliptic partial differential equations of second order*, 2nd ed., Springer-Verlag, Heidelberg, 1983.

4. T. Kusano and M. Naito, Positive entire solutions of superlinear elliptic equations, *Hiroshima Math. J.* **16** (1986), 361–366.

5. W.M. Ni, On the elliptic equation $\Delta u + Ku^{(n+2)/(n-2)} = 0$, its generalization and applications in geometry, *Indiana Math. J.* **31** (1982), 493–529.

6. W.M. Ni and J. Serrin, Nonexistence theorems for singular solutions of quasilinear partial differential equations, *Comm. Pure Appl. Math.* **39** (1986), 379–399.

7. Tadie, Weak and classical positive solutions of some elliptic equations in $\mathbb{R}^n$, $n \geq 3$. Radially symmetric cases, *Quart. J. Math.* **45** (1994), 397–406.

Matematisk Institut, Universitetsparken 5, 2100 Copenhagen, Denmark

E-mail: tad@math.ku.dk

G.R. THOMSON and C. CONSTANDA

On stationary oscillations in bending of plates

1. Preliminaries

In what follows, Greek and Latin subscripts and superscripts take the values 1,2 and 1,2,3, respectively, unless otherwise stated, $x = (x_1, x_2)$ and $x = (x_1, x_2, x_3)$ are generic points with respect to orthogonal Cartesian coordinates in $\mathbb{R}^2$ and $\mathbb{R}^3$, the superscript T denotes matrix transposition, and the convention of summation over repeated indices is understood.

Let S be a domain in $\mathbb{R}^2$ bounded by a C^2-curve ∂S. We consider a homogeneous and isotropic plate occupying a region $\bar{S} \times [-h_0/2, h_0/2]$ in $\mathbb{R}^3$, where $h_0 \ll \operatorname{diam} S$ is the constant thickness. The Lamé constants of the elastic material are denoted by λ and μ, and ρ is the (constant) mass density.

We assume that the body forces F acting on the plate are of the form

$$F(x,t) = \operatorname{Re}\left[f(x)e^{-i\omega t}\right], \tag{1}$$

where f is a complex-valued (3×1)-vector function, and that the boundary conditions prescribed on ∂S are separable with respect to the space and time variables in the same way. The body performs stationary oscillations of frequency ω and it is reasonable to assume that the displacements V can be written as

$$V(x,t) = \operatorname{Re}\left[u(x)e^{-i\omega t}\right], \tag{2}$$

where u is an unknown complex-valued (3×1)-vector function.

In accordance with the Mindlin-type plate model proposed in [1], we expect u to be of the form

$$u_\alpha(x_1, x_2, x_3) = x_3 v_\alpha(x_1, x_2), \quad u_3(x_1, x_2, x_3) = v_3(x_1, x_2). \tag{3}$$

This kinematic assumption takes into account transverse shear deformation, which is neglected in the classical Kirchhoff model.

Substituting (1), (2) and (3) into the equations of motion of three-dimensional elasticity and averaging over the thickness of the plate leads to the system of equations [2]

$$A^\omega(\partial_x)v(x) = \left[A(\partial/\partial x_1, \partial/\partial x_2) + \begin{pmatrix} \rho\omega^2 h^2 & 0 & 0 \\ 0 & \rho\omega^2 h^2 & 0 \\ 0 & 0 & \rho\omega^2 \end{pmatrix}\right] v(x) = H(x),$$

where

$$A(\xi_1, \xi_2) = \begin{pmatrix} h^2\mu\Delta + h^2(\lambda+\mu)\xi_1^2 - \mu & h^2(\lambda+\mu)\xi_1\xi_2 & -\mu\xi_1 \\ h^2(\lambda+\mu)\xi_1\xi_2 & h^2\mu\Delta + h^2(\lambda+\mu)\xi_2^2 - \mu & -\mu\xi_2 \\ \mu\xi_1 & \mu\xi_2 & \mu\Delta \end{pmatrix},$$

$h^2 = h_0^2/12$, $\Delta = \xi_1^2 + \xi_2^2$, and H is a combination of the body forces and moments, and of the forces and moments acting on the faces. If H is sufficiently smooth, we can reduce (4) to a homogeneous system by constructing a so-called Newtonian potential [3]. Consequently, for simplicity, from now on we consider the homogeneous system

$$A^\omega(\partial_x)v(x) = 0. \tag{5}$$

We also consider the boundary stress operator $T(\partial_x) = T(\partial/\partial x_1, \partial/\partial x_2)$ defined by [2]

$$T(\xi_1, \xi_2) = \begin{pmatrix} h^2(\lambda + 2\mu)\nu_1\xi_1 + h^2\mu\nu_2\xi_2 & h^2\mu\nu_2\xi_1 + h^2\lambda\nu_1\xi_2 & 0 \\ h^2\lambda\nu_2\xi_1 + h^2\mu\nu_1\xi_2 & h^2\mu\nu_1\xi_1 + h^2(\lambda + 2\mu)\nu_2\xi_2 & 0 \\ \mu\nu_1 & \mu\nu_2 & \mu\nu_\alpha\xi_\alpha \end{pmatrix},$$

where $\nu = (\nu_1, \nu_2)^{\mathrm{T}}$ is the unit outward normal to ∂S.

In what follows, S^+ denotes the domain enclosed by ∂S, $S^- = \mathbb{R}^2 \setminus (S^+ \cup \partial S)$, and we assume that

$$\lambda + \mu > 0, \quad \mu > 0, \tag{6}$$

$$\omega > h^{-1}(\mu/\rho)^{1/2}. \tag{7}$$

The conditions (6) ensure that the system (5) is elliptic [1].

2. Boundary value problems

We introduce the class $\mathcal{A}^\omega$ of (3×1)-vector functions v in S^- that, as $R = |x| \to \infty$, admit an asymptotic expansion of the form

$$\begin{aligned}
&v_1(R, \theta) \\
&= R^{-1/2}\{m_0^{(1)} + m_0^{(2)} - m_0^{(3)} + (m_1^{(1)} + m_1^{(2)} + 2m_2^{(1)} + 2m_2^{(2)} - m_2^{(3)}) \cos\theta \\
&\qquad + (m_3^{(1)} + m_3^{(2)} - m_3^{(3)}) \sin\theta + (m_0^{(1)} + m_0^{(2)} + m_0^{(3)}) \cos 2\theta \\
&\qquad + (m_4^{(1)} + m_4^{(2)} + m_4^{(3)}) \sin 2\theta + (m_2^{(1)} + m_2^{(2)} + m_2^{(3)} - m_5^{(1)} - m_5^{(2)}) \cos 3\theta \\
&\qquad + (m_3^{(1)} + m_3^{(2)} + m_3^{(3)}) \sin 3\theta\} + O(R^{-3/2}),
\end{aligned}$$

$$\begin{aligned}
&v_2(R, \theta) \\
&= R^{-1/2}\{m_4^{(1)} + m_4^{(2)} - m_4^{(3)} + (m_3^{(1)} + m_3^{(2)} - m_3^{(3)}) \cos\theta \\
&\qquad + (m_1^{(1)} + m_1^{(2)} + 2m_5^{(1)} + 2m_5^{(2)} + m_2^{(3)}) \sin\theta - (m_4^{(1)} + m_4^{(2)} + m_4^{(3)}) \cos 2\theta \\
&\qquad + (m_0^{(1)} + m_0^{(2)} + m_0^{(3)}) \sin 2\theta - (m_3^{(1)} + m_3^{(2)} + m_3^{(3)}) \cos 3\theta \\
&\qquad + (m_2^{(1)} + m_2^{(2)} + m_2^{(3)} - m_5^{(1)} - m_5^{(2)}) \sin 3\theta\} + O(R^{-3/2}),
\end{aligned}$$

$$\begin{aligned}
&v_3(R, \theta) \\
&= R^{-1/2}\{m_6^{(1)} + m_6^{(2)} + (m_7^{(1)} + m_7^{(2)}) \cos\theta + (m_8^{(1)} + m_8^{(2)}) \sin\theta \\
&\qquad + (m_9^{(1)} + m_9^{(2)}) \cos 2\theta + (m_{10}^{(1)} + m_{10}^{(2)}) \sin 2\theta\} + O(R^{-3/2}).
\end{aligned}$$

Here the $m_l^{(j)}$, $l = 0, \ldots, 10$, are functions of R and θ such that, as $R \to \infty$,

$$m_l^{(j)} = O(1), \quad \frac{\partial m_l^{(j)}}{\partial R} - ik_j m_l^{(j)} = O(R^{-1}), \quad \frac{\partial m_l^{(j)}}{\partial \theta} = O(1), \quad \frac{\partial^2 m_l^{(j)}}{\partial \theta^2} = O(R),$$

where

$$k_1^2 + k_2^2 = \frac{k^2(\lambda + 3\mu)}{\lambda + 2\mu}, \quad k_1^2 k_2^2 = \frac{\mu k^2 (k^2 h^2 - 1)}{h^2(\lambda + 2\mu)}, \quad k_3^2 = k^2 - \frac{1}{h^2}, \tag{8}$$

with $k^2 = \rho\omega^2/\mu$. If (6) and (7) are satisfied, the k_j^2 are positive and distinct [2].

It can be shown that every function $v \in \mathcal{A}^\omega$ satisfies the radiation conditions for a solution of system (5) in S^- formulated in [4].

Let $\mathcal{P}, \mathcal{Q}, \mathcal{R}$, and $\mathcal{S}$ be (3×1)-vector functions prescribed on ∂S. We consider the following interior and exterior Dirichlet and Neumann boundary value problems:

($\mathrm{D}^{\omega+}$) Find $v \in C^2(S^+) \cap C^1(\bar{S}^+)$ which satisfies (5) in S^+ and $v|_{\partial S} = \mathcal{P}$.

($\mathrm{N}^{\omega+}$) Find $v \in C^2(S^+) \cap C^1(\bar{S}^+)$ which satisfies (5) in S^+ and $Tv|_{\partial S} = \mathcal{Q}$.

($\mathrm{D}^{\omega-}$) Find $v \in C^2(S^-) \cap C^1(\bar{S}^-) \cap \mathcal{A}^\omega$ which satisfies (5) in S^- and $v|_{\partial S} = \mathcal{R}$.

($\mathrm{N}^{\omega-}$) Find $v \in C^2(S^-) \cap C^1(\bar{S}^-) \cap \mathcal{A}^\omega$ which satisfies (5) in S^- and $Tv|_{\partial S} = \mathcal{S}$.

The exterior problems ($\mathrm{D}^{\omega-}$) and ($\mathrm{N}^{\omega-}$) have at most one solution for any ω satisfying (7) [4].

We introduce the single and double layer potentials

$$(V^\omega\varphi)(x) = \int_{\partial S} D^\omega(x,y)\varphi(y)\,ds(y), \quad (W^\omega\varphi)(x) = \int_{\partial S} P^\omega(x,y)\varphi(y)\,ds(y),$$

where

$$D^\omega(x,y) = (\mathrm{adj}\ A^\omega)(\partial_x)[t(x,y)E_3], \quad P^\omega(x,y) = \left[T(\partial_y)D^\omega(y,x)\right]^{\mathrm{T}},$$

$$t(x,y) = \sum_{j=1}^{3} b_j H_0^{(1)}(k_j|x-y|),$$

E_3 is the identity (3×3)-matrix, and $H_0^{(1)}$ is the Hankel function of the first kind of order zero. The constants b_j are chosen so that

$$(\det A^\omega(\partial_x))\, t(x,y) = h^4\mu^2(\lambda + 2\mu)(\Delta + k_1^2)(\Delta + k_2^2)(\Delta + k_3^2)t(x,y) = -\delta(|x-y|),$$

where δ is the Dirac delta distribution and the k_j are defined by (8).

Theorem 1. (i) $V^\omega\varphi \in \mathcal{A}^\omega$, $W^\omega\varphi \in \mathcal{A}^\omega$.

(ii) *If* $\varphi \in C(\partial S)$, *then* $V^\omega\varphi$ *and* $W^\omega\varphi$ *are analytic and satisfy* (5) *in* $S^+ \cup S^-$.

(iii) *If* $\varphi \in C^{0,\alpha}(\partial S)$, $\alpha \in (0,1)$, *then the direct values* $V_0^\omega\varphi$ *and* $W_0^\omega\varphi$ *of* $V^\omega\varphi$ *and* $W^\omega\varphi$ *on* ∂S *exist (the latter as principal value), the functions*

$$\mathcal{V}^{\omega+}(\varphi) = (V^\omega\varphi)|_{\bar{S}^+}, \quad \mathcal{V}^{\omega-}(\varphi) = (V^\omega\varphi)|_{\bar{S}^-}$$

are of class $C^\infty(S^+) \cap C^{1,\alpha}(\bar{S}^+)$ *and* $C^\infty(S^-) \cap C^{1,\alpha}(\bar{S}^-)$, *respectively, and*

$$T\mathcal{V}^{\omega+}(\varphi) = (W_0^{\omega*} + \tfrac{1}{2}I)\varphi, \quad T\mathcal{V}^{\omega-}(\varphi) = (W_0^{\omega*} - \tfrac{1}{2}I)\varphi,$$

on ∂S, *where* $W_0^{\omega*}$ *is the adjoint of* W_0^ω *and* I *is the identity operator.*

(iv) *If* $\varphi \in C^{1,\alpha}(\partial S)$, $\alpha \in (0,1)$, *then the functions*

$$\mathcal{W}^{\omega+}(\varphi) = \begin{cases} (W^\omega\varphi)|_{S^+} & \text{in } S^+, \\ (W_0^\omega - \frac{1}{2}I)\varphi & \text{on } \partial S, \end{cases} \quad \mathcal{W}^{\omega-}(\varphi) = \begin{cases} (W^\omega\varphi)|_{S^-} & \text{in } S^-, \\ (W_0^\omega + \frac{1}{2}I)\varphi & \text{on } \partial S, \end{cases}$$

are of class $C^\infty(S^+) \cap C^{1,\alpha}(\bar{S}^+)$ *and* $C^\infty(S^-) \cap C^{1,\alpha}(\bar{S}^-)$, *respectively, and the equality* $T\mathcal{W}^{\omega+}(\varphi) = T\mathcal{W}^{\omega-}(\varphi)$ *holds on* ∂S.

Proof. Parts (ii),(iii), and (iv) follow from the behaviour of these functions in the neighbourhood of ∂S. The singularities that occur coincide with those of the corresponding functions in the static bending of plates, which have been fully investigated in [1]. Part (i) follows from the far-field pattern of the potentials.

3. Existence of solutions

If we seek the solutions of $(\mathrm{D}^{\omega+})$, $(\mathrm{N}^{\omega+})$, $(\mathrm{D}^{\omega-})$, and $(\mathrm{N}^{\omega-})$ in the form $\mathcal{W}^{\omega+}(\varphi)$, $\mathcal{V}^{\omega+}(\varphi)$, $\mathcal{W}^{\omega-}(\varphi)$, and $\mathcal{V}^{\omega-}(\varphi)$, respectively, then, by Theorem 1, these problems reduce to the singular integral equations

$$(W_0^\omega - \tfrac{1}{2}I)\varphi = \mathcal{P}, \qquad (\mathcal{D}^{\omega+})$$

$$(W_0^{\omega*} + \tfrac{1}{2}I)\varphi = \mathcal{Q}, \qquad (\mathcal{N}^{\omega+})$$

$$(W_0^\omega + \tfrac{1}{2}I)\varphi = \mathcal{R}, \qquad (\mathcal{D}^{\omega-})$$

$$(W_0^{\omega*} - \tfrac{1}{2}I)\varphi = \mathcal{S}. \qquad (\mathcal{N}^{\omega-})$$

The Fredholm Alternative is applicable to $(\mathcal{D}^{\omega+})$, $(\mathcal{N}^{\omega-})$ and $(\mathcal{N}^{\omega+})$, $(\mathcal{D}^{\omega-})$, which are mutually adjoint, since the index of each of these equations is zero [1].

In what follows, a subscript zero denotes a homogeneous equation (problem).

Theorem 2. (i) $(\mathcal{D}_0^{\omega-})$ *has non-trivial* $C^{1,\alpha}$*-solutions if and only if* ω *is an eigenfrequency of* $(\mathrm{N}_0^{\omega+})$. *These solutions coincide with the boundary values of the linearly independent eigenfunctions of* $(\mathrm{N}_0^{\omega+})$.

(ii) $(\mathcal{N}_0^{\omega-})$ *has non-trivial* $C^{0,\alpha}$*-solutions if and only if* ω *is an eigenfrequency of* $(\mathrm{D}_0^{\omega+})$. *These solutions coincide with the boundary stresses generated by the linearly independent eigenfunctions of* $(\mathrm{D}_0^{\omega+})$.

Theorem 3. (i) *If ω is not an eigenfrequency of* $(\mathrm{D}_0^{\omega+})$*, then* $(\mathrm{D}^{\omega+})$ *is solvable for any* $\mathcal{P} \in C^{1,\alpha}(\partial S)$*. If ω is an eigenfrequency of* $(\mathrm{D}_0^{\omega+})$*, then* $(\mathrm{D}^{\omega+})$ *is solvable if and only if*

$$\int\limits_{\partial S} \mathcal{P}^{\mathrm{T}}(y) T(\partial_y) v^{(k)}(y)\, ds(y) = 0, \quad k = 1, \ldots, \mu,$$

where $\{v^{(k)}\}_{k=1}^{\mu}$ *is a complete system of eigenfunctions for* $(\mathrm{D}_0^{\omega+})$*. The solution is given by* $\mathcal{W}^{\omega+}(\varphi)$ *with* $\varphi \in C^{1,\alpha}(\partial S)$*.*

(ii) *If ω is not an eigenfrequency of* $(\mathrm{N}_0^{\omega+})$*, then* $(\mathrm{N}^{\omega+})$ *is solvable for any* $\mathcal{Q} \in C^{0,\alpha}(\partial S)$*. If ω is an eigenfrequency of* $(\mathrm{N}_0^{\omega+})$*, then* $(\mathrm{N}^{\omega+})$ *is solvable if and only if*

$$\int\limits_{\partial S} \mathcal{Q}^{\mathrm{T}}(y) v^{(k)}(y)\, ds(y) = 0, \quad k = 1, \ldots, \mu,$$

where $\{v^{(k)}\}_{k=1}^{\mu}$ *is a complete system of eigenfunctions for* $(\mathrm{N}_0^{\omega+})$*. The solution is given by* $\mathcal{V}^{\omega+}(\varphi)$ *with* $\varphi \in C^{0,\alpha}(\partial S)$*.*

(iii) $(\mathrm{D}^{\omega-})$ *has a unique solution for any* $\mathcal{R} \in C^{1,\alpha}(\partial S)$*. If ω is not an eigenfrequency of* $(\mathrm{N}_0^{\omega+})$*, then the solution is given by* $\mathcal{W}^{\omega-}(\varphi)$ *with* $\varphi \in C^{1,\alpha}(\partial S)$*. If ω is an eigenfrequency of* $(\mathrm{N}_0^{\omega+})$*, then the solution is given by a linear combination of* $\mathcal{W}^{\omega-}(\varphi)$*, with* $\varphi \in C^{1,\alpha}(\partial S)$*, and* $\mathcal{V}^{\omega-}(\psi_*^{(k)})$*,* $k = 1, \ldots, \nu$*, where* $\{\psi_*^{(k)}\}_{k=1}^{\nu}$ *is a complete system of* $C^{0,\alpha}$*-eigenfunctions for* $(\mathcal{N}_0^{\omega+})$*.*

(iv) $(\mathrm{N}^{\omega-})$ *has a unique solution for any* $\mathcal{S} \in C^{0,\alpha}(\partial S)$*. If ω is not an eigenfrequency of* $(\mathrm{D}_0^{\omega+})$*, then the solution is given by* $\mathcal{V}^{\omega-}(\varphi)$ *with* $\varphi \in C^{0,\alpha}(\partial S)$*. If ω is an eigenfrequency of* $(\mathrm{D}_0^{\omega+})$*, then the solution is given by a linear combination of* $\mathcal{V}^{\omega-}(\varphi)$*, with* $\varphi \in C^{0,\alpha}(\partial S)$*, and* $\mathcal{W}^{\omega-}(\psi_*^{(k)})$*,* $k = 1, \ldots, \nu$*, where* $\{\psi_*^{(k)}\}_{k=1}^{\nu}$ *is a complete system of* $C^{1,\alpha}$*-eigenfunctions for* $(\mathcal{D}_0^{\omega+})$*.*

These results are proved by means of Theorem 2 and the Fredholm Alternative.

References

1. C. Constanda, *A mathematical analysis of bending of plates with transverse shear deformation*, Pitman Res. Notes Math. Ser. **215**, Longman, Harlow, 1990.
2. P. Schiavone and C. Constanda, Oscillation problems in thin plates with transverse shear deformation, *SIAM J. Appl. Math.* **53** (1993), 1253-1263.
3. V.D. Kupradze et al., *Three-dimensional problems of the theory of elasticity and thermoelasticity*, North-Holland, Amsterdam, 1979.
4. C. Constanda, Small oscillations in bending of plates with transverse shear deformation, *Strathclyde Math. Research Report* **9** (1996).

Department of Mathematics, University of Strathclyde, Glasgow, U.K.

E-mail: g.r.thomson@strath.ac.uk, c.constanda@strath.ac.uk

T. TIIHONEN

A non-local problem arising from heat radiation on non-convex surfaces

1. Introduction

Radiative heat exchange plays an important role in many situations. In general, it has to be taken into account whenever the temperature on a visible surface of the system is high enough, or when other heat transfer mechanisms are not present (for example, in vacuum). In non-convex geometries, the radiative heat exchange between different surfaces leads to a non-local boundary condition on the radiating part of the boundary.

We consider heat transfer in a system that consists of a union of finitely many opaque, conductive and bounded objects which have diffuse and gray surfaces and are surrounded by a perfectly transparent and non-conducting medium (such as vacuum). We denote the union of conductive objects by Ω and note that in general Ω is not a connected set. We assume that the boundary $\partial\Omega$ of Ω can be represented as $\partial\Omega = \Gamma \cup \Sigma$, where Σ denotes the part of the boundary on which radiative heat transfer takes place. For simplicity, on Γ we assume linear Newton-type boundary conditions as we concentrate mainly on questions relating to radiating boundaries.

Σ can be naturally partitioned into disjoint components, $\Sigma = \Sigma_0 \cup_i \Sigma_i$, so that the points of Σ_0 do not 'see' any other points of Σ and the points of Σ_i do not see any points on $\Sigma \setminus \Sigma_i$.

Under the above assumptions, the stationary heat equation for Ω combined with heat radiation on Σ results in a coupled system for the absolute temperature T and radiosity λ (intensity of total radiation leaving the surface). Namely, we have

$$-k\Delta T + v \cdot \nabla T = f \quad \text{in } \Omega, \tag{1}$$

where k is the coefficient of heat conductivity, v denotes the convection velocity multiplied by the heat capacity and f is the internal heat source. On the boundary part Γ we have

$$k\partial T/\partial n + \alpha(T - T_0) = 0 \quad \text{on } \Gamma \tag{2}$$

for some positive α. Finally, on Σ_i we have

$$k\partial T/\partial n + (I - K_i)\lambda_i = K_i^\infty \lambda^\infty \quad \text{on } \Sigma_i. \tag{3}$$

Here K_i^∞ specifies the amount of outside radiation that reaches Σ_i and λ_i is the radiosity of the surface Σ_i. The latter depends on the surface temperature through the relation

$$\lambda_i = (1-\epsilon)(K_i\lambda_i + K_i^\infty\lambda^\infty) + \epsilon\sigma T^4, \tag{4}$$

which says that the radiosity is the sum of the reflected radiation and the Stefan-Boltzmann radiation of the surface itself. The operator K_i (see [1] or [2]) defined by

$$(K_i\lambda)(z) = \pi^{-1}\int_{\Sigma_i} n_z \cdot (s-z)\; n_s \cdot (z-s)\|s-z\|^{-4}\Xi(z,s)\lambda(s)ds \tag{5}$$

describes the self-illumination on non-convex surfaces. Here n denotes the unit outer normal and $\Xi(z,s) = 1$ if $\overline{zs} \cap \Omega = \emptyset$, otherwise $\Xi(z,s) = 0$. As the Stefan-Boltzmann law is physically meaningful only for positive temperatures, we monotonize it by replacing σT^4 with $h(T) = \sigma |T|^3 T$.

Writing (1)–(4) in variational form, we get the system

$$\int_\Omega (k\nabla T \nabla \phi + v \cdot \nabla T \phi) + \int_\Gamma \alpha T \phi + \sum_i \int_{\Sigma_i} (I - K_i)\lambda_i \phi = \langle f, \phi \rangle_{V' \times V} \quad \forall \phi \in V, \quad (6)$$

$$-\int_{\Sigma_i} \epsilon h(T)\psi_i + \int_{\Sigma_i} (I - (1-\epsilon)K_i)\lambda_i \psi_i = \langle g, \psi_i \rangle_{W_i' \times W_i} \quad \forall \psi_i \in W_i \quad (7)$$

for the radiosity λ_i and absolute temperature T. The above system is well defined (in the three-dimensional case) if we set $V = H^1(\Omega) \cap_i L^5(\Sigma_i)$ and $W_i = L^5(\Sigma_i)$. Then the radiosity λ_i will be in $W_i' = L^{5/4}(\Sigma_i)$ and T in V. In this paper we recall the main results that imply that the above problem is well posed and has a solution. The proofs can be found in [3] and [4].

2. Radiation on non-convex surfaces

In this section we recall some properties of the operators K_i and the corresponding kernel. Unless stated otherwise, in what follows we consider Ω to be a three-dimensional body. In order to simplify the notation, we denote the integral kernel in (5) by $w(z,s)$.

The kernel and the operators K_i have the following properties.

Lemma 1. *Let Σ_i be a piecewise $C^{1,\delta}$-surface, so that the set of non-smooth points of Σ_i has zero surface measure. Then for any $s \in \Sigma_i$ for which the normal is defined, the integral $\int_{\Sigma_i} w(s,z)dz$ exists and its value is less than or equal to one.*

Lemma 2. *The operator K_i is non-negative, that is, it has a non-negative kernel. K_i maps $L^p(\Sigma_i)$ into itself compactly and $\|K_i\| \le 1$. If there exists a constant $k < 1$ such that $\int_{\Sigma_i} w(z,s)ds \le k$ for any $z \in \Sigma_i$, then $\|K_i\| \le k$.*

Lemma 3. *If Σ_i is not an enclosure, then $\|K_i\|_{L^p} < 1$ for any p, $1 < p < \infty$.*

Lemma 4. *The operator $I - (I - E)K_i$ from $L^p(\Sigma_i)$ into itself is invertible with non-negative inverse whenever $0 < \epsilon_0 \le \epsilon \le 1$.*

Here E is the operator induced by multiplication by ϵ. Lemma 4 implies that λ_i can be obtained from (4) and the problem (6), (7) can be written in terms of T alone by solving for λ_i from (7). If we introduce the operators

$$G_i = (I - K_i)(I - (I - E)K_i)^{-1}E,$$

then $(I - K_i)\lambda_i = G_i h(T) + \hat{g}$, and the problem reads

$$\int_\Omega k\nabla T \nabla \phi + v \cdot \nabla T \phi + \int_\Gamma \alpha T \phi + \sum_i \int_{\Sigma_i} G_i h(T) \phi = \langle \tilde{f}, \phi \rangle_{V' \times V}. \quad (8)$$

The operator G_i has the following properties.

Lemma 5. *The operator G_i from $L^2(\Sigma_i)$ into itself is symmetric and positive semi-definite. Moreover, if $\|K_i\|_2 < 1$, then G_i is positive definite.*

Lemma 6. *The operator G_i can be written as $G_i = I - H_i$, where H_i is non-negative and $\rho(H_i) \leq 1$. Moreover, for $1 < p < \infty$, $\|H_i\| \leq 1$ as a mapping from $L^p(\Sigma_i)$ into itself. Strict inequality is obtained when Σ_i is not an enclosure.*

Lemma 6 implies the (semi-) coerciveness of the non-local term modelling radiative heat transfer. In particular, it can be used to show that $\int_{\Sigma_i} G_i(h(T))T \geq 0$ for any T (see Lemma 7 in the next section). However, this does not imply the monotonicity of the radiation term. In fact, non-monotonicity is an intrinsic property of the system.

3. Existence results

We write

$$\tilde{a}(T,\phi) = a_0(T,\phi) + \sum_i a_i(T,\phi), \tag{9}$$

where

$$a_0(T,\phi) = \int_\Omega k\nabla T \nabla\phi + v\cdot\nabla T\phi + \int_\Gamma \alpha T\phi,$$
$$a_i(T,\phi) = \int_{\Sigma_i} G_i h(T)\phi.$$

Hence, our problem (8) takes the form

$$\tilde{a}(T,\phi) = \langle \tilde{f},\phi\rangle_{V'\times V} \quad \forall\phi\in V. \tag{10}$$

We have some partial coerciveness results in the case where the domain Ω is connected.

Lemma 7. *Suppose that Ω is connected, $\nabla\cdot v = 0$, $v\cdot n \geq 0$ on $\partial\Omega\setminus\Gamma$, $2\alpha + v\cdot n > 0$ on Γ and either* (i) *Γ has positive surface measure, or* (ii) *there exists at least one Σ_i with positive surface measure which is not an enclosure. Then the form a_0 is coercive in $H^1(\Omega)$, or $a_0 + a_i$ is coercive in $H^1(\Omega)\cap L^5(\Sigma_i)$.*

First we consider two important cases when the problem is coercive.

In two dimensions, the trace theorem implies that the boundary values of $H^1(\Omega)$-functions are in $L^p(\Sigma)$ for all finite p. Thus, the non-linear and non-local boundary terms are well defined in a two-dimensional setting for standard H^1-functions. Moreover, the mapping from $H^1(\Omega)$ to $L^5(\Sigma)$ is compact. Hence, the boundary terms can be considered in some sense as small perturbations of the elliptic main part. Of course, in 2D the definition of the kernel w must be changed appropriately to keep the physical properties.

Theorem 1. *Suppose that Ω is a connected piecewise $C^{1,\delta}$-domain in R^2 which satisfies one of the conditions in Lemma 7. Then* (10) *has a solution for every* $\tilde{f} \in (H^1(\Omega))'$.

Proof. The proof follows from Lemma 6 and a result in [4]. Essential ingredients are the coerciveness of the problem in $H^1(\Omega)$, by Lemma 7, and the compactness of the non-monotone part of G_i.

The second case when coerciveness can be achieved is the situation where the radiating surfaces emit part of the radiation out of the system. This makes the non-local operator contractive and results in coerciveness with respect to the $L^5(\Sigma)$-norm. More precisely, the following assertion holds.

Theorem 2. *Suppose that Ω is a connected piecewise $C^{1,\delta}$-domain with no enclosures. Then the problem* (10) *has a solution for every* $\tilde{f} \in V'$.

Proof. We make use of pseudomonotonicity. To prove that the non-local part is compact, we have to show that $h(T_n)$ converges weakly in $L^{5/4}(\Sigma)$ if T_n converges weakly in $H^1(\Omega) \cap L^5(\Sigma)$. The rest follows from the compactness of K and Lemma 7 (see [3] for details).

In the general case (that is, a 3D geometry with enclosures) we have neither coerciveness, nor monotonicity. Hence, we use the technique of sub- and supersolutions. We denote by V^+ the cone of the non-negative elements of V:

$$V^+ = \{v \in V |\ v \geq 0\}.$$

Theorem 3. *Let Ω be a three-dimensional domain with a piecewise $C^{1,\delta}$-boundary that satisfies the Lipschitz condition $\delta > 0$, and suppose that $\tilde{f} \in V'$, that the velocity v satisfies the conditions in Lemma* 7, *and that there exist two functions $\phi \leq \psi$, $\phi, \psi \in H^1(\Omega) \cap_i L^5(\Sigma_i)$, such that*

$$\tilde{a}(\phi, w) \leq \langle \tilde{f}, w \rangle \quad \forall w \in V^+,$$
$$\tilde{a}(\psi, w) \geq \langle \tilde{f}, w \rangle \quad \forall w \in V^+.$$

Then (10) *has a solution T. Moreover, $\phi \leq T \leq \psi$ in Ω.*

Proof. We sketch the general scheme; the detailed proof can be found in [3]. We can show that the sequence T_n defined by $T_1 = \psi$ and

$$a_0(T_n, w) + \sum_i \int_{\Sigma_i} h(T_n) w = \sum_i \int_{\Sigma_i} H_i h(T_{n-1}) w + \langle \tilde{f}, w \rangle \quad \forall w \in V, \quad n > 1,$$

is a decreasing sequence of supersolutions bounded below by ϕ. Hence, it has a limit, which is a solution of our problem.

Finally, we discuss briefly the extension of the above results to the parabolic case. Using the notation (9), we can write the parabolic problem formally as

$$\int_{\Omega} T_t w + \tilde{a}(T, w) = \langle f, w \rangle_{V' \times V} \quad \forall w \in V, \text{ for a.a. } t \in [0, \tau], \tag{11}$$

with initial condition $T(0) = T_0 \in L^2(\Omega)$. In the absence of non-linear terms on Σ_i, the natural space for the solution would be $X = L^2(0, \tau : H^1(\Omega)) \cap H^1(0, \tau : (H^1(\Omega))')$. However, in the 3D case, the non-linear term is not well defined for functions in X.

In order to be able to work with standard spaces, we implicitly assume some additional regularity for the solutions. Namely, we assume that there exist sub- and supersolutions $\phi \leq \psi$, $\phi, \psi \in X \cap L^\infty([0, \tau] \times (\cup_i \Sigma_i))$. Then we can define a truncation by setting $[T] = \min(\max(T, \phi), \psi)$.

Theorem 4. *Suppose that* $f \in L^2(0, \tau : (H^1(\Omega))')$ *and* $T_0 \in L^2(\Omega)$ *are such that the problem* (11) *has sub- and supersolutions* $\phi \leq \psi$ *such that* $\phi(0) \leq T_0 \leq \psi(0)$ *and* $\phi, \psi \in X \cap L^\infty([0, \tau] \times (\cup_i \Sigma_i))$. *Then, if* Ω *satisfies the assumptions in Theorem* 3, *the problem* (11) *has a solution* $T \in X$.

References

1. R. Siegel and J.R. Howell, *Thermal radiation heat transfer*, Hemisphere, New York, 1981.
2. E.M. Sparrow and R.D. Cess, *Radiation heat transfer*, Hemisphere, New York, 1978.
3. T. Tiihonen, Non-local problem arising from heat radiation on non-convex surfaces, *European J. Appl. Math.* (to appear).
4. T. Tiihonen, Stefan-Boltzmann radiation on non-convex surfaces, *Math. Methods Appl. Sci.* **20** (1997), 47–57.
5. E. Zeidler, *Nonlinear functional analysis, II B: Nonlinear monotone operators*, Springer-Verlag, Berlin, 1988.

Department of Mathematics, University of Jyväskylä, Box 35, FIN-40351 Jyväskylä, Finland

P.E. TOVSTIK

Asymptotic methods in the problem of large axisymmetric deflections of a thin elastic shell of revolution

1. Introduction

The large axisymmetric deflections of a thin elastic shell of revolution under axial loading are studied. Two-dimensional equations of Kirchhoff–Love type are used. This problem is very sensitive to this type of equations, which is why the non-linear three-dimensional equations of the theory of elasticity are used to derive approximate two-dimensional equations. For the case of axial force applied to the shell edges two problems are solved. The limiting value of force and the value of force for post-buckling deformations are found.

2. The shell deformation and the equilibrium equations

We study the compression of a thin elastic shell of revolution with non-negative Gaussian curvature by an axial force P (for compression $P < 0$). We assume that the shell deformations are axisymmetric. The question of bifurcation to non-symmetric modes is not examined here.

Fig. 1 illustrates the shell generatrix before and after deformation. The curve on the right side of Fig. 1 shows the typical axial force P—shell shortening Δ dependence. The interesting points on this curve are A and B. A corresponds to the maximal compressing force and is called *the limiting point*.

B corresponds to large axisymmetric deflections (see also [1]–[3]). The form of the deformed generatrix is such that one of its parts ($s_1 \leq s_0 < s_*$) is close to the initial form while the other ($s_* < s_0 \leq s_2$) is close to the mirror-image form. If not the force P but the shortening Δ is given, then the position of the point s_* may be arbitrary.

We expand all unknown values and functions in power series in the small geometric parameter μ. Let C be the dimensionless axial force. Its expansion is

$$C = \frac{P}{2\pi R E h \mu^2} = C^0 + \mu C^1 + \cdots,$$
$$\mu^4 = \frac{h^2}{12(1-\nu^2)R^2}, \tag{2.1}$$

where E is Young's modulus and h, R and ν are the shell thickness, its typical linear dimension and Poisson's ratio. In the case A, the limiting point satisfies $C^0 \neq 0$; in the case B of mirror reflection, $C^0 = 0$. So to find correctly the force P in the case B it is necessary to use more accurate initial equations than in the case A.

This work was partly supported by the RFFI grant 96-01-00411.

In this problem the strain is comparatively large, of order μ. Therefore, we take into account the main non-linear terms in the stress-strain relations, which are also of order μ compared to the linear ones.

The neutral surface before deformation (see Fig. 1) is described by

$$r_0' = \cos\theta_0, \quad k_{10} = \theta_0', \quad k_{20} = \sin\theta_0/r_0,$$
$$(\ldots)' \equiv d(\ldots)/ds_0, \tag{2.2}$$

where s_0 is the length of the generatrix, $r_0(s_0)$ is the distance between the current point and the axis of symmetry, $\theta_0(s_0)$ is the angle between the shell normal and the axis, and k_{10} and k_{20} are the main curvatures of the neutral surface. We denote the same values after deformation by s, r, θ, k_1, k_2. Formulae similar to (2.2) are hold.

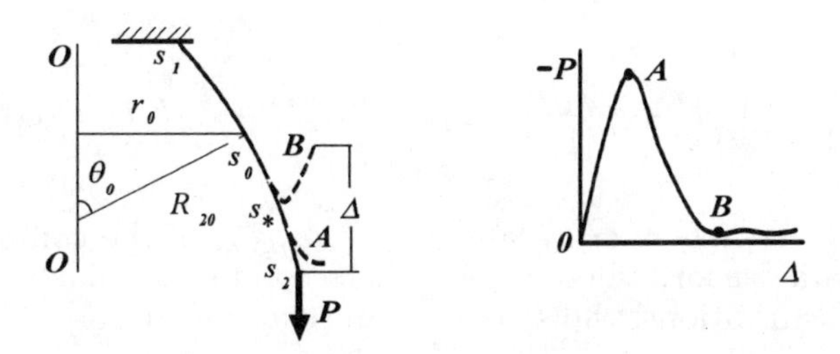

Fig. 1. Shell of revolution under axial compression.

The stretching deformations ε_1 and ε_2 of the neutral surface and the changes in its curvatures κ_1 and κ_2 are

$$\begin{aligned} \varepsilon_1 &= s' - 1, \\ \varepsilon_2 &= (r - r_0)/r_0, \\ \kappa_1 &= \theta'/(1+\varepsilon_1) - \theta_0', \\ \kappa_2 &= \sin\theta/r - \sin\theta_0/r_0. \end{aligned} \tag{2.3}$$

From (2.2) and (2.3) it follows that $(r_0\varepsilon_2)' = (1+\varepsilon_1)\cos\theta - \cos\theta_0$.

The equilibrium equations are

$$\begin{aligned} &(r_0 U)' = T_2, \qquad U = T_1\cos\theta + Q_1\sin\theta, \\ &(r_0 V)' = 0, \qquad V = T_1\sin\theta - Q_1\cos\theta, \\ &(r_0 M_1)' - M_2\cos\theta - r_0(1+\varepsilon_1)Q_1 = 0, \end{aligned} \tag{2.4}$$

where T_j, Q_1, and M_j are the stress-resultants and stress couples related to the unit length of neutral surface *before* deformation.

3. The stress-strain relations

All previous formulae are exact. Errors occur when we add the so-called *stress-strain relations* between T_j, M_j and ε_1, ε_2, κ_1, κ_2. The simplest linear elasticity relations are

$$T_i = K(\varepsilon_i + \nu\varepsilon_j), \quad M_i = D(\kappa_i + \nu\kappa_j), \quad j \neq i, \tag{3.1}$$

where $K = Eh/(1-\nu^2)$, and $D = Eh^3/[12(1-\nu^2)]$. To derive more exact relations, we assume that the stress-strain state satisfies the estimates

$$\{\varepsilon_1, \varepsilon_2\} \sim \mu, \quad \kappa_2 \sim 1, \quad \{\theta', \kappa_1\} \sim 1/\mu, \quad \frac{dy}{ds_0} \sim y/(\mu R), \tag{3.2}$$

where y is any unknown function, the shell material is homogeneous, isotropic and non-linearly elastic, and the elastic potential Φ may be expanded in a series of the form

$$\Phi = \frac{E\nu}{2(1+\nu)(1-2\nu)} I_1^2 + \frac{E}{2(1+\nu)} I_2 + E(\alpha_1 I_1^3 + \alpha_2 I_1 I_2 + \alpha_3 I_3) + \cdots \tag{3.3}$$

in the invariants $I_1 = \varepsilon_{ii}$, $I_2 = \varepsilon_{ij}\varepsilon_{ji}$ and $I_3 = \varepsilon_{ij}\varepsilon_{jk}\varepsilon_{ki}$ of the components ε_{ij} of the Cauchy–Green strain tensor, where α_k are the second order elastic modules.

Under these assumptions, the stress-strain relations are [4]

$$\begin{aligned}
T_1 &= K(\varepsilon_1 + \nu\varepsilon_2 + a_1\varepsilon_1^2 + 2a_2\varepsilon_1\varepsilon_2 + a_2\varepsilon_2^2) + Da_3\kappa_1^2 + O(Eh\mu^3), \\
T_2 &= K(\varepsilon_2 + \nu\varepsilon_1 + a_2\varepsilon_1^2 + 2a_2\varepsilon_1\varepsilon_2 + a_1\varepsilon_2^2) + Da_4\kappa_1^2 + O(Eh\mu^3), \\
M_1 &= D(\kappa_1 + \nu\kappa_2 + a_5\varepsilon_1\kappa_1 + a_6\varepsilon_2\kappa_1) + O(Eh^2\mu^3), \\
M_2 &= D(\kappa_2 + \nu\kappa_1 + a_7\varepsilon_1\kappa_1 + a_8\varepsilon_2\kappa_1) + O(Eh^2\mu^3),
\end{aligned} \tag{3.4}$$

where the constants a_j depend on ν and linearly on the α_k. The relative error in (3.4) is of order μ^2. In the next section we obtain a simplified version of (2.4) with the same degree of accuracy.

4. Simplification of system (2.4)

In the neighbourhood of the shell edges and of the point s_* (see Fig. 1) the estimates (3.2) hold. We introduce dimensionless variables of order 1 by

$$\begin{aligned}
\{s_0, r_0\} &= R\{s_0^\circ, r_0^\circ\}, \quad \{\varepsilon_1, \varepsilon_2\} = \mu\{\varepsilon_1^\circ, \varepsilon_2^\circ\}, \\
\{M_1, M_2\} &= EhR\mu^3\{M_1^\circ, M_2^\circ\}, \\
\{T_1, Q_1, U, V\} &= Eh\mu^2\{T_1^\circ, Q_1^\circ, U^\circ, V^\circ\}, \quad T_2 = Eh\mu T_2^\circ.
\end{aligned} \tag{4.1}$$

For simplicity, in what follows we omit the superscript $^\circ$.

We rewrite equations (2.4) and (3.4) in the form

$$\begin{gathered}
\mu(r_0 U)' = \varepsilon_2 + \mu\nu(U\cos\theta + V\sin\theta) + \mu c_1\varepsilon_2^2 + \mu^3 c_2 \theta'^2, \\
\mu(r_0\varepsilon_2)' = (1-\mu\nu\varepsilon_2)\cos\theta - \cos\theta_0, \\
\mu(r_0 M_1)' = r_0(1-\mu\nu\varepsilon_2)(U\sin\theta - V\cos\theta) + \mu\nu M_1\cos\theta, \\
\mu\theta' = M_1 + \mu(\theta_0' - \nu\kappa_2^0) - 2\mu^2 c_2\varepsilon_2\theta',
\end{gathered} \tag{4.2}$$

where $V = C/r_0$ and $\kappa_2^0 = (\sin\theta - \sin\theta_0)/r_0$. Dependence on non-linear elasticity takes effect only through the terms with coefficients c_1 and c_2, where

$$\begin{gathered}
c_1 = 3/2 + 3\alpha_1\nu_1^3 + 3\alpha_2\nu_1(1+2\nu^2) + 3\alpha_3(1-2\nu^3), \\
c_2 = -2\nu + \nu_2(1+\nu)[3\alpha_1\nu_1^3 + \alpha_2\nu_1(1-4\nu+6\nu^2) - 3\alpha_3\nu(\nu_1+2\nu^2)],
\end{gathered} \tag{4.3}$$

with $\nu_1 = 1-2\nu$ and $\nu_2 = (1-\nu)^{-1}$.

5. Asymptotic solution of system (4.2)

We assume that the edge $s_0 = s_1$ is clamped and the edge $s_0 = s_2$ may move freely in the radial direction. Then

$$\varepsilon_2 = 0, \quad \theta = \theta_0 \quad \text{at } s_0 = s_1, \quad U = M_1 = 0 \quad \text{at } s_0 = s_2, \tag{5.1}$$

and we rewrite (4.2) in the form

$$\mu\mathbf{x}' = \mathbf{F}(\mathbf{x}, C, s, \mu), \quad \mathbf{x} = \{x_1, x_2, x_3, x_4\} = \{U, \varepsilon_1, M_1, \theta\}. \tag{5.2}$$

In the case of the limiting point for the boundary conditions (5.1), the main deformations are concentrated near the edge $s_0 = s_2$. Therefore, we seek the solution of (4.2) (or (5.2)) in the form

$$\mathbf{x} = \mathbf{x}^0(\xi) + \mu\mathbf{x}^1(\xi) + O(\mu^2), \quad C = C^0 + \mu C^1 + O(\mu^2), \tag{5.3}$$

where $\xi = (s_0 - s_2)/\mu$ is a stretched independent variable. The exactness of system (4.2) allows us to construct correctly only the first two terms in (5.3).

By equating the coefficients of μ^0 and μ^1 with zero and solving the corresponding systems, we get the approximate expression of the limiting force $P = P^l$:

$$\begin{gathered}
P^l = 2\pi REh\mu^2(C^0 + \mu C^1 + O(\mu^2)), \\
C^1 = j_1 + j_2 k_1 + j_3\nu + f_1\alpha_1 + f_2\alpha_2 + f_3\alpha_3, \\
f_1 = \nu_1^3 j_4 + (1+\nu)\nu_1^3\nu_2 j_5, \\
f_2 = \nu_1(1+2\nu^2)j_4 + \nu_2(1+\nu)\nu_1(1-4\nu+6\nu^2)j_5, \\
f_3 = \nu_1^3 j_4 - 3\nu\nu_2(1+\nu)(\nu_1+2\nu^2)j_5.
\end{gathered} \tag{5.4}$$

Here C^0 and j_k depend only on the angle $\gamma = \theta_0(s_2)$ (see Table 1).

γ	C^0	j_1	j_2	j_3	j_4	j_5
15°	−0.0271	−0.0216	0.7429	0.0666	−0.0024	−0.0002
30	−0.1078	−0.0439	1.0362	0.1751	−0.0260	−0.0024
45	−0.2402	−0.0336	1.2330	0.2807	−0.0989	−0.0085
60	−0.4221	−0.0323	1.3637	0.3430	−0.2374	−0.0172
75	−0.6535	−0.1487	1.4423	0.3142	−0.4133	−0.0209

Table 1.

For the case $\gamma > \pi/2$ we get $C^0(\pi - \gamma) = C^0(\gamma)$ and $C^1(\pi - \gamma) = -C^1(\gamma)$.

From (5.4) and Table 1 it follows that for the shallow shells (with small values of γ) the effect of the non-linear terms with factors c_1 and c_2 decreases.

In a similar way, the final expression for the axial force $P = P^m$ of the mirror-reflected state is found as

$$P^m(s_*) = 2\pi REh\mu^3(C^1(s_*) + \mu C^2(s_*) + O(\mu^2)),$$
$$C^1 = \frac{a_1 + \rho\gamma' a_2}{\sqrt{\rho}}, \tag{5.5}$$
$$C^2 = \frac{(\rho\hat{\gamma}m_1)'}{\sin\gamma},$$

where $\rho = r_0(s_*)$, $\gamma = \theta_0(s_*)$, $\hat{\gamma} = \gamma$ for $\gamma < \pi/2$ and $\hat{\gamma} = \pi - \gamma$ for $\gamma > \pi/2$, and $m_1 = k_{10} + \nu k_{20}$. The coefficients a_1 and a_2 depend only on γ (see [5] and Table 2).

γ	a_1	a_2	γ	a_1	a_2
10°	0.02003	0.58212	50°	0.16071	1.39607
20	0.05469	0.83023	60	0.17201	1.58250
30	0.09440	1.03142	70	0.15681	1.78332
40	0.13216	1.21549	80	0.10409	2.00831

Table 2.

If $\gamma > \pi/2$, then $C^1(\pi - \gamma) = -C^1(\gamma)$ and $C^2(\pi - \gamma) = C^2(\gamma)$.

As an example, we consider a conic shell for which

$$\theta_0 = \gamma = \text{const}, \quad \rho = s_* \cos\gamma, \quad C^1 = a_1/\sqrt{\rho}, \quad C^2 = 0.$$

We note that if the wide edge is reflected, then $P^m > 0$, and vice versa. The absolute value of the force P^m increases when s_* decreases and the angle γ increases.

6. Discussion

Let $\alpha_k = 0$. In some works (see [6] and [7]) a different initial system is used. This system has the same form (2.4) but the stress resultants T_1 and T_2 and the stress couples M_1 and M_2 are related to the unit of length *after* deformation. At the same time, the stress-strain relations (3.1) are used. As a result, in system (2.4) the values T_1, T_2, M_1, and M_2 are replaced respectively by $(1+\varepsilon_2)T_1$, $(1+\varepsilon_1)T_2$, $(1+\varepsilon_2)M_1$, and $(1+\varepsilon_1)M_2$.

We solve this new system by the same asymptotic method as in Section 5 and calculate the force P^m. For definiteness we take a conic shell with parameters $\gamma = 45°$ and $\nu = 0.3$. Our formula (5.5) gives $C^1 = 0.148$, while the new system gives $C^1 = 0.209$. This difference does not decrease with the shell thickness. As shown by Udin [8], in the case of the new system the potential energy does not exist.

References

1. A.V. Pogorelov, *Geometric theory of shell stability*, Moscow, Nauka, 1966 (Russian).
2. G.A. Kriegsmann and C.G. Lange, On large axisymmetrical deflection states of spherical shells, *J. Elasticity* **10** (1980), 179–192.
3. A.Yu. Evkin and A.V. Korovaitsev, Asymptotic analysis of the post-buckling stress-strain state of shells of revolution under strong flexure, *Izv. Ross. Akad. Nauk Mekh. Tverd. Tela* **27** (1992), 125–133 (Russian).
4. P.E. Tovstik, The post-buckling axisymmetrical deflections of thin shells of revolution under axial loading, *Technische Mech.* **16** (1996), 117–132.
5. P.V. Koroteeva, P.E. Tovstik and S.P. Shuvalkin, On the value of the axial force for post-buckling axisymmetric deflections of a shell of revolution, *Vestnik S.-Peterburg. Univ. Mat. Mekh. Astronom.* **1995**, no. 2, 58–61 (Russian).
6. N.V. Valishvili, *Computer design of shells of revolution*, Moscow, Mashinostroenie, 1976 (Russian).
7. E.I. Grigoluk and V.I. Shalashilin, *Problems of nonlinear deformations*, Nauka, Moscow, 1988 (Russian).
8. A.C. Udin, On some non-linear equations of the axisymmetric deformations of shells of revolution, *Trudy Severo-Kavkaz. Nauchn. Tsentra Vyssh. Shkoly* **4** (1973), 93–98.

Department of Theoretical and Applied Mechanics, St. Petersburg University,
2 Bibliotechnaya Sq., Peterhof, 198904 St. Petersburg, Russia

E-mail: pet@pet.niimm.spb.su

V. TRETYNYK

Discrete symmetries and the reduction of the two-body Dirac equations with interactions

We consider two-particle relativistic equations with scalar, pseudoscalar, vector and oscillator-like interactions. These basic equations describing the steady states of the system are [1], [2]

$$L_i\psi = [\alpha_1 p_1 + \alpha_2 p_2 + (\beta_1 + \beta_2)m + U_i]\psi = E_i\psi, \quad i = 1,2,3,4, \tag{1}$$

where

$$\begin{aligned} U_1 &= \beta_1\beta_2 f(x) && \text{(scalar)},\\ U_2 &= -\alpha_1\alpha_2\beta_1\beta_2 f(x) && \text{(pseudoscalar)},\\ U_3 &= \tfrac{1}{2}(1-\alpha_1\alpha_2)f(x) && \text{(vector)},\\ U_4 &= \tfrac{1}{2}im\omega x(\beta_1\alpha_1 + \beta_2\alpha_2) && \text{(oscillator-like)}, \end{aligned}$$

$$\psi = \psi(x), \quad f(x) = f(-x), \quad x = x_1 - x_2, \quad p_1 = -i\frac{\partial}{\partial x_1}, \quad p_2 = -i\frac{\partial}{\partial x_2},$$

and α_1, α_2, β_1 and β_2 are 4×4 matrices satisfying the commutation relations

$$\{\alpha_i,\alpha_j\} \equiv \alpha_i\alpha_j + \alpha_j\alpha_i = 2\delta_{ij}, \quad \{\beta_i,\beta_j\} = 2\delta_{ij},$$
$$[\alpha_i,\beta_j] \equiv \alpha_i\beta_j - \beta_j\alpha_i = 0, \quad i,j = 1,2.$$

For interactions depending on $x_1 - x_2$ we can make a canonical transformation to total and relative coordinates and momenta and consider equations (1) in the centre of mass frame. Moreover, making the similarity transformation

$$\psi \to \psi' = \beta_2\psi, \quad L_i \to L_i' = \beta_2 L_i \beta_2,$$

we reduce equations (1) to the form

$$L_1'\psi' = \{2[\hat\beta_0,\hat\beta_1]p + 2\hat\beta_0 m + (2\hat\beta_0^2 - 1)f(x)\}\psi' = E_1\psi', \tag{2a}$$

$$L_2'\psi' = \{2[\hat\beta_0,\hat\beta_1]p + 2\hat\beta_0 m - (2\hat\beta_1^2 + 1)f(x)\}\psi' = E_2\psi', \tag{2b}$$

$$L_3'\psi' = \left\{2[\hat\beta_0,\hat\beta_1]p + 2\hat\beta_0 m + \tfrac{1}{2}[1 + (2\hat\beta_0^2 - 1)(2\hat\beta_1^2 + 1)]f(x)\right\}\psi' = E_3\psi', \tag{2c}$$

$$L_4'\psi' = \{2[\hat\beta_0,\hat\beta_1]p + 2\hat\beta_0 m + im\omega x\hat\beta_1\}\psi' = E_4\psi', \tag{2d}$$

where $p = p_1 - p_2$; the matrices $\hat\beta_0$, $\hat\beta_1$ of (2a)–(2d) satisfy the Kemmer-Duffin-Petiau algebra

$$\hat\beta_a\hat\beta_b\hat\beta_c + \hat\beta_c\hat\beta_b\hat\beta_a = g_{ab}\hat\beta_c + g_{bc}\hat\beta_a,$$
$$a,b,c = 0,1, \quad g_{00} = -g_{11} = 1.$$

Our interest in equations (2a)–(2d) resides in using discrete symmetries [5] such as parity, time-inversion and complex conjugation transformations to reduce these equations to sets of uncoupled sub-systems that can be solved independently.

Definition. We say that (2a)–(2d) are invariant under the transformation generated by an operator

$$Q = MR, \tag{3}$$

where M is a numerical matrix and R is either a discrete symmetry operator or a composition of such operators, if

$$[L_i', Q] = 0, \quad i = 1, 2, 3, 4.$$

We concentrate our attention on searching for symmetry operators (3) of equations (2a)–(2d).

Proposition 1. *Equations* (2a)–(2d) *are invariant under the transformation*

$$Q\psi(x) = MR\psi(x) = M\psi(-x), \tag{4}$$

where

$$M = 2\hat{\beta}_0^2 - 1 \tag{5}$$

and R is a space inversion operator.

We now use the algorithm for reducing discrete-invariant systems of differential equations originally proposed in [3]. The main idea of this algorithm is to transform the corresponding discrete symmetry operator to a diagonal form.

Proposition 2. *The similarity transformation*

$$\begin{gathered} Q \to Q' = U_1^+ Q U_1^-, \quad \psi' \to \psi'' = U_1^+ \psi', \\ L_i' \to L_i'' = U_1^+ L_i' U_1^- = \begin{pmatrix} L_i^+ & 0_2 \\ 0_2 & L_i^- \end{pmatrix}, \\ U_1^\pm = I_2 \otimes I_2 R_+ \mp I_2 \otimes i\sigma_2 R_-, \quad R_\pm = \tfrac{1}{2}(1 \pm R), \end{gathered}$$

where Q is defined in (4), (5), *L_i' are defined in* (2), *$A \otimes B$ is the direct product of the matrices A and B, I_2 and 0_2 are the unit and zero 2×2 matrices, and σ_2 is the Pauli matrix, reduces each of the equations* (2a)–(2d) *to two uncoupled sub-systems.*

We can represent these sub-systems in the form

$$\begin{aligned} L_i^+ \psi^+ &= E_i \psi^+, \\ L_i^- \psi^- &= E_i \psi -, \end{aligned} \tag{6}$$

where ψ^+, ψ^- are two-component functions. The explicit forms of $L_i^{\pm}$ are

$$\begin{aligned}
L_1^{\pm} &= K^{\pm} \pm Rf(x), \\
L_2^{\pm} &= K^{\pm} - (2+\sigma_3)f(x), \\
L_3^{\pm} &= K^{\pm} + \left[R_{\pm} \pm \tfrac{1}{2}(1+\sigma_3)R\right] f(x), \\
L_4^{\pm} &= K^{\pm} \mp \tfrac{1}{2}im\omega(1+\sigma_3)x, \\
K^{\pm} &= \pm(-i\sigma_1+\sigma_2)R_{\mp}p \pm (i\sigma_1+\sigma_2)R_{\pm}p + 2m\sigma_2 R_{\pm}.
\end{aligned}$$

We see that we can continue this process until our equations are reduced to one-component uncoupled sub-systems.

Indeed, it can be easily verified that equations (6) are invariant under the transformation

$$Q\psi(x) = \sigma_3 RC = \sigma_3\psi^*(-x),$$

where C is the operator of complex conjugation and σ_3 is the Pauli matrix.

In an analogous way we construct a diagonalizing unitary operator for Q:

$$\begin{aligned}
U_2^+ &= (C_+ - i\sigma_2 C_-)(R_+ - i\sigma_2 R_-), \\
U_2^- &= (R_+ + i\sigma_2 R_-)(C_+ + i\sigma_2 C_-), \\
C_{\pm} &= \tfrac{1}{2}(1 \pm C),
\end{aligned}$$

and the corresponding uncoupled systems

$$\tilde{L}_i\tilde{\psi}_i = E_i\tilde{\psi}_i, \quad i = 1,2,3,4 \text{ (no summation on } i).$$

Hence, the initial 4-component equations (2) reduce to four 1-component uncoupled equations. In the case of equation (2a) they have the form

$$\begin{aligned}
[-2imR_+C + f(x)R]\psi_1 &= E_1\psi_1, \\
[-2imR_+C + f(x)R + 2iRp]\psi_2 &= E_1\psi_2, \\
[-f(x) + 2imR_-C + 2ip]\psi_3 &= E_1\psi_3, \\
[-f(x) + 2imR_-C]\psi_4 &= E_1\psi_4;
\end{aligned}$$

for equation (2b):

$$\begin{aligned}
[-2imR_+C + f(x)(2+R)]\psi_1 &= E_2\psi_1, \\
[-2imR_+C + 2iRp + f(x)(2-R)]\psi_2 &= E_2\psi_2, \\
[2imR_-C + 2ip + f(x)(2+R)]\psi_3 &= E_2\psi_3, \\
[2imR_-C + f(x)(2-R)]\psi_4 &= E_2\psi_4;
\end{aligned}$$

for equation (2c):

$$\begin{aligned}
\big[-2imR_+C + f(x)(R_+ + \tfrac{1}{2}(R+1))\big]\psi_1 &= E_3\psi_1,\\
\big[-2imR_+C + 2iRp + f(x)(R_+ + \tfrac{1}{2}(R-1))\big]\psi_2 &= E_3\psi_2,\\
\big[2imR_-C + 2ip + f(x)(R_- - \tfrac{1}{2}(R+1))\big]\psi_3 &= E_3\psi_3,\\
\big[2imR_-C + f(x)(R_- - \tfrac{1}{2}(R-1))\big]\psi_4 &= E_3\psi_4;
\end{aligned}$$

for equation (2d):

$$\begin{aligned}
[im\omega(C_+R_- + C_-R_+)x - 2imR_+C]\psi_1 &= E_4\psi_1,\\
[-im\omega(C_-R_- + C_+R_+)x - 2imR_+C + 2iRp]\psi_2 &= E_4\psi_2,\\
[-im\omega(C_+R_- + C_-R_+)x + 2imR_-C + 2ip]\psi_3 &= E_4\psi_3,\\
[im\omega(C_-R_- + C_+R_+)x + 2imR_-C]\psi_4 &= E_4\psi_4.
\end{aligned}$$

Thus, we have shown how the use of discrete symmetries enables us to reduce systems of differential equations to uncoupled sub-systems.

References

1. F. Dominguez-Adame and B. Mendez, *Canad. J. Phys.* **69** (1991), 780–785.
2. F.A.B. Continho, W. Clockle, Y. Nogami and F. Toyama, *Canad. J. Phys.* **66** (1988), 769–772.
3. J. Niederle and A.G. Nikitin, *J. Math. Phys.* (to appear).
4. W.I. Fushchych and A.G. Nikitin, *Symmetries of equations of quantum mechanics*, Allerton Press, New York, 1994.
5. L. Gendenstein and I.V. Krive, Supersymmetry in quantum mechanics, *Uspekhi Fiz. Nauk* **146** (1985), 554–589 (Russian).

Department of Physics, Ukrainian State University of Food Technology,
34 Volodymyrs'ka St., 252034 Kyiv, Ukraine

E-mail: viola@apmat.freenet.kiev.ua

N. VIRCHENKO

On some new types of multiple integral equations and their applications

The method of multiple integral (or series) equations is one of the most effective modern analytic techniques for solving mixed boundary value problems [1]. This method has been widely used of late.

A system of lN-integral equations is a set of integral equations of the form

$$\int_{\Delta_i^j} \sum_{k=1}^{l} a_{ik}^j(\tau)\,\phi_k(\tau)\,K_i^j(\tau,x)\,d\tau = f_i^j(x) \tag{1}$$

$$(x \in I_i^j, \quad i=\overline{1,N}, \quad j=\overline{1,l}),$$

where Δ_i^j, I_i^j are segments of the real axis, $\phi_k(\tau)$ are unknown functions on the set

$$\Delta = \bigcup_{i=1}^{N} \bigcup_{j=1}^{l} \Delta_i^j,$$

$K_i^j(\tau,x)$ is the i-kernel of j-equation, $f_j^j(x)$ are known functions given on I_i^j, and $a_{ik}^j(\tau)$ are known weight functions. It should be noted that (1) may be formulated in matrix form if on every I_i^j we define a vector function f as a column matrix with l components. The set of N-matrix integral operators is defined on Δ and the unknown vector-column $\phi(\tau)$ must satisfy N-matrix integral relations. It is easy to generalise (1) to the n-dimensional case.

If $l=1$ and $N=2$ in (1), we have the pair of integral equations

$$\int_{\Delta_1} a_{11}(\tau)\,\phi(\tau)\,K_1(\tau,x)\,d\tau = f_1(x), \quad x \in I_1,$$

$$\int_{\Delta_2} a_{21}(\tau)\,\phi(\tau)\,K_2(\tau,x)\,d\tau = f_2(x), \quad x \in I_2.$$

When $l=1$ and $N=3$, we get triples of integral equations. N-tuples of integral equations have the form

$$\int_0^\infty a_i(\tau)\,\phi(\tau)\,K_1(\tau,x)\,d\tau = f_i(x)$$

$$(e_i < x < e_{i+1}, \quad i=\overline{1,N}, \quad e_1=0, \ e_{N+1}=\infty).$$

We consider pairs of integral equations with the generalised Legendre function $P_k^{m,n}(z)$.

It is well known [2] that the generalised Legendre functions $P_k^{m,n}(z)$ and $Q_k^{m,n}(z)$ are linearly independent solutions of the differential equation

$$(1-z^2)\frac{d^2u}{dz^2}-2z\frac{du}{dz}+\left[k(k+1)-\frac{m^2}{2(1-z)}-\frac{n^2}{2(1+z)}\right]u=0,$$

where k, m and n may be complex. When $m=n=\mu$ and $k=\nu$, we have the usual Legendre functions $P_\nu^\mu(z)$ and $Q_\nu^\mu(z)$ [3].

Let $|z-1|<2$, let $k+(m+n)/2$ be a positive integer, and let $k-(m-n)/2 \notin \mathbb{Z}$, $m<1$. Then

$$P_k^{m,n}(z)=\frac{1}{\Gamma(1-m)}\frac{(z+1)^{n/2}}{(z-1)^{m/2}}\,{}_2F_1\Big(k-\frac{m-n}{2}+1,-k-\frac{m-n}{2};1-m;\frac{1-z}{2}\Big),$$

where ${}_2F_1(a,b;\,c;\,z)$ is the hypergeometric function.

We establish two new integral representations for the generalised Legendre function $P_k^{m,n}(z)$.

Theorem 1. *If* $\mathrm{Re}(1/2-m)>0$, $k+(m+n)/2\neq -1,-2,\dots$, $k-(M-n)/2\notin\mathbb{Z}$, $m\neq 1,2,\dots$, *and* $z=\cosh\alpha$, *then*

$$P_k^{m,n}(\cosh\alpha)=\frac{2^{(n-m+1)/2}\sinh^m\alpha}{\sqrt{\pi}\,\Gamma(\frac{1}{2}-m)}$$

$$\times\int_0^\alpha\frac{\cosh\big(k+\frac{1}{2}\big)t}{(\cosh\alpha-\cosh t)^{m+1/2}}\,{}_2F_1\Big(\frac{n-m}{2},-\frac{m+n}{2};\tfrac{1}{2}-m;\frac{\cosh\alpha-\cosh t}{1+\cosh\alpha}\Big)dt. \quad (2)$$

Theorem 2. *If* $|\mathrm{Re}\,m|<1/2$, $\mathrm{Re}(-k-(m+n)/2)>0$, $\mathrm{Re}(1+k-(3m-n)/2)>0$, $\chi=\mathrm{Re}(k+(m+n)/2)$, $m\notin\mathbb{Z}$, $|(1-\cosh\alpha)/(\cosh t-\cosh\alpha)|<1$, *and* $z=\cosh\alpha$, *then*

$$P_k^{m,n}(\cosh\alpha)$$

$$=\frac{\sqrt{2\pi}\sec\pi\Big(k-\frac{m-n}{2}\Big)\sinh^{-m}\alpha(\cosh\alpha+1)^{(m-n)/2}}{2^{m-n}\Gamma(\frac{1}{2}+m)\Gamma\Big(1+k-\frac{3m-n}{2}\Big)\Gamma\Big(-k-\frac{m+n}{2}\Big)}\int_\alpha^\infty\frac{\sinh\Big(k+\frac{n-m+1}{2}\Big)t}{(\cosh t-\cosh\alpha)^{1/2-m}}$$

$$\times\int_\alpha^\infty\frac{\sinh\Big(k+\frac{n-m+1}{2}\Big)t}{(\cosh t-\cosh\alpha)^{1/2-m}}\,{}_2F_1\Big(n-m,\tfrac{1}{2}-m;\,1+k-\frac{3m-n}{2};\,\frac{1-\cosh\alpha}{\cosh t-\cosh\alpha}\Big)dt. \quad (3)$$

The proof of (2) and (3) can be found in [1].

We consider the pair of integral equations

$$\int_0^\infty f(\tau)\,P^{m,n}_{-1/2+i\tau}(\cosh\alpha)\,d\tau = g(\alpha) \quad (0<\alpha<a), \tag{4}$$

$$\int_0^\infty \omega(\tau)\,f(\tau)\,P^{m,n}_{-1/2+i\tau}(\cosh\alpha)\,d\tau = h(\alpha) \quad (a<\alpha<\infty), \tag{5}$$

where $f(\tau)$ is an unknown function, $f(\tau)=O(\tau^{\delta})$ as $\tau\to 0$, $f(\tau)=O\big(\tau^{-m-\varepsilon-1/2}\big)$ as $\tau\to\infty$, $\delta>0$, $\varepsilon>0$,

$$\omega(\tau)=\frac{2^{m-n-2}\Gamma\Big(\dfrac{1-m+n}{2}+i\tau\Big)\Gamma\Big(\dfrac{1-m-n}{2}-i\tau\Big)\Gamma\Big(\dfrac{1-m-n}{2}+i\tau\Big)}{\pi\Gamma(2i\tau)\Gamma(-2i\tau)\Gamma^{-1}\Big(\dfrac{1-m-n}{2}-i\tau\Big)},$$

and $g(\alpha)$, $h(\alpha)$ are given functions.

Theorem 3. *If $|\operatorname{Re} m|>1/2$, n and m satisfy the conditions for the existence of the Γ-function of $w(\tau)$, $g(\alpha)=O((a-\alpha)^{-m-1/2+\beta})$ as $\alpha\to a$, $g(\alpha)=O(\alpha^{m+\mu-2})$ as $\alpha\to 0$, $\beta>0$, $\mu>0$ and $h(\alpha)=O((\alpha-a)^{-1/2-m+\gamma})$ as $\alpha\to a$, $\gamma>0$, then the solution of (4), (5) exists, is unique and can be written as*

$$f(\tau)=f_1(\tau)+f_2(\tau),$$

where

$$f_1(\tau)=\frac{2^{(m-n+1)/2}}{\sqrt{\pi}\,\Gamma(\frac12+m)}\Big[(\cos\tau a)(\cosh a+1)^{(n-m)/2}$$

$$\times\int_0^a(\cosh a-\cosh\alpha)^{m-1/2}(\cosh\alpha+1)^{(m-n)/2}$$

$$\times{}_2F_1\Big(\frac{m-n}{2},\frac{1+m-n}{2};\tfrac12+m;\frac{\cosh a-\cosh\alpha}{1+\cosh a}\Big)\sinh^{1-m}\alpha g(\alpha)d\alpha$$

$$+\tau\int_0^a\sin\tau t(\cosh t+1)^{(n-m)/2}dt\int_0^t(\cosh t-\cosh\alpha)^{m-1/2}(\cosh\alpha+1)^{(m-n)/2}$$

$$\times{}_2F_1\Big(\frac{m-n}{2},\frac{1+m-n}{2};\tfrac12+m;\frac{\cosh t-\cosh\alpha}{1+\cosh t}\Big)\sinh^{1-m}\alpha g(\alpha)d\alpha\Big],$$

$$f_2(\tau)=a_1(\tau)+a_2(\tau),$$

$$a_1(\tau) = \int_a^{\infty} h(\alpha)\, P^{m,n}_{-1/2+i\tau}(\cosh\alpha)\sinh\alpha\, d\alpha,$$

$$a_2(\tau) = \frac{2^{m-n+1/2}}{\sqrt{\pi}\,\Gamma\left(\frac{1}{2}+m\right)}\Bigg[\cos\tau a(\cosh a+1)^{(n-m)/2}$$

$$\times \int_0^a (\cosh a - \cosh\alpha)^{m-1/2}(\cosh\alpha+1)^{(m-n)/2}$$

$$\times_2 F_1\left(\frac{m-n}{2}, \frac{1+m-n}{2}; \tfrac{1}{2}+m; \frac{\cosh a - \cosh\alpha}{1+\cosh a}\right)^* \sinh^{1-m}\alpha\, {}^*\Phi(\alpha)d\alpha$$

$$+\tau\int_0^a \sin\tau t^*(\cosh t+1)^{(n-m)/2}\, dt^* \int_0^1 (\cosh t - \cosh\alpha)^{m-1/2}(\cosh\alpha+1)^{(m-n)/2}$$

$$\times_2 F_1\left(\frac{m-n}{2}, \frac{1+m-n}{2}; \tfrac{1}{2}+m; \frac{\cosh t - \cosh\alpha}{1+\cosh t}\right)\sinh^{1-m}\alpha\Phi(\alpha)d\alpha\Bigg],$$

$$\Phi(\alpha) = \frac{2^{n-m}\sinh^m\alpha}{\Gamma^2\left(\frac{1}{2}-m\right)}\int_a^{\infty} h(y)\sinh^{m+1}y\, dy \int_0^{\min(\alpha,y)} \frac{(\cosh y - \cosh s)^{-m-1/2}}{(\cosh K\alpha - \cosh s)^{m+1/2}}$$

$$\times_2 F_1\left(\frac{n-m}{2}, -\frac{m+n}{2}; \tfrac{1}{2}-m; \frac{\cosh\alpha - \cosh s}{1+\cosh\alpha}\right)$$

$$\times\ {}_2F_1\left(\frac{n-m}{2}, -\frac{m+n}{2}; \tfrac{1}{2}-m; \frac{\cosh y - \cosh s}{1+\cosh y}\right)ds.$$

To prove this theorem, we use the generalised integral Mehler-Fock transform [4], the first integral representation (2) for $P^{m,n}_k(z)$, the solution of the integral equation with the hypergeometric function (see [5] and [6]), and the properties of ${}_2F_1$ [3].

As an example, we consider the mixed boundary value problem for a space wedge in toroidal coordinates ($0 < \alpha < \infty$, $0 < \phi < \psi$, $-\pi < \beta \leq \pi$):

$$\frac{\partial}{\partial\alpha}\left(\frac{\sinh\alpha}{\cosh\alpha - \cos\beta}\frac{\partial u}{\partial\alpha}\right) + \left(\frac{\sinh\alpha}{\cosh\alpha - \cos\beta}\frac{\partial u}{\partial\alpha}\right)$$

$$+\frac{\sinh^{-1}\alpha}{\cosh\alpha - \cos\beta}\frac{\partial^2 u}{\partial\phi^2} - \frac{\cosh\alpha}{\cosh\alpha - \cos\beta}\frac{\partial u}{\partial t} = 0,$$

$$u|_{t=0} = f_1(\alpha,\beta,\phi) \quad (0<\alpha<\infty,\ 0<\phi<\psi,\ -\pi<\beta\leq\pi),$$

$$u|_{\phi=0} = 0 \quad (0<t<+\infty),$$

$$\left.\frac{\partial u}{\partial\phi}\right|_{\phi=\psi} = 0 \quad (0<\alpha<\alpha_0,\ 0<t<+\infty),$$

$$u|_{\phi=\psi} = f_2 \quad (\alpha_0 < \alpha < \infty, 0 < t < +\infty).$$

Substituting

$$u = v\sqrt{2(\cosh\alpha - \cos\beta)}$$

and using the method of partial solutions and the asymptotic behaviour of the generalised Legendre functions, we reduce the solution of this problem to the solution of a pair of integral equations of the form

$$\begin{gathered}\int_0^\infty \phi(\tau)\, P^{m,n}_{-1/2+i\tau}(\cosh\alpha)\, d\tau = 0 \quad (0 < \alpha < \alpha_0),\\ \int_0^\infty \omega(\tau)\, \phi(\tau)\, P^{m,n}_{-1/2+i\tau}(\cosh\alpha)\, d\tau = f_2 \quad (\alpha_0 < \alpha < +\infty),\end{gathered} \tag{6}$$

where $\phi(\tau)$ is an unknown function such that $\phi(\tau) = O(\tau^\delta)$ as $\tau \to 0$, $\phi(\tau) = O(\tau^{-m-\varepsilon-1/2})$ as $\tau \to \infty$, $\delta > 0$, $\varepsilon > 0$, and $\omega(\tau)$ is a given weight function. The integral equations (6) are a particular case of (4), (5).

References

1. N. Virchenko, *Pairs (triples) of integral equations*, Vyshcha Shkola, Kyiv, 1989.
2. L. Kuipers and B. Meulenbeld, On a generalization of Legendre's associated differential equation, *Proc. Konink. Nederl. Akad. Wetensch.* **60** (1957), 337–350.
3. H. Bateman and A. Erdelyi, *Higher transcendental functions*, vol. 1, New York, 1953.
4. F. Götze, Verallgemeinerung einer Integraltransformation von Mehler-Fock durch dem von Kuipers und Meulenbeld eingeführten Kern $P_k^{m,n}(z)$, *Proc. Konink. Nederl. Akad. Wetensch. A* **68** (1965), 396–404.
5. N. Virchenko, Pairs of integral equations with hypergeometric functions, *Dokl. Akad. Nauk. UkrainSSR* **1986**, no. 2, 3–5.
6. M. Saigo, A certain boundary value problem for the Euler-Darboux equation, *Math. Japon.* **24** (1979), 377–385.

Department of Mathematics, Technical State University of Ukraine (Kyiv Polytechnic Institute), 11 Polarna Street, ap. 18, 254201 Kyiv, Ukraine

D.B. VOLKOV-BOGORODSKY

On integration of Fuchs-class equations with four singular points in connection with a conformal mapping problem

1. Introduction

Integration of an ordinary differential equation of Fuchs class with a large number of singular points (more than 3) is known to be a difficult problem [1]–[3]. It has a great theoretical and applied significance and is related, in particular, to the problem of conformal mapping of circular polygons [4]–[6]. One of the basic difficulties that arise here is the relation between the so-called auxiliary parameters of the equation and the geometric parameters of the circular polygon [2]–[10].

In this paper we consider Fuchs-class equation with four real-valued singular points and the associated conformal mapping on to a circular quadrangle. A method for the integration of the equation and for constructing the mapping proposed in [11] is presented. The method includes: 1) the determination of the auxiliary parameters; 2) representations for the solution of the Fuchs-class equation in the form of Burmann-Lagrange series by combinations of exponentials and elliptic functions; 3) the associated representation for the direct and inverse conformal mappings; 4) an exact algorithm for obtaining the coefficients of these representations; 5) the determination of the monodromy group and transfer group. The method has been numerically realised and applied to some specific problems [11].

2. Fuchs-class equation and Schwarz equation

Consider an ordinary second order differential equation in the complex z-plane, of the form:

$$\Xi''(z) + \tfrac{1}{2}\, R(z)\,\Xi(z) = 3D0, \tag{1}$$

where $R(z)$ is a rational function with four singular points 0, τ, 1 and ∞, namely

$$R(z) = 3D\frac{1-\alpha_1^2}{2\,z^2} + \frac{1-\alpha_2^2}{2\,(z-\tau)^2} + \frac{1-\alpha_3^2}{2\,(z-1)^2} + \frac{1}{z\,(z-1)^2}\left[\frac{\alpha_1^2+\alpha_2^2+\alpha_3^2-\alpha_4^2-2}{2} + \frac{\gamma_1}{z-\tau}\right], \tag{2}$$

the parameters α_k^2, $k \in \{1,2,3,4\}$, and γ_1 are arbitrary real numbers, and τ belongs to the interval $(0,1)$. This is a Fuchs-class equation (see [4] and [5]) with four singular points. An arbitrary Fuchs-class equation of second order with four real singular points can be reduced to the form (1), (2) by means of well-known transformations [4].

This work has been supported by the Russian Foundation for Basic Research, Grant No. 95-01-01367a.

A Fuchs-class equation is known to be closely related to a Schwarz equation (see [4] and [5]), which is a non-linear third order equation of the form

$$\{f;z\} = 3D\frac{f'''}{f'} - \tfrac{3}{2}\left(\frac{f''}{f'}\right)^2 = 3DR(z). \tag{3}$$

This relationship is shown as follows. Let $\Xi(z)$ be a solution of equation (1); then the function $f(z)$ defined by the integral

$$f(z) = 3D\int_{z_0}^{z} \Xi^{-2}(z)\,dz + A \tag{4}$$

satisfies the Schwarz equation (3). We can provide the normalisation $f(z_0) = 3DA$ by choosing appropriate values of A and $z_0 \in \overline{\mathcal{C}}$; the latter cannot coincide with the zeros of $\Xi(z)$ and its non-integrable singularities.

The other way round, the function $f(z)$ from (3) generates two canonical linearly independent solutions $\Xi(z)$ and $\Xi^*(z)$ of the Fuchs equation (1),

$$\Xi(z) = 3D\big[f'(z)\big]^{-1/2}, \quad \Xi^*(z) = 3D\big[(I\circ f)'\big]^{-1/2}, \quad f(z) = 3D\frac{\Xi^*(z)}{\Xi(z)},$$

where $I(\xi) = 3D - 1/\xi$ and $I \circ f$ denotes the composition $I(f)$.

It can be proved that a solution of (3) performs a conformal mapping of the upper half-plane $\mathcal{H}$ on to a circular quadrangle whose vertices with inner angles $\pi\alpha_k$ (in absolute value) are images of the singular points of (1).

Conversely, let us take a circular quadrangle in the complex w-plane and construct its conformal mapping on to $\mathcal{H}$. First of all we need to determine a Fuchs-class equation associated with the mapping. We know the indices α_k in this equation because they are simply related to the angles of the quadrangle as mentioned above; besides, we may define three vertex prototypes as 0, 1 and ∞. Hence, the prototype τ of the fourth vertex and the auxiliary parameter γ_1 remain unknown in (1), (2). Of course, the quantities γ_1 and τ are, in principle, uniquely determined by the geometry of the quadrangle. The purpose of this paper is to obtain an explicit algorithm for the determination of γ_1 and τ through the geometric parameters of the quadrangle.

3. Geometric parameters of the domain

The coefficient $R(z)$ in equations (1) and (3) contains 6 real parameters: the indices α_k and the quantities γ_1 and τ. Since equation (3) is invariant with respect to linear-fractional transformations, the domain is determined (modulo linear-fractional transformations) by 6 parameters. Four of them are angles simply related to the indices α_k in (2) as mentioned above. Hence, we have to define two additional geometric parameters denoted by ν_1 and ν_2, and to clarify their relation to γ_1 and τ.

To introduce ν_1, we consider two opposite arcs $\mathcal{L}_1$ and $\mathcal{L}_3$ of the quadrangle. If the extensions of these arcs intersect at an angle θ_1, then $\nu_1 = 3D\theta_1/\pi$. If their extensions do not intersect, then they generate a domain bounded by two circles;

this domain can be transformed into a ring with radii ρ' and ρ by a linear-fractional function; in this case, $\nu_1 = 3D = C4(i/\pi)\ln(\rho'/\rho)$. The parameter ν_2 is similarly defined by means of the arcs $\mathcal{L}_2$ and $\mathcal{L}_4$.

4. Reducing a Fuchs-class equation to a Lamé-type equation

We make the substitution $z = 3D\tau\ \mathrm{sn}^2\,\zeta$, which corresponds to mapping $\mathcal{H}$ on to a rectangle when the vertices of G are transformed into the vertices of the rectangle. We also make the substitution $\Xi(z) = 3D\big(2\tau\,\mathrm{sn}\,\zeta\,\mathrm{cn}\,\zeta\,\mathrm{dn}\,\zeta\big)^{1/2}\ U(\zeta)$, where sn, cn and dn are the Jacobi elliptic functions with modulus $k = 3D\sqrt{\tau}$. Then (1) is turned into a Lamé-type equation, namely

$$U''(\zeta) + \left[\sum_{k=3D1}^{4}\left(\tfrac{1}{4} - \alpha_k^2\right)\mathcal{P}_k(\zeta) + \gamma_0\right]U(\zeta) = 3D0, \tag{5}$$

with a doubly periodic coefficient, which is an analogue of the rational function $R(z)$; here $\mathcal{P}_1 = 3D\mathcal{P}(\zeta) - \mathcal{P}(K)$, $\mathcal{P}_2 = 3D\mathcal{P}_1(\zeta - K)$, $\mathcal{P}_3 = 3D\mathcal{P}_1(\zeta - K - iK')$, and $\mathcal{P}_4 = 3D\mathcal{P}_1(\zeta - iK')$, where $\mathcal{P}$ is the Weierstrass function and γ_0 is connected to γ_1 by a simple relation.

The function $U(\zeta)$ can be considered to be defined on a doubly periodic structure, that is, on a torus with periods $\tau_1 = 3D2K$ and $\tau_2 = 3D2iK'$, where K and K' are elliptic integrals of the first kind.

5. The representation of the solution of a Lamé-type equation

We have found two representations for the coefficient of equation (5) in the form of series converging in the periodic rectangle, and thus we have rewritten the equation in the form [11]

$$U''(\zeta) + \omega_s^2 \sum_{n=3D-\infty}^{+\infty} c_{sn} i^n \exp\left[in\omega_s\left(\zeta - \frac{K+iK'}{2}\right)\right]U(\zeta) = 3D0, \quad s \in \{1,2\}. \tag{6}$$

As a consequence, we have obtained two canonical solutions of this equation in the form containing the above geometric parameters ν_1 and ν_2:

$$U_s(\zeta) = 3D \sum_{p=3D-\infty}^{+\infty} U_{sp}\, i^p \exp\left[i\left(p + \frac{\nu_s}{2}\right)\omega_s\left(\zeta - \frac{K+iK'}{2}\right)\right], \quad s \in \{1,2\}.$$

The coefficients c_{sn} in equation (6) are real numbers. For $n \neq 0$ they are determined by the series $q_s^{|n/2|}\sum_{k=3D0}^{\infty} q_s^k\, c_{sn}^{(k)}$. Here $c_{sn}^{(k)}$ are related to α_k by a simple rule [11], the parameters $q_1 = 3D\exp(-\pi\,K'/K)$ and $q_2 = 3D\exp(-\pi\,K/K')$ depend on τ only, and the auxiliary parameter γ_0 appears only in the coefficients c_{s0}:

$$c_{10} = 3D\frac{K^2}{\pi^2}\left[\gamma_0 - \frac{E}{K}\sum_{k=3D1}^{4}\left(\tfrac{1}{4} - \alpha_k^2\right)\right], \quad c_{20} = 3D\frac{K'^2}{\pi^2}\left[\frac{E}{K}\sum_{k=3D1}^{4}\left(\tfrac{1}{4} - \alpha_k^2\right) - \gamma_0\right]. \tag{7}$$

The quantities U_{sp} have been derived in the form of a similar series:

$$U_{sp} = 3DU_{s0}\, q_s^{|p/2|} \sum_{k=3D0}^{\infty} q_s^k\, U_{sp}^{(k)}, \quad U_{s0}^{(0)} = 3D1, \quad U_{s0}^{(k)} = 3D0, \quad k \neq 0. \qquad (8)$$

The second canonical solution is obtained by substituting $-\nu_s$ for ν_s in the above formulae. The coefficients U_{sp} satisfy the infinite system of linear equations

$$\left[c_{s0} - \left(p + \frac{\nu_s}{2}\right)^2\right] U_{sp} + \sum_{n=3D-\infty,\, n\neq 0}^{+\infty} U_{s,p-n} c_{sn} = 3D0, \quad p = 3D0, \pm 1, \pm 2, \ldots, \qquad (9)$$

which can be solved by means of the representations (8).

6. The relationship between ν_1, ν_2 and the parameters τ, γ_1

It is important to emphasise that all the geometric and analytic parameters τ, γ_1, α_k and ν_s are completely separated in the expressions of the coefficients c_{sn}, U_{sn} and the solution itself. It is this fact that allows us to derive an explicit algorithm for solving system (9) for the coefficients of the canonic solutions:

$$p(p+\nu_s)\, U_{sp}^{(k)} = 3D \sum_{n=3D1}^{p} \sum_{m=3D0}^{k} U_{s,p-n}^{(k=-m)}\, c_{sn}^{(m)}$$
$$+ \sum_{n=3D1}^{k} \left[c_{s0}^{(n)}\, U_{sp}^{(k-n)} + \sum_{m=3D0}^{k-n} \left(U_{s,-n}^{(k-n-m)}\, c_{s,p+n}^{(m)} + U_{s,p+n}^{(k-n-m)}\, c_{s,-n}^{(m)}\right)\right],$$

$$p(p-\nu_s)\, U_{s,-p}^{(k)} = 3D \sum_{n=3D1}^{p} \sum_{m=3D0}^{k} U_{s,-p+n}^{(k-m)}\, c_{s,-n}^{(m)}$$
$$+ \sum_{n=3D1}^{k} \left[c_{s0}^{(n)}\, U_{s,-p}^{(k-n)} + \sum_{m=3D0}^{k-n} \left(U_{sn}^{(k-n-m)}\, c_{s,-p-n}^{(m)} + U_{s,-p-n}^{(k-n-m)}\, c_{sn}^{(m)}\right)\right],$$

$$c_{s0}^{(k)} = 3D - \sum_{n=3D1}^{k} \sum_{m=3D0}^{k-n} \left(U_{s,-n}^{(k-n-m)}\, c_{sn}^{(m)} + U_{sn}^{(k-n-m)}\, c_{s,-n}^{(m)}\right),$$

as well as for finding a transcendental equation that determines τ:

$$\frac{\pi^2}{K^2}\left[\frac{\nu_1^2}{4} + \sum_{k=3D1}^{\infty} q_1^k\, c_{10}^{(k)}\right] + \frac{\pi^2}{K'^2}\left[\frac{\nu_2^2}{4} + \sum_{k=3D1}^{\infty} q_2^k\, c_{20}^{(k)}\right] = 3D0,$$

where $(\ln q_1)\,(\ln q_2) = 3D\pi^2$. The auxiliary constant γ_0 is found from (5) in terms of the coefficients $c_{s0} = 3D\nu_s^2/4 + \sum_{k=3D1}^{\infty} q_s^k\, c_{s0}^{(k)}$.

Thus, we have obtained an explicit algorithm for solving the posed problem; it includes the solution of the auxiliary parameter problem [1]–[4], [7]–[9] and the determination of the fourth vertex prototype, as well as the construction of the conformal mapping itself. This algorithm has been numerically realised and implemented for a number of specific applied problems [11].

References

1. O.A. Ladyzhenskaya, On the life and scientific activity of Vladimir Ivanovich Smirnoff, *Uspekhi Mat. Nauk* **42** (1987), no. 6, 3–23 (Russian).
2. A.B. Venkov, On the auxiliary coefficients of a Fuchs equation of second order with real singular points, *Zap. Nauchn. Sem. LOMI* **129** (1983), 17–29 (Russian).
3. P.G. Zograf and L.A. Takhtadzhan, On Liouville's equation, auxiliary parameters and the geometry of Teichmüller spaces for Riemann surfaces of genus 0, *Mat. Sb.* **132** (1987), 147–166 (Russian).
4. V.V. Golubev, *Lectures on the analytic theory of differential equations*, Gostekhizdat, 1950 (Russian).
5. W. Koppenfels and F. Stallmann, *Praxis der konformen Abbildung*, Springer-Verlag, Berlin-Göttingen-Heidelberg, 1959.
6. P.Ya. Polubarinova-Kochina, An application of the theory of differential equations to some problems of ground water (number of singular points above 3), *Izv. Akad. Nauk SSSR Ser. Mat.* **5–6** (1939), 579–602 (Russian).
7. V.A. Fock, Über die konforme Abbildung eines Kreisvierecks mit verschwindenden Winkeln auf eine Halbebene, *J. Reine Angew. Math.* **161** (1929), 137–151.
8. E.N. Bereslavskii, On a method of conformal mapping of circular quadrangles, *Ukraïn. Mat. Zh.* **32** (1980), 197–201 (Russian).
9. E.D. Pergamentseva, Some plane electrostatic problems connected with conformal mapping of circular quadrangles, *Differentsial'nye Uravneniya* **5** (1969), 935–944 (Russian).
10. A.R. Tsitskishvili, On constructing analytic functions mapping conformally a half-plane on to circular polygons, *Differentsial'nye Uravneniya* **21** (1985), 646–656 (Russian).
11. D.B. Volkov, On the integration of a Fuchs-class equation with four singular points (in connection with the problem of conformal mapping), *Comm. Appl. Math.*, CC AS USSR, 1991 (Russian).
12. V.I. Vlasov and D.B. Volkov, On the inversion problem for a Fuchs-class equation, *Differentsial'nye Uravneniya* **22** (1986), 1854–1864 (Russian).

Computing Center of the Russian Academy of Sciences, 40 Vavilova St.,
117967 Moscow, Russia

V.V. ZALIPAEV and M.M. POPOV

Asymptotic analysis of wave scattering on a periodic boundary

1. Introduction

This paper deals with one of the important aspects of grating theory in optics—the scattering of a plane wave by a periodical structure in the high frequency approximation (the period of grating is far greater than the wavelength). The well-known traditional numerical methods face difficulties in calculating characteristics of the reflected field with high frequency oscillations since the order of the corresponding system of complex linear equations becomes very high and the calculations are quite time-consuming. Under these conditions it is natural to employ the asymptotic methods of the geometric theory of diffraction (GTD), that is, ray-optical analysis (see [1] and [2]), to investigate the problem of short wave scattering by a diffraction grating. The methods we refer to are the uniform asymptotic theory of diffracted waves and the summation of multiple diffracted fields (MDF).

In [3] and [4] the authors have developed a new asymptotic theory of scattering by an echelette grating and smooth periodic (sinusoidal) boundary in the high frequency approximation based on GTD and the method of summation of MDF. These were the first examples of GTD applied to scattering by diffraction gratings.

In this paper we demonstrate briefly the main points of the asymptotic analysis of short wave scattering by an echelette and sinusoidal grating which are common for both structures. It is worth noting that the results of the numerical computations for the Rayleigh coefficients have been presented in [4]. It has been shown that the agreement between the ray-optical asymptotic solution and the exact values obtained on the basis of the rigorous integral equation method is acceptable at the boundary of the resonance domain and excellent in the short wave domain. The calculation time for the computer code based on the asymptotic analysis is very small compared with the classical numerical methods.

2. Formulation of the problem

The following time-harmonic diffraction problem is considered in the paper. In Fig. 1, the periodic surface S is an echelette grating with right angle of the wedge, or a sinusoidal grating illuminated by a plane electromagnetic wave $u_i(x,y)$. We assume that it is a perfectly conductive grating and that the problem is two-dimensional. The total wave field $u(x,y)$ regarded as a scalar complex function satisfies the Helmholtz equation $\Delta u + k^2 u = 0$ outside S, where k is the wave number, and Dirichlet or Neumann boundary conditions on S corresponding to the cases of the transverse electric (TE) or transverse magnetic (TM) polarisation. The solution must satisfy the Floquet condition and the radiation condition. The latter can be presented as the Rayleigh expansion of the diffracted field for $y > 0$ in the form of a series of diffracted outgoing plane waves in the upper half-plane, that is,

$$u_d = \sum_{n=-\infty}^{\infty} R_n \exp[ik(x\cos\varphi_n + y\sin\varphi_n)], \quad \cos\varphi_n = 2\pi n/kd + \cos\chi, \qquad (1)$$

where d is the grating period. Here R_n is called the Rayleigh coefficient and φ_n is the diffraction (observation) angle, $0 < \varphi_n < \pi$. For the homogeneous diffracted plane waves ($|\cos\varphi_n| < 1$) propagating in the upper half-plane, the following energy balance criterion holds: $\Sigma e_n = 1$, where $e_n = |R_n|^2 \sin\varphi_n / \sin\chi$ is the efficiency of the diffraction orders.

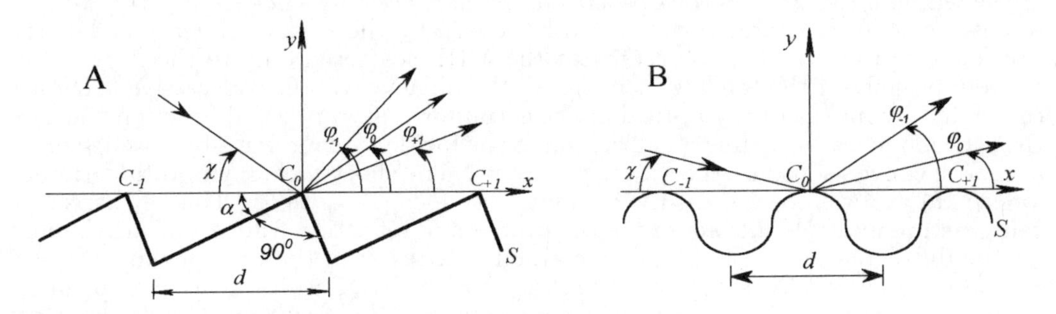

Fig. 1. Scattering of a plane wave by echelette (A) and sinusoidal grating (B).

Up to this point we have discussed the general formulation of the problem. We now confine ourselves to considering a special case. The wavelength is assumed to be small compared to the grating period ($kd \gg 1$), and thus we deal with a high frequency approximation. In this paper we briefly describe a ray-optical analysis based on the summation of MDF, which enables us to derive an asymptotic expansion for the Rayleigh coefficients of homogeneous plane waves in the form

$$R_n = R_n^{(0)} + R_n^{(1)}/p + O(p^{-2}), \tag{2}$$

where, owing to the short wave approximation of the problem, p is a large parameter. We have $p = \sqrt{kd}$ for echelette grating and $p = (k\rho/2)^{1/6}$ for sinusoidal grating, $k\rho \gg 1$ (ρ is the radius of curvature at C_0). In the case of scattering by echelette, the grazing angle χ is arbitrary, but for sinusoidal grating $\chi = (k\rho)^{-1/2}$ is considered to be sufficiently small. Hence, this is the case of grazing scattering.

The idea of deriving an asymptotic solution is based on the formula [1]

$$R_n = \frac{\sqrt{2\pi} e^{i\pi/4}}{kd \sin\varphi_n} \Phi(\varphi_n). \tag{3}$$

This is the relation between the Rayleigh coefficient R_n and $\Phi(\varphi_n)$, the radiation pattern of the grating element (the condition $\sin\varphi_n \neq 0$ is implied throughout the paper). The asymptotic formulae for $\Phi(\varphi_n)$ can be derived from the expansion for the diffracted field represented as an infinite sum of cylindrical waves centred at each point C_m, $m = 0, \pm 1, \pm 2, \ldots$, of the grating (Fig. 1) (see [3] and [4]).

3. Multiple diffracted fields

The asymptotic solution of the problem is constructed by means of the method of summation of MDF scattered by a grating element, that is, we find the diffracted field as an infinite sum of MDF: $u_d(M) = \sum_{n=1}^{\infty} u_n(M)$, where $u_n(M)$ is the nth order diffracted field ($M(x, y)$ is the observation point). For the echelette $u_n(M)$, we represent the MDF scattered forward and backward by the wedges of grating. For the sinusoidal grating $u_n(M)$, we represent the MDF scattered only in the forward direction, because of the grazing incidence of the plane wave. The successive MDF are calculated by means of ray-optical analysis (uniform asymptotic diffraction theory). Here $u_1(M)$ gives the primary diffraction of the incident plane wave by a wedge or by a smooth convex contour (the grating element). The short wave asymptotic formulae for $u_1(M)$ are well-known results of ray theory (see [1] and [2]). The key aspect of constructing multiple diffractions is the problem of reflection from a grating element for the diffracted penumbral field, generated as a result of the previous diffraction.

Here we omit the asymptotic expansion for the MDF in the scattering problem for the echelette, but it is worth remarking that, if the observation point is far from the grating, the MDF are expressed via the multiple integrals

$$I_n(\gamma) = \frac{1}{(\pi i)^{n/2}} \int_0^{\infty} dz_1 \int_0^{\infty} dz_2 \cdots \int_0^{\infty} dz_{n-1} \int_0^{\infty} dz_n \exp[i\phi(z_1, \ldots, z_n, \gamma)], \quad (4)$$

$$J_n(\gamma) = \frac{1}{(\pi i)^{n/2}} \int_0^{\infty} dz_1 \int_0^{\infty} dz_2 \cdots \int_0^{\infty} dz_{n-1} \int_{-\infty}^{0} dz_n \exp[i\phi(z_1, \ldots, z_n, \gamma)], \quad (5)$$

$$\phi(z_1, \ldots, z_n, \gamma) = 2z_1^2 - 2z_1 z_2 + 2z_2^2 - 2z_2 z_3 \ldots + z_n^2 - 2z_n\gamma,$$

where $\gamma = \varphi\sqrt{kd/2}$ (the MDF are described properly in terms of polar coordinates r and φ with respect to the centre C_m (see Fig. 1A)).

The principal term of the primary diffraction field scattered by a smooth convex contour in the problem of sinusoidal grating is given by the Fok-Babich asymptotic penumbral formulae [2] (the vicinity of a shadow boundary at a finite distance from the tangency point). The principal term of the short wave asymptotic expansion for MDF far from grating is described properly in terms of polar coordinates r and φ with respect to the centre C_m, and is given by the formula

$$u_n(M) = u_i(C_m) e^{ikr} \sqrt{\frac{d}{r}} P_n(\gamma, \gamma_c) + O(p^{-1}), \quad (6)$$

$$P_n(\gamma, \gamma_c) = -\frac{\Omega^{n-1}}{(\pi)^{n/2}} \int_{-\infty}^{0} dx_1 \int_0^{\infty} dx_2 \cdots \int_0^{\infty} dx_n \exp[i\psi(\gamma_c, x_1, \ldots, x_n, \gamma)], \quad (7)$$

$$\psi(\gamma_c, x_1, \ldots, x_n, \gamma) = 2e^{-i\pi/4} \sqrt{\frac{d}{\rho}} \gamma_c x_1 - \sum_{m=1}^{n-1} (x_m - x_{m+1})^2 + 2e^{3i\pi/4} \sqrt{\frac{d}{\rho}} \gamma x_n,$$

with $\gamma_c = \chi\sqrt{k\rho/2}$, $\gamma = \varphi\sqrt{k\rho/2}$, and $\Omega = \exp[ikd(1 - \cos\chi)]$. This is the Fresnel diffraction part, which is independent of polarisation. The next term of the expansion for $u_n(M)$ is expressed via the same integral, but the integration with respect to x_m, $m = 1, \ldots, n$, should contain the ratio of Airy functions [3].

4. Asymptotic formulae for the Rayleigh coefficients

Using the asymptotic formulae for the MDF, we are able to construct an expansion for $\Phi(\varphi_n)$ and, ultimately, for the Rayleigh coefficients. These are expressed in the form

$$J(\gamma,\Omega)=\sum_{n=1}^{\infty}\Omega^n J_n(\gamma),\quad I(\gamma,\Omega)=\sum_{n=1}^{\infty}\Omega^n I_n(\gamma),\quad P(\gamma,\gamma_c)=\sum_{n=2}^{\infty}P_n(\gamma,\gamma_c),\tag{8}$$

with $\Omega=\exp[ikd(1\pm\cos\chi)]$ for both types of grating.

First, we must stress that the MDF have the same structure for both the echelette and the sinusoidal grating. For this reason, all the series in (8) can be summed up by means of one single method. Using the Wiener-Hopf technique to solve the integral equation

$$\Psi(\tau)=e^{-\tau^2}+\frac{\Omega}{\sqrt{\pi}}\int_0^{\infty}e^{-(\tau-s)^2}\Psi(s)\,ds,\tag{9}$$

we find closed form results for $J(\gamma,\Omega)$ and $I(\gamma,\Omega)$:

$$\begin{aligned}J(\gamma,\Omega)&=\frac{\Omega}{\sqrt{\pi}}\int_{-\infty}^{0}e^{-2i\tau\gamma e^{i\pi/4}}\Psi(\tau)\,d\tau=-W(\varphi,q),\\ I(\gamma,\Omega)&=\frac{\Omega}{\sqrt{\pi}}\int_{0}^{\infty}e^{-2i\tau\gamma e^{i\pi/4}}\Psi(\tau)\,d\tau=W(\varphi,q),\end{aligned}\tag{10}$$

where $W(\varphi,q)=\exp[U(\varphi\sqrt{kd},q)]-1$, $\Omega=\exp(2\pi iq)$ and $\gamma=\varphi\sqrt{kd/2}$. Furthermore, $U(s,q)$ is the Weinstein function, well known in waveguide theory [5] and given by

$$U(s,q)=\frac{1}{2\pi i}\int_{-\infty}^{\infty}\frac{\ln(1-e^{2\pi iq-t^2/2})}{t-se^{i\pi/4}}\,dt.\tag{11}$$

For $P(\gamma,\gamma_c)$ we have

$$P=-\frac{\Omega}{\pi}\int_{-\infty}^{0}dt_1\int_0^{\infty}dt_2\exp\left(2\gamma_c e^{-i\pi/4}t_1\sqrt{\frac{d}{\rho}}\right)\Psi(t_1,t_2)\exp\left(2\gamma e^{3i\pi/4}t_2\sqrt{\frac{d}{\rho}}\right),\tag{12}$$

where $\Psi(\tau,t)$ is the solution of (9) with inhomogeneous term $\exp[-(\tau-t)^2]$, and $\gamma=\varphi\sqrt{k\rho/2}$.

As a result, in the transverse electric case (the Dirichlet boundary condition) in which an incident plane wave illuminates only one face of the wedge, for the echelette grating we get

$$R_n^{(0)}=\frac{i}{2kd\sin\varphi_n}[V(\varphi_n,\chi)+T(\varphi_n,\chi)],\tag{13}$$

where

$$\begin{aligned}T(\varphi_n,\chi)&=V(0,\chi)W(\varphi_n,q_1),\quad 0\le\varphi_n\le\alpha,\\ T(\varphi_n,\chi)&=V(\pi,\chi)W(\pi-\varphi_n,q_2),\quad \alpha+\pi/2\le\varphi_n\le\pi,\end{aligned}\tag{14}$$

$$T(\varphi_n,\chi) = -[V(0,\chi)W(\varphi_n - 2\alpha, q_1) + V(\pi,\chi)W(2\alpha - \varphi_n, q_2) \\ + e^{i\delta_1}V_1(\varphi_n,\chi) + e^{i\delta_1}V(\pi,\chi)W(\varphi_n - 2\alpha, q_2)], \quad \alpha \le \varphi_n \le 2\alpha, \quad (15)$$

$$T(\varphi_n,\chi) = -[V(0,\chi)W(\varphi_n - 2\alpha, q_1) + V(\pi,\chi)W(2\alpha - \varphi_n, q_2) + e^{i\delta_2}V_2(\varphi_n,\chi) \\ + e^{i\delta_2}V(0,\chi)W(2\alpha - \varphi_n, q_1)], \quad 2\alpha \le \varphi_n \le \alpha + \pi/2, \quad (16)$$

with $\delta_{1,2} = kd[\cos(\varphi_n - 2\alpha) \pm \cos\chi]$ and $q_{1,2} = kd(1 \pm \cos\chi)/2\pi$. The function $V(\varphi,\chi)$ is the Keller diffraction coefficient defined by the well-known formula

$$V = \tfrac{2}{3}\big[\cot\tfrac{1}{3}(\varphi + \chi - 2\pi) - \cot\tfrac{1}{3}(\varphi - 2\alpha - \chi + \pi) \\ - \cot\tfrac{1}{3}(\varphi + \chi) + \cot\tfrac{1}{3}(\varphi - 2\alpha - \chi)\big].$$

The functions $V_1(\varphi,\chi)$ and $V_2(\varphi,\chi)$ are analogous to $V(\varphi,\chi)$. In the TM case, where both sides of the wedge are illuminated by a plane wave (quasi-normal incidence, $\pi/2 - \alpha < \chi < \pi - 2\alpha$), the formulae for $T(\varphi_n,\chi)$ look similar. For the sinusoidal grating, the principal term in the Rayleigh coefficient has the form

$$R_n^{(0)} = \frac{\exp\big[U(\chi\sqrt{kd}, q) + U(\varphi_n\sqrt{kd}, q)\big]}{ikd\sin\varphi_n(\varphi_n + \chi)}, \quad (17)$$

where $q = kd(1 - \cos\chi)/2\pi$, which holds for the sector $0 < \varphi_n < c(k\rho)^{-1/3}$, $c = \text{const}$. Above this sector, R_n becomes negligible. Thus, this asymptotic technique yields analytic expressions for the Rayleigh coefficients for both scattering problems, and provides a physical interpretation for the diffraction phenomenon in terms of MDF.

References

1. V.A. Borovikov and B.Y. Kinber, *Geometrical theory of diffraction*, IEE Electromagnet. Waves Ser. **37**, London, 1994.
2. V.M. Babich and N.Ya. Kirpichnikova, *The boundary-layer method in diffraction problems*, Springer-Verlag, Berlin-Heidelberg-New York, 1979.
3. M.M. Popov and V.V. Zalipaev, Short-wave grazing scattering of a plane wave by a smooth periodic boundary, *J. Soviet Math.* **55** (1991), 1685–1705.
4. V.V. Zalipaev, Calculation of the wave field scattered by an echelette grating in the short-wave approximation, in *Proc. Fifteenth URSI Internat. Symp. on Electromagnetic Theory*, St. Petersburg, 1995, 305–307.
5. L.A. Weinstein, *The theory of diffraction and the factorisation method*, Golem Press, Boulder, CO, 1969.

St. Petersburg Branch of the Steklov Mathematical Institute, Russian Academy of Sciences, Fontanka 27, 191011 St. Petersburg, Russia

B. ZHANG and S.N. CHANDLER-WILDE

An integral equation method for electromagnetic scattering by an inhomogeneous layer on a perfectly conducting plate

1. Introduction

We consider a time harmonic electromagnetic plane wave incident on a layer of some inhomogeneous, isotropic, conducting or dielectric material sitting on a perfectly conducting plate in $\mathbb{R}^3$. The medium above the layer consists of some homogeneous dielectric material. We assume throughout that the material is invariant in one coordinate direction parallel to the plate. Thus, in effect, the problem geometry is two-dimensional. Furthermore, we assume that the magnetic permeability is a fixed positive constant. The material properties of the media are then characterized completely by an index of refraction, which is assumed to be a bounded measurable function in the layer and a positive constant above the layer. The scattering problem consists in studying the electromagnetic field distributions. In this paper, an integral equation method is used to settle the question of existence and uniqueness of solution to the problem with TM polarization, in particular, for the case where the inhomogeneity extends to infinity without decaying. Uniqueness of solution is proved by introducing a radiation condition for the problem, which is a generalisation of the Rayleigh expansion condition for one-dimensional periodic gratings [1]. Existence of solution is established by reformulating the problem as an equivalent Lipmann-Schwinger integral equation over an infinite strip in $\mathbb{R}^2$ and then employing general results developed in [2] on the solvability of integral equations on unbounded domains.

Integral equation methods have played a very important role in the study of wave scattering problems by finite obstacles or local inhomogeneities (see, for example, [3] and [4] and the references quoted therein). Recently, the integral equation method has been used to study scattering by periodic structures [5], by an impedance half-plane in a homogeneous medium [6], and by a one-dimensional rough surface in a homogeneous medium [7] (see also the references quoted in the above works).

The present paper is most closely related in terms of results and methods of argument to [6] and to previous work on plane wave scattering by an inhomogeneous periodic layer (see [1] and [8]). Our uniqueness and existence results for the case of a dielectric layer include those of [1] for a periodic dielectric layer as a special case, but are obtained without an *a priori* assumption of quasiperiodicity of the scattered field.

The outline of this paper is as follows. In the next section, the scattering problem and the radiation condition are introduced. The existence and uniqueness results for the problem are given in Section 3.

This work was supported by the UK Engineering and Physical Science Research Council under Grant GR/K24408.

2. The scattering problem and radiation condition

Let $\mathbb{R}^3_+ = \{(x_1, x_2, x_3) \in \mathbb{R}^3 | x_3 > 0\}$ and let us assume that $\mathbb{R}^3_+$ is filled with an inhomogeneous, isotropic, conducting or dielectric medium of electric permittivity ϵ, magnetic permeability μ and electric conductivity σ. Suppose that the medium is non-magnetic, that is, the magnetic permeability μ is a fixed constant in $\mathbb{R}^3_+$, and that the fields are source-free. Then the electromagnetic wave propagation is governed by the time-harmonic Maxwell equations

$$\nabla \times E - i\omega\mu H = 0, \tag{1}$$
$$\nabla \times H + (i\omega\epsilon - \sigma)E = 0, \tag{2}$$

where E and H are the electric field and magnetic field, respectively, and $\omega > 0$ is the angular frequency. Suppose also that the inhomogeneous medium has a perfectly conducting boundary. Then the corresponding boundary condition is that the tangential component of the electric field E vanishes, that is,

$$\nu \times E = 0, \tag{3}$$

where ν is the unit normal at the boundary. In this paper we assume that the medium is invariant in the x_3-direction, that is, $\epsilon = \epsilon(x)$ and $\sigma = \sigma(x)$, with $x = (x_1, x_2) \in \mathbb{R}^2$. Also, we restrict ourselves to the TM polarization case, that is, the electric field E is assumed to point along the x_3-axis. Let $E = (0, 0, u)$, where $u = u(x)$ is a scalar function. Then it follows from the Maxwell equations (1), (2) and the boundary condition (3) that u satisfies

$$\Delta u + k^2 u = 0 \quad \text{in } U, \tag{4}$$
$$u = 0 \quad \text{on } \Gamma, \tag{5}$$

where Δ is the Laplacian in $\mathbb{R}^2$ and $k^2 = \omega^2\mu\epsilon[1 + i\sigma/(\omega\epsilon)]$, so that $\mathrm{Im}(k^2) \geq 0$. Hereafter, for $h \in \mathbb{R}$ we define $\Gamma_h = \{x \in \mathbb{R}^2 | x_2 = h\}$ and $U_h = \{x \in \mathbb{R}^2 | x_2 > h\}$, write U and Γ for U_0 and Γ_0, respectively, and set $E_h = U \backslash \bar{U}_h$ for $h > 0$.

Throughout this paper we make the following additional assumptions on k.

(A1) $k \in L_\infty(U)$.

(A2) There are two positive constants B and k_+ such that $k(x) = k_+$ for all $x \in U_B$.

Assumption (A2) implies that the medium is homogeneous and non-conducting (that is, ϵ is a positive constant and $\sigma = 0$) above a finite layer E_B.

In order to prove uniqueness of solution for the problem, we also require that one or other of the following assumptions (A3) and (A4) is satisfied.

(A3) There is a positive constant λ_1 such that $\mathrm{Im}(k^2(x)) \geq \lambda_1$ for almost all $x \in E_B$.

(A4) There are positive constants λ_2 and λ_3, with $\lambda_3 < 1/(4B^3)$, such that $\lambda_2 \,\mathrm{Im}(k^2) + \partial_2 \,\mathrm{Re}(k^2) \geq -\lambda_3$ in a distributional sense in U, where $\partial_j = \partial/\partial x_j$, $j = 1, 2$.

Clearly, (A3) is satisfied if and only if the finite layer E_B has finite conductivity everywhere, more precisely, if and only if $\sigma(x) \geq \sigma_0 > 0$ for almost all $x \in E_B$.

Assumption (A1) is sufficient to ensure that $\partial_2 \operatorname{Re}(k^2)$ exists in a distributional sense. Since $\partial_2 \operatorname{Re}(k^2)$ is to be understood in a distributional rather than a classical sense, (A4) permits jumps in $\operatorname{Re}(k^2)$ in the x_2-direction, provided these are jumps where $\operatorname{Re}(k^2)$ increases. In particular, when the layer E_B is dielectric (that is, $\sigma = 0$ and $\operatorname{Im}(k^2) = 0$), (A4) is satisfied if and only if $\operatorname{Re}(k^2(x)) = \pi(x) - \lambda_3 x_2$ for almost all $x \in U$, with $\partial_2 \pi \geq 0$. Moreover, $\partial_2 \pi \geq 0$ in a distributional sense if and only if $\pi(x)$ is monotonically increasing as x_2 increases, more precisely, if and only if $\operatorname{ess\,inf}_{x \in U}[\pi(x + e_2 h) - \pi(x)] \geq 0$ for all $h > 0$, where $e_2 = (0, 1)$.

We remark that the radiation condition which we intend to impose ensures that the scattered field does not contain a downwards propagating component; however (in line with the usual radiation condition for plane wave incidence on periodic gratings), it does not rule out guided waves localised in the inhomogeneous layer. It is not surprising then that a condition such as (A3) or (A4) is required to ensure uniqueness. In particular, although it is not clear that the bound on λ_3 in (A4) is the best possible one, a condition of this form is certainly appropriate given that, in the case when the medium is purely dielectric with $k \in C(\overline{U})$ and $\partial_1(k^2) = 0$ and $\partial_2(k^2) = -\lambda_3$ in E_B, elementary separation of variable arguments establish the existence of a guided wave for $\lambda_3 > \alpha^3/B^3$, where $-\alpha \approx -1.987$ is the largest negative zero of the Airy function $\operatorname{Ai}(z) + \operatorname{Bi}(z)/\sqrt{3}$.

Let $u^{\mathrm{i}}(x) = \exp(ik_+ x \cdot \alpha)$ be the time-harmonic incoming plane wave incident from U_B on the finite inhomogeneous layer E_B, where $x \in \mathbb{R}^2$, $\alpha = (\cos\theta, -\sin\theta) \in \mathbb{R}^2$, and $\theta \in (0, \pi)$ is the incident angle. We are interested in finding the total field u satisfying (4), (5).

In order to determine the physical solution u, a radiation condition as x_2 tends to infinity has to be imposed on the scattered field $u^{\mathrm{s}} = u - u^{\mathrm{i}}$, that is, the scattered field u^{s} should behave as an outgoing wave as $x_2 \to \infty$. Specifically, the scattered field u^{s} is required to satisfy a radiation condition of the form

$$u^{\mathrm{s}}(x) = \int_{\Gamma_b} \frac{\partial \Phi(x, y)}{\partial y_2} \phi(y) ds(y), \quad x \in U_b, \tag{6}$$

for some $\phi \in L_\infty(\Gamma_b)$ with some $b > B$, where $\Phi(x, y) = (i/4) H_0^{(1)}(k_+|x - y|)$ for $x, y \in \mathbb{R}^2$, $x \neq y$, is a fundamental solution for $\Delta + k_+^2$, and $H_0^{(1)}$ is the Hankel function of the first kind of order zero. This radiation condition is a generalisation of the Rayleigh expansion condition for one-dimensional periodic gratings.

Our problem of scattering of a time-harmonic plane wave by an inhomogeneous layer can now be formulated as the following boundary value problem.

Problem (P). Find $u \in C(\overline{U})$ such that

(i) u satisfies the Helmholtz equation (4) in a distributional sense, and the Dirichlet condition (5);

(ii) u^{s} satisfies the radiation condition (6);

(iii) u is bounded in E_A for every $A > 0$.

In the next section, the existence of a unique solution to Problem (P) is established by means of integral equation methods.

3. Existence and uniqueness of solutions

Let $G(x,y) = \Phi(x,y) - \Phi(x,y')$, where $y = (y_1, y_2)$ and $y' = (y_1, -y_2)$, be the Dirichlet Green's function for $\Delta + k_+^2$ in U. The following equivalence theorem between Problem (P) and an integral equation formulation is proved in [9] by using Green's second theorem in combination with the radiation condition (6).

Theorem 1. *Let u be a solution of Problem* (P). *Then*

$$u(x) = u^{\mathrm{i}}(x) + u^{\mathrm{r}}(x) + \int_{E_B} u(y)[k_+^2 - k^2(y)]G(x,y)dy, \quad x \in \overline{U}, \tag{7}$$

where u^{i} is the incident wave from the previous section and $u^{\mathrm{r}}(x) = -\exp(ik_+ x \cdot \alpha')$ with $\alpha' = (\cos\theta, \sin\theta)$.

Conversely, if u satisfies (7) *and $u \in BC(\overline{E_B})$, the Banach space of functions bounded and continuous on $\overline{E_B}$, then u satisfies Problem* (P).

With the help of Theorem 1 we can prove the following assertion on the existence of a unique solution for Problem (P).

Theorem 2. *If either* (A3) *or* (A4) *holds, then Problem* (P) *has exactly one solution. Furthermore, for any $K > 0$ there exists a constant $C > 0$ depending only on K, k_+, B, and λ_1 (when* (A3) *is satisfied), or on K, k_+, B, λ_2, and λ_3 (when* (A4) *is satisfied), such that, provided $\|k\|_\infty \le K$,*

$$|u(x)| \le C(1 + x_2)^{1/2}, \quad x \in \bar{U}. \tag{8}$$

Theorem 2 is proved in [9]. In the proof of uniqueness an essential role is played by the following basic inequality for the solution to the problem (4), (5): for some $a > B$ and non-negative constants C_j, $j = 1,2,3,4$,

$$\int_{E_B(A)} |u|^2 dx \le C_1 \int_{\Gamma_a(A)} \{|\partial_2 u|^2 - |\partial_1 u|^2 + k_+^2 |u|^2\} ds$$
$$+ C_2 \operatorname{Re} \int_{\Gamma_a(A)} \bar{u}\partial_2 u \, ds - C_3 \operatorname{Im} \int_{\Gamma_a(A)} \bar{u}\partial_2 u \, ds + C_4 R_A \tag{9}$$

for all $A > 0$, where $\Gamma_a(A) = \{x \in \Gamma_a \| x_1| \le A\}$, $E_B(A) = \{x \in E_B \| x_1| \le A\}$, and $|R_A| \le C\left[\int_{\gamma(A)} + \int_{\gamma(-A)}\right](|u|^2 + |\nabla u|^2)ds$ with $\gamma(b) = \{(b, x_2) \,|\, 0 \le x_2 \le a\}$. In the case when (A3) is satisfied, the inequality (9) can be easily proved using Green's first theorem. In the case when (A4) is satisfied, the inequality (9) is derived by first applying the transformation $v(x) = \eta(x_2)u(x)$, $x \in \bar{U}$, to the problem (4), (5), where $\eta(t) = \exp\left[\int_0^t \theta(s)ds\right]$ for $t \ge 0$ with $\theta(s) = (\alpha^{1/2} - \alpha)/(4a) + \alpha s(3a - s)/(8a^3)$ and $\alpha < 1$ being such that $4a^3\lambda_3 < \alpha$ for some $a > B$ (which, by (A4), is possible), and then employing a distributional approach to the transformed problem for v to deal with the difficulty of lack of regularity of k ($k \in L_\infty$). Existence of solution for Problem (P) is established by using Theorem 1 and showing that the integral equation

$$\psi(x) = \phi(x) + \int_{\Omega} \psi(y)w(y)G(x,y)dy, \quad x \in \Omega, \tag{10}$$

has a unique solution, where $\psi(x) = u(x)$, $\phi(x) = u^{\mathrm{i}}(x)+u^{\mathrm{r}}(x)$ and $w(x) = k_+^2 - k^2(x)$ for $x \in \Omega$ with $\Omega = E_B$. The existence of a unique solution to the integral equation (10) is proved by employing general results on the solvability of integral equations on unbounded domains (see [2], Theorem 1) along with the uniqueness result for Problem (P) and Theorem 1.

Finally, we should mention that the exponent 1/2 in the bound (8) is the best possible one in the general case, in that it is proved in [9] that, for a particular choice of k, this rate of growth in the solution can be achieved.

References

1. A. Kirsch, An inverse problem for periodic structures, in *Inverse scattering and potential problems in mathematical physics* (R.E. Kleinman, R. Kress and E. Martensen, eds.), P. Lang, Frankfurt, 1995, 75–93.
2. S.N. Chandler-Wilde and B. Zhang, On the solvability and numerical treatment of a class of second kind integral equations on unbounded domains (in this volume).
3. D. Colton and R. Kress, *Integral equation methods in scattering theory,* Wiley, New York, 1983.
4. D. Colton and R. Kress, *Inverse acoustic and electromagnetic scattering theory,* Springer, Berlin, 1992.
5. J.C. Nédélec and F. Starling, Integral equation methods in a quasi-periodic diffraction problem for the time-harmonic Maxwell's equations, *SIAM J. Math. Anal.* **22** (1991), 1679–1701.
6. S.N. Chandler-Wilde, The impedance boundary value problem for the Helmholtz equation in a half-plane, *Math. Methods Appl. Sci.* **20** (1997), 813–840.
7. S.N. Chandler-Wilde and R.C. Ross, Scattering by rough surfaces: The Dirichlet problem for the Helmholtz equation in a non-locally perturbed half-plane, *Math. Methods Appl. Sci.* **19** (1996), 959–976.
8. B. Szemberg, An acoustic scattering problem for a periodic, inhomogeneous medium, preprint no. 167 (1996), Institut für Angewandte Mathematik, Universität Erlangen-Nürnberg.
9. S.N. Chandler-Wilde and B. Zhang, Electromagnetic scattering by an inhomogeneous conducting or dielectric layer on a perfectly conducting plate, *Proc. R. Soc. Lond. A* (to appear).

Department of Mathematics and Statistics, Brunel University, Uxbridge UB8 3PH, UK

O.Yu. ZHARII

Singular integral equations of the theory of ultrasonic motors and their solutions

1. Introduction

In an ultrasonic motor, vibrational and wave motions of a solid (stator) are transformed into a progressive motion of another body (rotor) due to frictional forces in contact. This phenomenon is adequately modelled by mixed boundary value problems of elastodynamics.

The mathematical model of a so-called travelling wave ultrasonic motor based on well-grounded ideas of contact mechanics [1] was formulated in the paper [2]. In the works [3]–[5], analytic solutions of several particular problems were reported. In this paper we systematise and generalise the most important mathematical points of the previous considerations and present some new results concerning the structure of contact stresses in the ultrasonic motor.

For the solution of the problems formulated we use the method of singular integral equations. It allows one to treat a complex physical and mathematical problem analytically and to investigate all the important characteristics of the solution. We remark that in problems of this kind, the fields of contact stresses and velocities have unbounded derivatives. So, the knowledge of specific features of behaviour of these quantities is a prerequisite for the elaboration of effective methods for the numerical solution of those problems that cannot be solved analytically.

2. Origin of singular equations in contact problems

The conventional method of solution of dynamic contact problems (see [2]–[5]) is based on the representation of unknown contact stresses by a Fourier series with unknown coefficients. Let us take, for example, the expression of normal stresses (the notation coincides with that in [2]–[5]), that is,

$$\sigma_y \equiv f(s) = \frac{f_0}{2} + \sum_{n=1}^{\infty}(f_n \cos ns + f'_n \sin ns), \tag{1}$$

$$f(s) \equiv 0 \quad \text{for } s < \alpha,\ s > \delta.$$

We use this series as the boundary value of the normal stresses and find the general solution of the Lamé equations of motion. When we calculate the tangential velocities in contact, which enter the kinematic boundary conditions, we obtain, besides the series (1), a series of another kind. We transform the latter by means of

The author gratefully acknowledges the support provided by the Alexander von Humboldt Foundation.

the well-known formulae for the Fourier coefficients, namely

$$\begin{aligned}\sum_{n=1}^{\infty}&(f_n \sin ns - f_n' \cos ns)\\ &= \frac{1}{\pi}\sum_{n=1}^{\infty}\left(\sin ns \int_{\alpha}^{\delta} f(s_1)\cos ns_1 ds_1 - \cos ns \int_{\alpha}^{\delta} f(s_1)\sin ns_1 ds_1\right)\\ &= -\frac{1}{\pi}\int_{\alpha}^{\delta} f(s_1)ds_1 \sum_{n=1}^{\infty}\sin n(s_1 - s) = -\frac{1}{2\pi}\int_{\alpha}^{\delta} f(s_1)\cot\frac{s_1 - s}{2}ds_1, \quad (2)\end{aligned}$$

where we have used the equality

$$\sum_{n=1}^{\infty}\sin ns = \frac{1}{2}\cot\frac{s}{2}. \quad (3)$$

Now, making use of the formula

$$\cot\frac{s_1 - s}{2} = \frac{1 + \tan\frac{s_1}{2}\tan\frac{s}{2}}{\tan\frac{s_1}{2} - \tan\frac{s}{2}} \quad (4)$$

and transforming both the independent and dependent variables by writing

$$\eta = \tan\frac{s}{2}, \quad \eta_1 = \tan\frac{s_1}{2}, \quad \varphi(\eta) = f(s)\cos^2\frac{s}{2}, \quad \mathrm{A} = \tan\frac{\alpha}{2}, \quad \Delta = \tan\frac{\delta}{2}, \quad (5)$$

we reduce the singular integral in (2) to the standard form

$$\frac{1}{2}\int_{\alpha}^{\delta} f(s_1)\cot\frac{s_1 - s}{2}ds_1 = (\eta^2 + 1)\int_{\mathrm{A}}^{\Delta}\frac{\varphi(\eta_1)d\eta_1}{\eta_1 - \eta} + \eta\int_{\mathrm{A}}^{\Delta}\varphi(\eta_1)d\eta_1. \quad (6)$$

The first integral on the right is now the standard principal value integral, and the value of the non-singular integral is known from the equilibrium conditions of the original problem [2].

3. Closed-form solutions of the singular integral equations

After satisfying the boundary conditions, we obtain linear equations containing both the unknown function φ and the Cauchy principal value integral of it:

$$a\varphi(\eta) + \frac{b}{\pi i}\int_{\alpha}^{\beta}\frac{\varphi(\eta_1)\,d\eta_1}{\eta_1 - \eta} = g(\eta), \quad \alpha < \eta < \beta, \quad (7)$$

where g is a prescribed function. The general solution of (7) is [6]

$$\begin{aligned}\varphi(\eta) = \frac{a}{a^2 - b^2}g(\eta) &+ \frac{b}{\pi i(a^2 - b^2)(\eta - \alpha)^{1-m}(\beta - \eta)^m}\\ &\times \int_{\alpha}^{\beta}\frac{(\zeta - \alpha)^{1-m}(\beta - \zeta)^m}{\zeta - \eta}g(\zeta)d\zeta + \frac{C}{(\eta - \alpha)^{1-m}(\beta - \eta)^m}, \quad (8)\end{aligned}$$

where C is an arbitrary constant and the quantity m is determined as

$$m = \frac{1}{2\pi i} \ln \frac{a+b}{a-b}, \quad 0 \leq \operatorname{Re} m < 1. \tag{9}$$

In standard situations, the limits of integration α and β are prescribed. At the expense of the proper choice of the constant C we may write solutions to (7) that are bounded at one of the points α or β. In general, it is impossible to obtain the solution bounded at both end-points of integration.

However, in contact problems, the end-points of integration are the ends of the contact area or the points separating domains with different types of boundary conditions. At these points the unknown function (one of the contact stresses) should be bounded. This follows from physical considerations. At the same time, the limits of integration (dimensions of the contact area and those of adhesion and slip zones) are unknown in advance.

The general procedure for solving the integral equations is as follows. First, before the calculation of the integrals in (8), we exclude the constant C demanding that the last integral in (6) should have a certain value. From the physical viewpoint, this value is equal to the prescribed stress resultant, known from the problem formulation. Then we calculate the integrals in (8). We can do this analytically for a rational function g (see [3] and [4]). It is the case for our problems. After that, we demand boundedness of the function φ at the end-points of integration. This gives two transcendental equations with respect to α and β. Finally, these equations are used for the further simplification of the expression of φ.

4. Resulting equations of particular problems

4.1. Problem of frictional contact. This problem was investigated in [4]. When the complete contact area is the slip zone (adhesion zones disappear), the contact problem is reduced to a single real integral equation, namely

$$\mu_1 \varphi_{\mathrm{fr}}(\eta) + \frac{\kappa}{\pi} \int_{\mathrm{A}}^{\Delta} \frac{\varphi_{\mathrm{fr}}(\eta_1)\, d\eta_1}{\eta_1 - \eta} = G_{\mathrm{fr}}(\eta), \quad \mathrm{A} < \eta < \Delta, \tag{10}$$

whose solution can be obtained in closed form because the right-hand side of (10) is a rational function. Its solution in terms of the original unknown function has the form

$$f(s) = 2A_0 \sin m\pi \sin^m \frac{s-\alpha}{2} \sin^{1-m} \frac{\delta - s}{2} \cos \frac{m\alpha + (1-m)\delta + s}{2}, \tag{11}$$

where the value of the parameter m belongs to the interval $m = 0.504 \operatorname{div} 0.540$. It depends on the frequency of excitation and on the value of the coefficient of dry friction. The two transcendental equations for the determination of α and δ contain only one trigonometrical function.

In the particular case of smooth contact ($\mu_1 = 0$), instead of (10) we arrive at the equation of the first kind

$$\frac{\kappa}{\pi} \int_{\mathrm{A}}^{\Delta} \frac{\varphi(\eta_1)\, d\eta_1}{\eta_1 - \eta} = G_{\mathrm{sm}}(\eta). \tag{12}$$

Due to the symmetry of the problem, we have $\mathrm{A} = -\Delta$, and the solution (11) takes the especially simple form

$$f(s) = 2A_0 \cos\frac{s}{2}\sqrt{\cos^2\frac{s}{2} - \cos^2\frac{\delta_{\rm sm}}{2}}, \quad \delta_{\rm sm} = 2\arcsin\sqrt{P/P_{\max}}. \tag{13}$$

In [5] this simplest problem was solved using another technique, namely the method of dual series equations. However, it turned out that this method was inapplicable to the problems of frictional and adhesive contact (see below). Therefore, the method of singular integral equations appeared to be more general than that of dual series equations.

4.2. Problem of adhesive contact. In the contact problem with complete adhesion considered in [3], we obtained a complex equation for the complex combination of the normal and tangential contact stresses:

$$\chi(\eta) + \frac{\kappa}{\pi i}\int_{\mathrm{A}}^{\Delta}\frac{\chi(\eta_1)\,d\eta_1}{\eta_1 - \eta} = G(\eta), \quad \chi(s) = \varphi(s) + i\psi(s), \quad \mathrm{A} < \eta < \Delta. \tag{14}$$

The solution obtained in [3] has been generalised to the case of non-zero tangential load, when $\mathrm{A} \neq -\Delta$. The solution of equation (14) exists in this case also in closed form:

$$f(s) + i\sqrt{\frac{\gamma_2}{\gamma_1}}g(s) = 2A_0 \sin m\pi \sin^m\frac{s-\alpha}{2}\sin^{1-m}\frac{\delta - s}{2}(\cos\Psi - iR\sin\Psi), \tag{15}$$

where $\Psi = \frac{1}{2}[m\alpha + (1-m)\delta + s]$ and the parameter m is now a complex quantity, $m = \frac{1}{2} + i\theta$. The contact stresses oscillate in the vicinity of the boundary of the contact boundary.

4.3. General problem with partial adhesion and slip. The general problem of dynamic contact when the contact area is subdivided into portions with adhesion and slip is reduced to a system of singular integral equations. After introducing a new auxiliary function $\phi(\eta)$, this system is written as

$$\begin{aligned} \mu_1\phi(\eta) - \frac{\kappa}{\pi}\int_{\mathrm{B}}^{\Gamma}\frac{\phi(\eta_1)d\eta_1}{\eta_1 - \eta} &= -(1+\mu_1^2)\varphi(\eta) + G_1(\eta) - \mu_1 G_2(\eta), \quad \mathrm{B} < \eta < \Gamma, \\ \mu_1\varphi(\eta) + \frac{\kappa}{\pi}\int_{\mathrm{A}}^{\Delta}\frac{\varphi(\eta_1)d\eta_1}{\eta_1 - \eta} &= -\phi(\eta) - G_2(\eta), \quad \mathrm{A} < \eta < \Delta. \end{aligned} \tag{16}$$

First, we invert the singular integral operators in these equations assuming that the functions on the right are known. We follow the procedure described at the end of Section 3. Then to the resulting integral expressions we apply the well-known asymptotic estimates for principal value integrals [7]: if $\chi(\eta) \sim C(\eta - \mathrm{B})^{-\epsilon}$ as $\eta \to \mathrm{B} + 0$ $(0 < \epsilon < 1)$, then

$$\frac{1}{\pi}\int_{\mathrm{B}}^{\Gamma}\frac{\chi(\zeta)\,d\zeta}{\zeta - \eta} \sim \begin{cases} \dfrac{1}{\sin\epsilon\pi}C(\mathrm{B} - \eta)^{-\epsilon} \text{ as } \eta \to \mathrm{B} - 0, \\ \cot\epsilon\pi C(\eta - \mathrm{B})^{-\epsilon} \text{ as } \eta \to \mathrm{B} + 0. \end{cases} \tag{17}$$

Asymptotic analysis allows us to study the detailed structure of the stress fields near the points of transition from adhesion to slip. Both the normal and tangential stresses are proved to be continuous inside the contact area. In the slip zones, both stresses have infinite derivatives near these points, while in the adhesion zone this is observed only for the tangential stresses. To the best of the author's knowledge, the existent literature does not contain any results like this, obtained on the basis of analytic solutions.

After some transformations, for the auxiliary function $\phi(s)$, we obtain the Fredholm equation of the second kind

$$\phi(s) + \int_{\beta}^{\gamma} \mathcal{K}(s, s_1)\phi(s_1)ds_1 = \mathcal{F}(s); \qquad (18)$$

we also obtain four transcendental equations for the determination of the end-points α and δ of the contact area, as well as for the determination of the location of the adhesion zone (β, γ). Then the unknown stress distributions $f(s)$ and $g(s)$ are found using explicit integral representations of these functions through $\phi(s)$.

5. Conclusions

The method of singular integral equations is the proper mathematical technique for solution of the mixed boundary value problems of elastodynamics arising in mathematical modelling of ultrasonic motors. In several particular cases it allows us to obtain analytic closed-form solutions. In all cases the method gives useful information about the solution for further numerical analysis.

The solutions obtained enable us to calculate all the characteristics of the motor that are important in applications, namely input voltage/velocity, load force/velocity relations, the maximal load force and the efficiency of the transformation of the wave energy into the energy of rotor motion. They form a reliable basis for the numerical analysis and design of ultrasonic motors.

References

1. K.L. Johnson, *Contact mechanics*, Cambridge University Press, Cambridge, 1987.
2. O.Yu. Zharii, An exact mathematical model of a travelling wave ultrasonic motor, *Proc. IEEE Ultrasonics Symp., 1994*, vol. 1, 545–548.
3. O.Yu. Zharii, Adhesive contact between the surface wave and a rigid strip, *J. Appl. Mech.* **62** (1995), 368–372.
4. O.Yu. Zharii, Frictional contact between the surface wave and a rigid strip, *J. Appl. Mech.* **63** (1996), 15–20.
5. O.Yu. Zharii and A.F. Ulitko, Smooth contact between the running Rayleigh wave and a rigid strip, *J. Appl. Mech.* **62** (1995), 362–367.
6. S.G. Mikhlin, *Integral equations*, OGIZ, Moscow, 1949.
7. N.I. Muskhelishvili, *Singular integral equations*, GIFML, Moscow, 1962.

Institut für Mechanik II, Technische Hochschule Darmstadt, Hochschulstrasse 1, 64289 Darmstadt, Germany